International Review of **Cytology** A Survey of **Cell Biology**

VOLUME 151

International Review of Cytology

A Survey of Cell Biology

Edited by

Kwang W. Jeon
Department of Zoology
The University of Tennessee
Knoxville, Tennessee

Jonathan Jarvik
Department of Biological Sciences
Carnegie Mellon University
Pittsburgh, Pennsylvania

VOLUME 151

ACADEMIC PRESS
A Division of Harcourt Brace & Company
San Diego   New York   Boston   London   Sydney   Tokyo   Toronto

Academic Press, Inc.
525 B Street, Suite 1900, San Diego, California 92101-4495

United Kingdom Edition published by
Academic Press Limited
24–28 Oval Road, London NW1 7DX

International Standard Serial Number: 0074-7696

International Standard Book Number: 0-12-364554-9

PRINTED IN THE UNITED STATES OF AMERICA

94 95 96 97 98 99   EB   9 8 7 6 5 4 3 2 1

CONTENTS

Inherited Genetic Defects: Analysis and Diagnosis at the Cellular Level in Preimplantation Embryos

Audrey L. Muggleton-Harris

Structure and Function of the Cyanelle Genome

Wolfgang Löffelhardt and Hans J. Bohnert

Diverse Distribution and Function of Fibrous Microtubule-Associated Proteins in the Nervous System

Thomas A. Schoenfeld and Robert A. Obar

Cholinesterases in Avian Neurogenesis

Paul G. Layer and Elmar Willbold

Role of Nuclear Trafficking in Regulating Cellular Activity

Carl M. Feldherr and Debra Akin

Phenolic Components of the Plant Cell Wall

Graham Wallace and Stephen C. Fry

CONTRIBUTORS

Numbers in parentheses indicate the pages on which the authors' contributions begin.

Debra Akin (183), *Department of Anatomy and Cell Biology, University of Florida, College of Medicine, Gainesville, Florida 32610*

Hans J. Bohnert (29), *Department of Biochemistry, The University of Arizona, BioSciences West, Tucson, Arizona 85721*

Carl M. Feldherr (183), *Department of Anatomy and Cell Biology, University of Florida, College of Medicine, Gainesville, Florida 32610*

Stephen C. Fry (229), *Centre for Plant Science, Division of Biological Sciences, University of Edinburgh, Edinburgh EH9 3JH, United Kingdom*

Paul G. Layer (139), *Technical University of Darmstadt, Institute for Zoology, 6100 Darmstadt, Germany*

Wolfgang Löffelhardt (29), *Institut für Biochemie und Molekulare Zellbiologie der Universität, Wien und Ludwig-Boltzmann-Forschungstelle für Biochemie, A-1030 Vienna, Austria*

Audrey L. Muggleton-Harris (1), *UMDS Division of Obstetrics & Gynaecology, St. Thomas' Hospital, London SE1 7EH, England*

Robert A. Obar (67), *Alkermes, Inc., Cambridge, Massachusettts 02139*

Thomas A. Schoenfeld (67), *Departments of Psychology and Biology, and the Neuroscience Program, Clark University, Worcester, Massachusetts 01610*

Graham Wallace (229), *Centre for Plant Science, Division of Biological Sciences, University of Edinburgh, Edinburgh EH9 3JH, United Kingdom*

Elmar Willbold (139), *Technical University of Darmstadt, Institute for Zoology, 6100 Darmstadt, Germany*

Inherited Genetic Defects: Analysis and Diagnosis at the Cellular Level in Preimplantation Embryos

Audrey L. Muggleton-Harris
UMDS Division of Obstetrics & Gynaecology, St Thomas' Hospital, London
SE1 7EH, England

I. Introduction

One of the major challenges of modern medicine is the prevention or cure of genetic diseases. There are more than 4000 known inherited disorders. However, with the recent advances in recombinant DNA technology, rapid progress has been made in the characterization of mutations and chromosome aberrations which give rise to genetic diseases. The genetic defects associated with human inherited diseases are systematically being identified. In some cases, an associated biochemical defect aids diagnosis. In other cases, the genetic lesion itself may be identified, or the disease phenotype may be associated with linked DNA markers, such as restriction enzyme polymorphisms. Prevention of a select number of genetic diseases has been achieved by genetic counseling and antenatal diagnosis. Since the development of amniocentesis and chorionic villus sampling (CVS), fetal cells can be obtained for assaying and identifying carriers or an affected fetus (Weatherall, 1991). Once a fetus has been identified as carrying the genetic defect, elective termination of pregnancy can be offered. Preimplantation diagnosis would prevent much of the physical and psychological trauma for parents who have to make such a decision.

II. Animal Models of Human Genetic Disease Used to Develop Methods for Preimplantation Diagnosis

The development of diagnostic techniques requires sufficient cells and/or embryos with the appropriate genotype. A mouse model for the Lesch-Nyhan syndrome (Lesch and Nyhan, 1964) was used to develop the biopsy

1

and analysis procedures for preimplantation embryos. Heterozygous (carrier) females have the hypoxanthine phosphoribosyl transferase, E.C. 2.4.2.8 (HPRT) deficiency allele on one of their X chromosomes (Hooper *et al.*, 1987). Thus when they are mated to a normal male, half of their female offspring are again carriers of the enzyme defect in the heterozygous condition. The technique of creating null enzyme or gene mutants depends on using selected negative mutant clones of embryonal mouse stem cells. The defect then incorporated in the germ line of mice is derived from transferred blastocysts into which a number of the selected embryonal cells have been injected (Robertson, 1987; Muggleton-Harris, 1992; Brown *et al.*, 1992).

Male mice embryos deficient in HPRT activity were diagnosed by biochemical microassay in a single cell isolated from the eight-cell preimplantation embryo. The sampled embryos were transferred to recipient mothers and examined on the 14th day of gestation to confirm the diagnosis. Of the embryos diagnosed as HPRT negative, all four grew into fetuses (Monk *et al.*, 1987). The same animal model was used to identify carrier female and affected male embryos by biopsing 5–10 extraembryonic trophectoderm cells from the blastocyst. The diagnosis was confirmed at 14 days' gestation, and live young were obtained from some of the biopsied embryos after transfer (Monk *et al.*, 1988). The sex of an embryo can also be diagnosed by the amount of HPRT activity in a single blastomere (Monk *et al.*, 1990). Both X chromosomes are active in female morulae and thus the blastomeres sampled from female preimplantation embryos have twice the X-coded HPRT activity of those from male embryos.

The polymerase chain reaction (PCR) (Saiki *et al.*, 1985, 1988; Mullis and Faloona, 1987) is capable of rapidly amplifying DNA sequences *in vitro*. A single DNA segment composed of a few hundred base pairs present in a human genome which has a complexity of 3 billion base pairs can be selectively amplified hundreds of millions of times.

In 1990 the polymerase chain reaction (discussed in detail later) was used to detect a deletion in the myelin basic protein gene (Gomez *et al.*, 1990). The mouse model used was the shiverer (shi/shi) mutant, which is characterized by tremors, hypomyelination of the central nervous system, and a shortened lifespan (Readhead and Hood, 1990). This mouse model is used to study dysmyelinating deficiencies in the central nervous system (CNS). Compact myelin contains a small number of very abundant proteins, for example, up to 90% of the protein mass of myelin is composed of proteolipid protein (PLP) and myelin basic protein (MBP). Both proteins are believed to be the key structural elements in the architecture of the compacted myelin sheath. The gene encoding myelin basic protein, a major constituent of central nervous system myelin, has been identified with mouse chromosome 18 and is largely deleted in shiverer mice. The normal MBP gene has seven exons which have been sequenced, while the mutant MBP gene has only exons I and II. The 5' breakpoint of the MBP gene deletion in shiverer mice has been sequenced.

Trophectoderm biopsy samples from the abembryonic area of the preimplantation mouse blastocyst were simultaneously analyzed for the presence of a normal and mutant allele of the MBP gene by PCR. The biopsied embryos were kept in culture during analysis of two to five biopsied cells. The cultured biopsied embryos were later introduced into a foster mother. Preimplantation diagnosis was completed in 7 hr. The identity of either amplification product was proved conclusively by direct sequence analysis of amplified products. Ninety-six percent of the recovered blastocysts survived the biopsy procedures, judged by re-formation of a blastocoele cavity *in vitro*. Fifty-nine percent of the biopsied embryos established pregnancy by day 6.5, compared with 88% of the nonbiopsied control blastocysts (Gomez *et al.*, 1990).

Sickle cell anemia is one of the most common human monogenic disorders (Weatherall, 1991) and is caused by a single base-pair mutation (β^S) in codon 6 of the human β-globin gene. The ability to determine the β-globin genotype in human preimplantation embryos before pregnancy would be a major advance in offering couples at risk the option to embark on a pregnancy in the certain knowledge their offspring would be unaffected by sickle cell anaemia. A transgenic mouse model created by Greaves *et al.* (1990) has provided a means to diagnose the Hb^{SA}-globin (hemoglobin S Antilles) transgene in biopsied trophectoderm cells and blastomeres of the preimplantation embryo. The overall success rate for diagnosis and confirmation of the presence or absence of the human β-globin sequence in the biopsied embryo was 70%. In individual experiments under optimal conditions, the presence of the transgene in biopsied cells was detected with 100% accuracy at the one-cell level. Fetuses and live young were born (Sheardown *et al.*, 1992). Once more PCR was used to amplify the DNA.

III. The Polymerase Chain Reaction

The fundamental principle of PCR and its applications to biological and medical science have been reviewed elsewhere (Erlich, 1989; Erlich *et al.*, 1991; Arnheim, 1990; Innis *et al.*, 1990; McPherson *et al.*, 1991). The first application of PCR was for prenatal diagnosis of sickle cell anemia (Saiki *et al.*, 1985). Since then, PCR has been applied to the prenatal diagnosis of many other genetic diseases using cells obtained from amniocentesis (second trimester of pregnancy) and chorionic villus sampling (first trimester of pregnancy) (Kazazian, 1989).

The ability of PCR to be so selective in its amplification of DNA is accompanied by excellent sensitivity; a single molecule of DNA present in a single sperm cell can be amplified and analyzed (Li *et al.*, 1988, 1990, 1991; Cui *et al.*, 1989). The DNA can be sequenced in a single haploid or diploid cell (Li *et al.*, 1988). Geneticists and reproductive scientists then proposed the idea that PCR amplification could be used to detect genetic disease in human embryos produced by *in vitro* fertilization (IVF), and thus diagnose the disease prior to implantation of the embryo (Li *et al.*, 1988; Handyside *et al.*, 1989; Cui *et al.*, 1989; Gomez *et al.*, 1990).

The first reports of progress toward this goal used mouse embryos. Preimplantation embryos obtained from mated females have been biopsied and the DNA analyzed for an inherited gene defect (Holding and Monk, 1989; Bradbury *et al.*, 1990). Normal pregnancies and live young were obtained following the biopsy and analysis for an inherited gene deletion (Gomez *et al.*, 1990; Readhead and Muggleton–Harris, 1991) and an inherited base pair substitution (Sheardown *et al.*, 1992).

PCR has been used to identify the sex of human preimplantation embryos, and a transfer of female embryos was attempted in eight couples known to be at risk of transmitting various X-linked diseases. Five of the eight women became pregnant, two with twins and three with singleton pregnancies, after a total of 13 treatment cycles. Six of the seven fetuses were confirmed as being female after chorionic villus sampling (Handyside *et al.*, 1990, 1992; Handyside and Delhanty, 1992; Hardy and Handyside, 1992). Pregnancy has also been achieved after embryo biopsy and coamplification of DNA from X and Y chromosomes simultaneously by multiplex PCR. This diminishes the likelihood of an incorrect diagnosis (Grifo *et al.*, 1992a). The multiplex PCR technique was used during this study to determine the sex of the embryos from a hemophilia carrier. Two blastomere embryo biopsies were performed after IVF, with the subsequent transfer of X-typed embryos.

Successful preimplantation diagnosis of genetic disease relies on being able to determine the genotype of one or perhaps a few cells using PCR. In addition, the manipulation of the embryo or unfertilized egg must not affect its normal development. Navidi and Arnheim (1991) have considered the accuracy of DNA analysis by PCR and suggest that since a single diploid cell or polar body contains only two DNA molecules representing each single-copy gene, the accuracy of genotype determination is much more sensitive to random fluctuations. Therefore, each of the two

DNA molecules of the target gene present in the diploid cell must be efficiently amplified to a detectable level, and if the cell is heterozygous for a gene, both alleles need to be identified. They conclude that blastomere and polar body DNA typing using a single cell is unacceptable under most circumstances, and that a combination of these, that is, two cells, must be analyzed by PCR before a decision is made. They suggest that a cell system, for example, sperm, oocytes, tissue-cultured cells, or buccal cells, would provide an appropriate method for working out the mathematical risk of which conditions are required at the single-cell level.

A DNA/PCR analysis of the first polar body of the human egg has been undertaken (Monk and Holding, 1990; Strom et al., 1990, Strom, 1992), and Verlinsky and Kuliev (1992) demonstrated that using multiple blastomeres from a single human embryo or combining polar body and blastomere analysis significantly lowers the level of misdiagnosis. The reliability of PCR detection of the sickle cell-containing region of the β-globin gene in a single human blastomere was examined by Pickering et al. (1992). They found that in contrast with a 97% positive confirmation in groups of cumulus cells and a 100% positive confirmation in whole, unmanipulated human embryos, a single blastomere gave a signal detection of only 45%. Recent data have shown that the signal from single human blastomeres has improved significantly (Pickering, 1992; Pickering et al., 1992; Pickering et al., 1993). However, the benefits of two or more samples for analysis cannot be overstated; thus the biopsy techniques and the selection of an appropriate stage of preimplantation development which can provide sufficient cell numbers are important.

PCR has been used to amplify a single-copy fragment of the β-globin gene from two to three human embryonic cells obtained from arrested human preimplantation embryos. Allele-specific priming of the PCR using nested primers was employed to detect β-thalassaemia mutations on 10 pg of DNA from individuals known to carry these mutations. False positive amplification was observed, and the authors caution against using this approach for clinical practice before this problem is overcome (Varawalla et al., 1991).

Recently PCR has been used to amplify the region of the most common mutation for cystic fibrosis (CF) in human blastomeres isolated from two eight-cell tripronucleate embryos. The nuclear status of the blastomeres was examined by light microscopy and by fluorescence microscopy using the fluorochrome Hoechst 33342 dye prior to PCR. The region around the ΔF508 mutation site was amplified by two PCRs with nested primers. After staining with Hoechst dye, the amplification was 91% and decreased to 71% after 20 hr of coloration (Liu et al., 1992).

IV. Biopsy Techniques and Stage of Biopsied Embryo

A. Polar Body

Micromanipulation of the early egg by removal of a polar body provides an easy method of obtaining a sample of DNA for analysis (Fig. 1). The sex of the egg can be identified and X-linked disorders avoided by replacing appropriately sexed embryos. Verlinsky and Kuliev (1992) have provided

Following fertilization

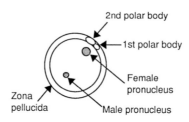

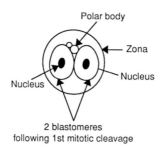

2 blastomeres
following 1st mitotic cleavage

4 blastomeres
following 2nd mitotic cleavage
(approx. 38–42 hr)

FIG. 1   Polar body biopsies can be undertaken at the fertilized egg stage. The polar body(s) can be seen trapped under the zona pellucida. Following the second mitotic cleavage, four blastomeres are available for biopsy.

data which show that removal of polar bodies produces no significant deleterious effects on human oocytes or embryos (Verlinsky *et al.*, 1990). Removal of the polar body and a blastomere to confirm the diagnosis, and subsequent culture of the manipulated embryo demonstrated that 16 out of 25 embryos developed to blastocyst *in vitro* (Verlinsky *et al.*, 1991). Preimplantation diagnosis was offered to five couples at risk for offspring with the ΔF508 mutation. Genetic analysis undertaken in 22 polar bodies and 15 corresponding blastomeres identified 21 embryos, of which 10 were transferred (Verlinsky *et al.*, 1992).

Polar body typing alone appears to be increasingly unacceptable when looking at autosomal recessive disease because if genotypes and errors in PCR efficiency. However, in the case of dominant diseases which are caused by genes lying extremely close to the centromere, there is less chance of misdiagnosis (Navidi and Arnheim, 1991; Trounson, 1992).

B. Early Cleavage

The ability to study biopsied human embryos *in vitro* and assess the results of micromanipulative processes on viability has obvious limitations. A true assessment of the viability of experimental biopsied embryos can only be obtained following their transfer back *in vivo*. Animal models provide sufficient numbers of eggs or embryos for a meaningful analysis of the effects of biopsy procedures and to establish pregnancy rates for the transferred embryos. The recipient animals can be sacrificed at various stages during pregnancy to obtain fetal material for confirmation of the original diagnosis, and the live young born provide information on birth rates, etc.

The mouse was used by Wilton and Trounson (1989) to develop methods for aspirating a single blastomere from a four-cell embryo. These authors found that the rate of fetal development of the biopsied embryo was not significantly different from that of the nonmanipulated embryo, although the implantation rate was significantly reduced. The implantation rate for biopsied embryos was 53.1 versus 81.8% for nonmanipulated embryos. When a single blastomere was removed from an eight-cell mouse embryo and the biopsied embryo cultured and replaced at the blastocyst stage, 31% of the transferred biopsied embryos were available for analysis on the 14th day of gestation (Monk *et al.*, 1987).

Takeuchi *et al.* (1992) have compared the usefulness of three micromanipulative methods at two different stages of preembryo development to assess the possible effects on postbiopsy survival and development using mouse preembryos. Four, seven- to eight-cell preembryos were biopsied using enucleation, aspiration, or extrusion of single blastomeres. After biopsy, preembryos were observed for *in vitro* and *in vivo* development. Few embryos died as the result of the biopsy trauma and high postbiopsy survival rates were associated with normal intrauterine and postnatal development. Expanded blastocyst formation rates from four-cell and eight-cell preembryos were 94.6% in the controls as opposed to 80.7% with nucleation, 19.1% with aspiration, and 83.1% with extrusion, respectively. Live birth rates at the four-cell stage were slightly lower in the enucleation group than in the blastomere/aspiration/extrusion groups or controls, but for the eight-cell stage there were no differences among the groups. No developmental abnormalities were found in body or organ weights in neonates or at 3 weeks of age, or in their subsequent ability to reproduce a second generation. Thus the authors concluded that the biopsy of mouse preembryos produces no loss of viability because of trauma and permits normal prenatal and postnatal development among surviving preembryos. It must be stressed that in this work only one blastomere or one nucleus was removed.

It does increasingly appear that at this stage of development it would be preferable for two nuclei or two blastomeres to be removed if one is to undertake controlled and effective preimplantation diagnosis. Removal of a quarter of the embryo from a four-cell or eight-cell stage human preimplantation embryo retards cleavage and the inner cell mass to trophectoderm ratio is strikingly reduced at the blastocyst stage in the biopsied group. This reduction was greater in embryos that reached the morula stage after day 4 (Tarin *et al.*, 1992). The conclusions of the authors were that more investigation is needed to assess whether the detrimental effects observed were the result of the biopsy method used in the study, or the result of a high sensitivity of human embryos at early stages to manipulation *in vitro* (Tarin *et al.*, 1992).

To assess any reduction in viability and development *in vitro* after biopsy of a quarter of the cells of human embryos on day 2 after insemination, Tarin and colleagues studied pyruvate uptake and cell number at the blastocyst stage. Biopsy did not have an adverse effect on either the proportion of embryos developing to blastocyst stage (32 of 64) in control and (31 of 65) biopsied groups, or on pyruvate uptake. The proportion of the biopsied embryos reaching morula stage on day 4 was significantly higher in biopsied groups; thus they were retarded in development. The normal ratio of numbers of inner cell mass to trophectoderm cells at the blastocyst was also affected when the retarded biopsied embryos reached the blastocyst stage (Tarin *et al.*, 1992). The effects of an abnormal ratio on *in vivo* viability and normal fetal development are not known for human embryos. This is one area which can be effectively explored using animal models. The viability of human embryos at the six-to-eight-cell stage following the biopsy of one cell is 24% (four established pregnancies) (Handyside *et al.*, 1990). Recently 10 out of 22 (45%) biopsied embryos implanted and 7 (32%) went on to become clinical pregnancies (Handyside *et al.*, 1992).

C. Blastocyst

Procedures have been developed for removing a number of cells from the abembryonic pole of the preimplantation animal blastocyst (Fig. 2). Results show that this procedure is not detrimental to the *in vitro* or *in vivo* development of the rabbit and mouse embryo (Gardner and Edwards, 1968; Gardner, 1972; Betteridge *et al.*, 1981; Monk *et al.*, 1988; Summers *et al.*, 1988; Gomez *et al.*, 1990; Readhead and Muggleton-Harris, 1991; Sheardown *et al.*, 1992). For primates, the marmoset blastocyst biopsy procedures resulted in 6 out of 15 normal offspring being born (Summers *et al.*, 1988).

"Spare" human preimplantation embryos which are donated for research after they no longer needed for patients in an IVF program can be used for *in vitro* studies and development of the biopsy procedures. Early cleavage embryos were cultured to the appropriate stage (Muggleton-Harris *et al.*, 1990) for undertaking zona drilling and thinning procedures using acid tyrodes. Blastomere biopsies were undertaken on fragmenting grade 3 embryos. The trophectoderm biopsy procedures used for the human blastocysts closely followed those developed for the mouse (Monk *et al.*, 1988; Gomez *et al.*, 1990; Sheardown *et al.*, 1992) (Fig. 3). The majority of the zona-thinned blastocysts re-formed their blastocoele cavities when replaced in culture, and all the biopsied blastocysts (21 out of 21) hatched and formed outgrowths *in vitro* (Muggleton–Harris and Findlay, 1991). Thus the removal of trophectoderm cells did not affect the *in vitro* development of the biopsied human blastocyst. Indeed the acid tyrode thinning of the zona plus the micromanipulation procedures improved the numbers of blastocysts which hatched and attached to a substrate *in vitro*.

To date there have been no transfers of biopsied human blastocysts. However, a comparative study of nonbiopsied early morula stage embryos with that of blastocysts demonstrated a similar rate of pregnancies (Bolton *et al.*, 1991; Bolton, 1992). Although the numbers in this study are small, a 25% pregnancy rate was recently obtained following the transfer of blastocysts (Boyer *et al.*, 1993; Janny and Menezo, 1993). The obvious advantages of a biopsy of three to ten cells from the abembryonic trophectoderm are (1) the biopsy procedures are simpler than those required for early cleavage cells, (2) sufficient cells are available for at least two attempts to analyze for the same or different genetic diseases, and (3) biopsy of tro-

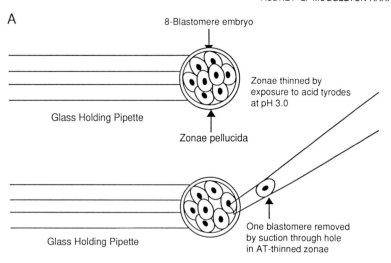

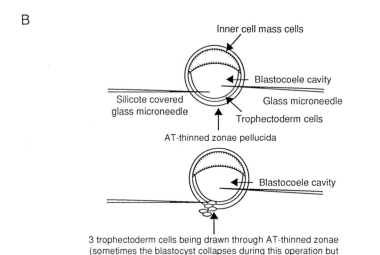

FIG. 2 (A). Biopsy procedures on an eight-cell embryo are shown. One blastomere has been removed and is within the pipette. (B). The procedures for blastocyst biopsy involve the use of two glass microneedles. The cells required for analysis are drawn through the zona-thinned blastocyst without harming the inner cell mass cells (embryo proper).

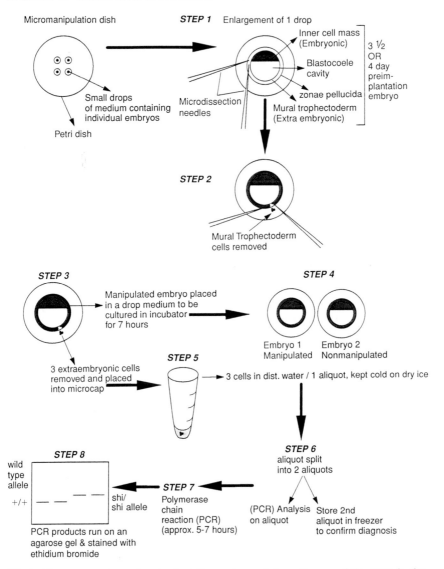

FIG. 3 Biopsy procedure of preimplantation embryo. A flow diagram of the steps in the biopsy and PCR analysis of trophectoderm cell samples. Procedures similar to those in steps 5–8 are followed when analyzing biopsied polar bodies or blastomeres. (Reproduced from C. Readhead and A. L. Muggleton-Harris, *Human Reproduction,* Vol. 6, No. 1, pp. 93–100, 1991.)

phectoderm may be ethically more acceptable than removal of one or two cells destined to form part of the embryo proper. The portion of the blastocyst that is sampled is destined to become placenta, and the effect of the biopsy procedures on implantation rates of mouse embryos is only slightly lower than nonmanipulated embryos in controls (Gomez *et al.*, 1990; Readhead and Muggleton-Harris, 1991).

This strategy may be used to screen human embryos obtained by IVF or lavage (Buster *et al.*, 1985; Brambati and Tului, 1990). Blastocysts recovered by uterine lavage after IVF appear to result in pregnancies more often than do earlier stage embryos (Simpson and Carson, 1992). However, in practice, lavage will probably require superovulation, which for unexplained reasons has not proven successful in terms of recovering embryos (Carson *et al.*, 1991).

V. Proliferation of Biopsied Cells

The possibility of increasing the original biopsied cell number by culturing the sample for 24 to 48 hr must be advantageous when an embryo at the early cleavage preimplantation stage is used for diagnosis. When a polar body and/or blastomere from a four-cell, or one blastomere from an eight-cell embryo is processed, there can only be one attempt at analysis. This severely limits the procedure should there be loss or contamination of the cell sample, or equivocal results. It was suggested that there is a definite requirement for removing two or more blastomeres, or to promote the replication of the biopsy blastomere, which would then permit duplicate analyses to be undertaken if required (Muggleton-Harris, 1990; Muggleton-Harris and Findlay, 1991).

A variety of techniques, methods, and conditions have been used to culture murine blastomeres from the preimplantation embryo (Wilton and Trounsen, 1989; Muggleton-Harris, 1990; Muggleton-Harris and Findlay, 1991). A fibronectin or gelatin substrate was used in conjunction with conditioned media to grow both mouse and human blastomeres. A finite number of replications of each blastomere was achieved: the first replication of individual blastomeres was noted within 18 to 48 hr (Muggleton-Harris and Findlay, 1991). Monolayer cell cultures derived from single mouse blastomeres cultured on extracellular matrix components have been obtained, with most single blastomeres undergoing several cell divisions within 6 days (Wilton and Trounson, 1989). Thus it is possible to achieve duplicate cell samples for PCR analysis or *in situ* hybridization.

VI. Culture of Preimplantation Biopsied Embryos

It has been found that if embryos were recovered following biopsy procedures and placed on an appropriate substrate, for instance fibronectin, in conditioned media, the embryo cavitated, hatched, and attached to the substrate and formed an outgrowth (Muggleton–Harris and Findlay, 1991). The zona thinning and drilling procedures and subsequent biopsy did not appear to harm the embryo. In fact, the embryo was shown in most cases to hatch through the slit in the zona that had been made following biopsy. The fibronectin or collagen substrate enhanced outgrowth, and earlier studies showed that the development of nonmanipulated cultured preimplantation embryos was improved when substrates were used (Muggleton-Harris et al., 1990; Muggleton-Harris, 1990; Muggleton-Harris and Findlay, 1991). It is interesting to note that others have reported that fibronectin, a glycoprotein growth factor, and laminin promotes in vitro attachments and outgrowth of mouse blastocysts (Armant et al., 1986), and that the relative expression of laminin and fibronectin receptors may determine the morphology and behavior of human trophoblast during implantation (Burrows et al., 1993).

VII. In Situ Hybridization, Fluorescent Techniques

Numerous refinements have been made to analyze the sex of preembryos and their chromosomes by in situ hybridization techniques over the past few years (Pinkel et al., 1986; West et al., 1988; Jones et al., 1987; Cremer et al., 1986). Improvements in sensitivity, in particular of fluorescent in situ hybridization (FISH), routinely allow the detection of single copy sequences (Grifo et al., 1990, 1992b). Using a multicolor FISH approach based on the combination of probe labeling and digital imaging microscopy, Reid et al. (1992) demonstrated the simultaneous visualization of probe sets specific for chromosomes 13, 18, 21, X, and Y. This approach enables one to evaluate aberrations of multiple chromosomes in a single hybridization experiment using metaphase chromosomes and interphase nuclei from a variety of cell types, including lymphocytes and amniocytes.

Dual FISH was used for simultaneous detection of X and Y chromosome-specific probes for sexing human preimplantation embryonic nuclei (Griffin et al., 1991, 1992a). The possibility of diagnosing males as females is virtually eliminated with this approach but it takes 24 hr to perform. Recently Griffin et al. (1992b) have reported a rapid dual FISH sexing strategy taking 6–7 hr which gives reliable results. Ten patients, each carrying X-linked disorders, have had preimplantation diagnosis of embryo sex in a total of 12 treatment cycles. In the first eight cycles biopsy was undertaken on day 2; in the final four, biopsy, spreading, FISH diagnosis, and transfer were undertaken on day 3. A pregnancy was achieved from these treatment cycles.

Diagnosis of the sex of day 3 human preimplantation embryos using single biopsied cells and dual fluorescent FISH probes on the interphase nuclei identified the sex of 26 out of 39 cells. Thirteen of the biopsied, typed embryos were transferred, with two pregnancies being achieved, resulting in one live birth of a normal female infant, and one miscarriage of a female fetus in the first trimester (Griffin et al., 1992b).

FISH can be used to distinguish whole chromosomes or specific regions of chromosomes. A series of human embryos developing in culture at various stages up to blastocyst have been examined with a probe specific for the centromeric region of chromosome 18. Fifty-four out of 59 embryos were considered diploid based on obtaining two hybridization signals. Some nuclei in individual embryos had one signal or none, and tetraploid nuclei were also widespread. Among the 5 remaining embryos, one five-cell was trisomic for chromosome 18, one four-cell was monosomic, and 3 other embryos were aneuploid mosaics and/or had multinucleated blastomeres (Handyside, 1992a). As suggested by Schrurs et al. (1993), preimplantation diagnosis of aneuploidy using FISH will require more than a single nucleus for an accurate diagnosis.

Diagnosis of Lipid Storage Diseases

The sphingolipidases are a group of lipid storage diseases, each of which is caused by a deficiency of a lysosomal enzyme, resulting in a block of the catabolic pathway of a specific sphingolipid (Schriver et al., 1989). Most sphingolipidases are inherited as autosomal recessive disorders without any available treatment. During the past few years a number of genetic mutations leading to these disorders have been mapped (Neufield, 1991), indicating that for each disease, the associated enzyme deficiency may be caused by different mutations. The diagnosis of lipid storage diseases in embryos at the preimplantation stage has used sensitive fluorescent or fluorogenic procedures (Epstein et al., 1991). The activities of several sphingolipid hydrolases were determined in extracts of murine embryos and also human oocytes and polyspermic embryos. These procedures indicated that Tay-Sachs, Gaucher's and Krabbe's diseases might be diagnosed in one blastomere, while for Niemann-Pick disease, two might be required.

VIII. Cytogenetic Studies to Diagnose Chromosome Abnormalities

Cytogenetic studies would be required to diagnose chromosome abnormalities and there are very few chromosome studies on preimplantation embryos (Roberts *et al.*, 1990; Geraedts *et al.*, 1992; Schmiady and Kentenich, 1993). This is due to difficulties obtaining high-quality chromosome preparations from blastomeres. As shown earlier (Muggleton-Harris, 1990; Muggleton-Harris and Findlay, 1991), single human blastomeres can be cultured, and they replicate within 18–24 hr; therefore, once a single blastomere is obtained, the possibilities of undertaking cytogenetic studies on more than one cell biopsied from early cleavage stages can be considered.

The techniques for processing individual cells to improve cytogenetic sexing of blastomeres by means of chromosome *c* banding have been carried out using mouse embryos. Chromosome spreads were obtained in 85.0% of 1 cell from 2 cell embryos and 75% of 2 cells from 4 cell embryos. Sexing was accomplished in 30.0% of 1 to 2 embryos and 52.5% of 2 to 4 embryos, with a loss due to poor technique of 12.5 and 15.0% respectively. Transfer of sexed embryos resulted in a low implantation rate; the two live fetuses obtained were the predicted sex (Gimenez *et al.*, 1993).

The sequential use of *c* banding and *in situ* hybridization techniques should improve the present status, and would allow sexing of interphase nuclei (Gimenez *et al.*, 1993). The diagnosis of a trisomy in the mouse preimplantation embryo was undertaken by Kola and Wilton (1991), who suggest that analysis of a single blastomere may be reliable enough to predict the genetic status of the embryo. However, performing two one-cell biopsies, or allowing the blastomere to replicate *in vitro* would allow a combination of methods—for example, FISH, PCR, and cytogenetics—to be used for preimplantation diagnosis.

The use of small numbers of cells obtained initially from mouse embryos, and cultured human cells obtained from cell lines or patients with known chromosomal abnormalities, should permit the development of reliable techniques for the preimplantation diagnosis of chromosomal abnormalities. Abnormal pronuclear human oocytes from an IVF program have provided details on morphology, chromosomal constitution, and sex using *in situ* hybridization (Balakier *et al.*, 1993). Also, parthenotes obtained by activating unfertilized human oocytes (Winston *et al.*, 1991a; Balakier and Casper 1991) can develop to early cleavage stages *in vitro*, and the use of parthenotes and arrested embryos can provide human cells on which these techniques can be developed and/or refined.

IX. Concluding Remarks

A. Human Embryo Quality

The quality of human embryos *in vitro* is variable and their capacity to develop to the blastocyst stage is limited. Only 10% of the embryos obtained after IVF are able to implant after transfer *in utero* (Plachot, 1992). Current evidence suggests that large numbers of cleaving human embryos may give unrepresentative single-cell biopsies because gross abnormalities of the nucleocytoplasmic ratio are evident in 5% of embryos from the two-to-four cell stage onward (Winston *et al.*, 1991b). Cytogenetic analyses suggest that errors of somy as well as ploidy also occur frequently (Plachot *et al.*, 1987). The implications for preimplantation diagnosis are clear.

The reason for such a high percentage of abnormalities is not known (Handyside, 1992b). However, the culture conditions within IVF laboratories for human embryos are not optimal. Most human embryos arrest at the four-to-eight-cell stage when not proceeding through the normal developmental phases *in vitro*. This period coincides with the activation of the human genome (Braude *et al.*, 1988). Efforts are being made to improve the media in which the preimplantation human embryo is routinely cultured (Muggleton-Harris and Findlay, 1991; Muggleton-Harris, 1990,1992; Menezo *et al.*, 1990; Plachot *et al.*, 1987; Fitzgerald and DiMattina, 1992; Weimer *et al.*, 1993). Within these studies a promising area has developed, involving the coculture of the embryo with feeder cells. The beneficial effects of a coculture system may involve provision of growth factors, or the protection from, or removal of, toxic compounds. The presence of reactive oxygen species and/or free radicals has been associated with the "two cell block" of cultured mouse embryos (Nasr-Esfahani *et al.*, 1990a,b).

Diagnosis of human embryo cells at preimplantation stages for inherited diseases will provide an effective alternative to intrauterine diagnosis and termination of pregnancy for couples at high risk (1 : 2 1 : 4) of transmitting a defect. Diagnosis of biopsied human cells of the preimplantation embryo and the transfer of the biopsied embryo back into the mother are being undertaken in a small number of clinics (Verlinsky *et al.*, 1991,1992;

Handyside *et al.*, 1990,1992; Handyside and Delhanty, 1992; Hardy and Handyside, 1992). Mistakes have been made. For example, the sex of an embryo was wrongly identified as female and the pregnancy was terminated after CVS procedures (Handyside and Delhanty, 1992). Another embryo was diagnosed as free of the predominant deletion (ΔF508) of the cystic fibrosis transmembrane regulator gene that causes cystic fibrosis, but the child was homozygous for the deletion and had cystic fibrosis (Verlinsky *et al.*, 1992). Therefore, there is an obvious need to constantly reevalute the techniques used, to ensure that both the biopsy and analysis procedures are at their most effective. Whenever possible, data on these procedures using cells (e.g., DNA from single sperm, buccal cells, or blastomeres) to validate laboratory procedures (Liu *et al.*, 1992; Holding *et al.*, 1993; Pickering *et al.*, 1992; Morsy *et al.*, 1992) and animal systems (Gomez *et al.*, 1990; Sheardown *et al.*, 1992) should be gathered before the procedures are used on human patients. Handyside and colleagues (1992) suggest that an analysis of a series of human pregnancies is required to assess methods adequately. However, I concur with the views of Trounson (1992) and Liebaers *et al.* (1992), who think that more experimentation on model systems is a reasonable procedure to adopt.

B. Recent Technical Innovations

Technical innovations are made every day, especially in the field of molecular biology. PCR has proved an invaluable tool for genetic analysis at the cellular level. However, it is also subject to artifacts caused by contamination, or mutations introduced by the infidelity of the DNA polymerase (Keohavong and Thilly, 1989) or recombination between amplified products. Methods to avoid contamination have been discussed in detail (Kwok and Higuchi, 1989; Longo *et al.*, 1990; Gomez *et al.*, 1990; Readhead and Muggleton-Harris, 1991; Arnheim, 1992; Pickering, 1992; Pickering *et al.*, 1993; Sheardown *et al.*, 1992).

A method for quantitative analysis of polymerase chain reaction products using primers labeled with biotin and a fluorescent dye allows immobilization and separation of the products, which can then be analyzed at a later date. This procedure circumvents electrophoretic separation and radioactive labeling. Exact quantitative analysis of reaction products is feasible during the logarithmic phase of amplification when *Taq* polymerase is not limiting, as it is during the plateau phase of the reaction (Landegren *et al.*, 1990). The theoretical plateau precludes the correlation of signal intensity with the number of cells and amount of DNA present in the cell sample preparation. The possible reasons for the PCR reaching a plateau are numerous but include thermal inactivation and limiting concentration of the DNA polymerase, a reduction in denaturation efficiency, inefficiency in primer annealing, and destruction of template due to DNA polymerase exonuclease activity (Sardelli, 1993). Amplification beyond the plateau will inevitably lead to amplification of nonspecific products. Thus, systematic studies to optimize the primers and PCR reaction conditions must be undertaken to avoid such problems.

Point mutations are not easily detected by PCR and confirmation is difficult unless the amplified product is denatured, and/or sequenced as in Gomez et al. (1990). An alternative method for detecting single base-pair mutations or substitutions is the oligonucleotide ligation assay (OLA) developed by Landegren et al. (1988a). In this assay, two oligonucleotides are designed to hybridize in exact juxtaposition on the target DNA sequences, permitting covalent joining by a DNA ligase. The successful ligation can be monitored by a variety of means, including visualization of the ligation product after size separation by gel electrophoresis or immobilizing one of the two oligonucleotides on a support and, after washing, detecting the covalent joining to the other appropriately labeled oligonucleotide. The OLA technique is characterized by a few false positives, since the chance that the two oligonucleotides will hybridize in immediate proximity in the absence of the target sequence is minimal. The sensitivity of the assay may be greatly increase by combining it with the target amplification by PCR. This measure permits the analysis of minute DNA samples or allows alternative detectable moieties such as fluorophores to be used.

Preliminary studies have demonstrated that the OLA technique can detect a primary defect in the synthesis of myelin caused by a proteolipid protein gene deficiency. The PLP gene has been mapped to the human X chromosome and has demonstrated a strong gene dosage effect, indicating that the PLP is linked to the amplified X chromosome. Indication of the PLP gene locus on the X chromosome in mouse and man provided evidence for implicating this gene in a group of sex-linked genetic myelin-deficiency disorders. Using molecular techniques, the characterization of the PLP gene has made it possible to identify the primary genetic defect of the jimpy mutant mouse, which in turn has established the jimpy mouse as a unique model system in which to study the function of PLP (Nave and Milner, 1989). Biopsy and analysis of blastomeres and trophectoderm cells of the jimpy mouse preimplantation embryo has enabled us to detect the PLP deficiency by OLA/PCR techniques at the single-cell level (A. L. Muggleton-Harris and C. Readhead, unpublished observations). The OLA method has also been automated and thus could be used in clinical laboratories to identify families with a history of inherited disease (Landegren et al., 1988b). Many genetic diseases, such as adenosine deaminase (ADA) deficiency, several forms of diabetes, Gaucher's disease (type 1 or 2), and others are due to single base-pair substitutions (Cooper and Krawczak, 1990; Weatherall, 1991).

A primer extension amplification (PEP) can produce on average 60 copies of at least 78% of the genome. Only an aliquot of the single cell PEP reaction needs to be taken for any one specific PCR test (Arnheim, 1992; Zhang et al., 1992). Also, amplification of small amounts of mRNA has been achieved using peripheral blood lymphocytes. The deletion and duplicate mutations responsible for Duchenne's and Becker's muscular dystrophy have been detected by amplification in 10 sections by reverse transcription and nested PCR (R. G. Roberts et al., 1990).

A new source of polymorphic DNA markers for sperm typing has been used to analyze microsatellite repeats in single cells (Hubert et al., 1992). The recombination fraction between CA repeat polymorphisms was determined after whole genome amplification of single sperm, followed by typing of two different aliquots, one for each polymorphic locus. Recently a CA repeat procedure (Pickering, 1992; Pickering et al., 1993) has been used in conjunction with probes for the human β-globin gene to analyze one to five cells biopsied from the trophectoderm of cultured human blastocysts. Clear, unequiviocal signals for both the CA repeat and sickle fragment are obtained by using nested primers. The biopsy procedure also allows five to ten cells to be removed without a significant effect on the viability (in vitro) of the embryo; thus two 1-cell samples can be obtained easily (Muggleton-Harris et al., 1993). The advantages of using a probe such as the CA repeat is that the presence of the signal confirms the presence of cells analyzed; thus, false negatives for the sickle fragment signal can be discounted. Also contamination from outside human sources can be controlled for.

A report on the successful PCR amplification of specific DNA sequences from routinely fixed chromosomal spreads could provide opportunities to reanalyze material that has been stored for 1–5 years. Thus, retrospective studies could be undertaken using up-to-date molecular analysis (Jonveaux, 1991). PCR and FISH techniques have been combined on single cells for the diagnosis of sex and the CF mutation (Wu et al., 1992).

The areas covered by this and other recent reviews (Simpson, 1992; Simpson and Carson, 1992; Kaufmann et al., 1992; Muggleton–Harris and Braude, 1993) demonstrate the exciting possibilities for the field of preimplantation diagnosis. The research is proceeding at a rapid rate and the future looks extremely encouraging for those couples "at risk" for inherited genetic defects. Successful implementation of many of the current techniques will make it possible to eliminate a mutant allele from a family in a single generation.

C. Legal and Ethical Issues

The legal and ethical issues arising from the new genetics have been discussed elsewhere (McLaren, 1984, 1986, 1987; Warnock, 1987; Johnson, 1989; Harvey, 1992; Robertson, 1992). Robertson suggests that even when the techniques are perfected, preimplantation genetic screening is likely to appeal to only a minority of women. Women undergoing IVF to treat infertility might choose to have their embryos screened if screening does not reduce the chances of achieving pregnancy. Some women at risk for a serious genetic defect might also choose to undergo IVF. However, "pursuing the goal of avoiding the birth of offspring with severe genetic handicaps is part of procreative liberty and parental discretion" and therefore "the means employed to achieve this goal—preimplantation genetic diagnosis and non transfer of embryos is also ethically acceptable. Because embryos are so rudimentary in develoment, they are not generally viewed as having interests or rights. Thus, they have no right to be placed in the uterus and may be discarded if they carry the gene for serious disease" (Robertson, 1992).

Acknowledgments

I wish to thank Carrie Victor-Smith for her expert word processing skills, and the Medical Research Council for supporting our work for this chapter with Grant G9216133.

References

Armant, D. R., Kaplan, H. A., and Lennarz, W. J. (1986). Fibronectin and laminin promoter *in vitro* attachment and outgrowth of mouse blastocysts. *Dev. Biol.* **116,** 519–523.

Arnheim, N. (1990). The polymerase chain reaction. *In* "Genetic Engineering, Principles and Methods" (Setlow, ed.), pp. 115–138. Plenum, New York.

Arnheim, N. (1992). Preimplantation genetic diagnosis—a rolling stone gathers no moss. *Hum. Reprod.* **7,** 1481–1483.

Balakier, H., and Casper, R. F. (1991). A morphologic study of unfertilised oocytes and abnormal embryos in human *in vitro* fertilisation. *J. In Vitro Fertil. Embryo Transfer* **8,** 73–79.

Balakier, H., Squire, J., and Casper, R. F. (1993). Characterisation of abnormal one pronuclear human oocytes by morphology, cytogenetics and in situ hybridisation. *Hum. Reprod.* **8,** 402–408.

Betteridge, K., Hare, W., and Singh, E. (1981). Approaches to sex selection in farm animals. *In* "New Technologies in Animal Breeding" (B. Brackett and G. S. Seidel, eds.), pp. 109–125. Academic Press, New York.

Bolton, V. N. (1992). Preimplantation diagnosis of genetic disease in human embryos. *Dev. Basis Inherited Disord., Br. Soc., Dev. Biol. Meet.* (Abstr.)

Bolton, V. N., Wren, M. E., and Parsons, J. H. (1991). Pregnancies after *in vitro* fertilization and transfer of human blastocysts. *Fertil. Steril.* **55**, 830–832.

Boyer, P., Mignot, T. M., Perez, J., Bulwa, S., Janssens, Y., and Zorn, J. R. (1993). Co-culture 1993. *Hum. Reprod.* **8.** Supplement 1, 47–48.

Bradbury, M. W., Isola, L. M., and Gordon, J. W. (1990). Enzymatic amplification of a Y chromosome repeat in a single blastomere allows identification of the sex of preimplantation mouse embryos. *Proc. Natl. Acad. Sci. U.S.A.* **87**, 4053–4057.

Brambati, B., and Tului, L. (1990). Preimplantation genetic diagnosis: a new simple uterine washing system. *Hum. Reprod.* **5**, 4,448–450.

Braude, P. R., Bolton, V. N., and Moore, S. (1988). Human gene expression first occurs between the four- and eight-cell stages of preimplantation development. *Nature (London)* **332**, 459–461.

Brown, D., Willington, M., Findlay, I., and Muggleton-Harris, A. L. (1992). Criteria which optimize the potential of murne embryonic stem cells for *in vitro* and *in vivo* developmental studies. *In Vitro Cell Dev. Biol.* **28A**, 773–778.

Burrows, T. D., King, A., and Loke, Y. W. (1993). Expression of intergrins by human trophoblast and differential adhesion to lamanin or fibronectin. *Hum. Reprod.* **8**, 475–484.

Buster, J. C., Bustillo, M., Rodi, I. A., Sydlee, W., Cohen, R. N. P., Hamilton, M., Simon, A., Thornycroft, I. H., and Marshall, J. R. (1985). Biologic and morphologic description of donated human ova recovered by non-surgical uterine lavage. *Am. J. Obstet. Gynecol.* **153**, 211–217.

Carson, S. A., Smith, A. L., Scoggan, J. L., and Buster, J. E. (1991). Superovulation fails to increase human blastocyst yield after uterine lavage. *Prenatal Diag.* **11**, 513–522.

Cooper, A., and Krawczak, C. (1990). The mutational spectrum of single base pair substitutions causing human genetic disease: patterns and predictions. *Hum. Genet.* **85**, 55–74.

Cremer, T., Landegent, J., Bruckner, A., Scholl, H. P., Schardin, M., Hager, H. D., Devillee, P., Pearson, P., and Ploeg van der, M. (1986). Detection of chromosome aberrations in the human interphase nucleus by visualisation of specific target DNAs with radioactive and non-radioactive in situ hybridisation techniques: diagnosis of trisomy 18 with probe L.1.84. *Hum. Genet.* **76**, 346–352.

Cui, X., Li, H., Goradia, T. M., Lange, K., Karazian, H. H., Gulas, D. J., and Arnheim, N. (1989). Single sperm typing: determination of genetic distance between G-gamma globin and parathyroid hormone loci. *Proc. Natl. Acad. Sci. U.S.A.* **86**, 9389–9393.

Epstein, M., Agmon, V., Fibach, E., Gatt, S., and Laufer, N. (1991). A model for preimplantation diagnosis of Gaucher disease in single mouse embryos. *Annu. Meet. Eur. J. Hum. Reprod. Embryol., 7th, Paris.* (Abstr.)

Erlich, H., ed. (1989). "PCR Technology: Principles and Applications for DNA Amplification." Stockton Press, New York.

Erlich, H., Gelfaud, D., and Sninsky, J. (1991). Recent advances in the polymerase chain reaction. *Science* **252**, 1643–1651.

Fitzgerald, L., and DiMattina, M. (1992). An improved medium for long term culture of human embryos overcomes the *in vitro* developmental block and increases blastocyst formation. *Fertil. Steril.* **57**, 641–647.

Gardner, R. L. (1972). Manipulation of development. *In* "Reproduction in Mammals 2: Embryonic and Fetal Development" (C. R. Austin and R. V. Short, eds.), pp. 110–133. Cambridge Univ. Press, Cambridge, England.

Gardner, R. L., and Edwards, R. G. (1968). Control of sex ratio at full term in the rabbit by transferring sexed blastocysts. *Nature (London)* **218**, 346–348.

Geraedts, J. P. M., Pieters, M. H. E. C., Loots, W. J. G., Coonen, E., and Hopman, A. H. N.

(1992). Detection of chromosomal abnormalities in single cells. *Int. Symp. Preimplantation Genet. Assisted Fert., Brussels.* (Abstr.)

Gimenez, C., Egozcue, J., and Vidal, F. (1993). Cytogenetic sexing of mouse embryos. *Hum. Reprod.* **8,** 470–475.

Gomez, C. M., Muggleton-Harris, A. L., Whittingham, D. G., Hood, L. E., and Readhead, C. L. (1990). Rapid preimplantation detection of mutant (Shiverer) and normal alleles of the mouse myelin basis protein gene (MBP) allowing selective implantation and birth of live young. *Proc. Natl. Acad. Sci. U.S.A.* **87,** 4481–4484.

Greaves, D. R., Fraser, P., Videl, M. A., Hedges, M. J., Ropers, D., Luzzalto, L., and Grosfeld, F. (1990). A transgenic model of sickle-cell disorder. *Nature (London)* **243,** 183–185.

Griffin, D. K., Handyside, A. H., Penketh, J. A., Winston, R. M. L., and Delhanty, J. D. A. (1991). Fluorescent in sity hybridisation to interphase nuclei of human preimplantation embryos with X and Y chromosome specific probes. *Hum. Reprod.* **6,** 101–105.

Griffin, D. K., Wilton, L. J., Handyside, A. H., Winston, R. M. L., and Delhanty, J. D. A. (1992a). Dual fluorescent in situ hybridisation for simultaneous detection of X and Y chromosome-specific probes for the sexing of human preimplantation embryonic nuclei. *Hum. Genet.* **89,** 18–22.

Griffin, D. K., Handyside, A. H., Wilton, L. J., Winston, R. M. L., and Delhanty, J. D. A. (1992b). Novel cytogenetic techniques diagnosing genetic disease in preimplantation human embryos. *Dev. Basis Inherited Disord., Br. Soc. Dev. Biol. Meet.* (Abstr.)

Grifo, J. A., Boyle, A., Fischer, E., Laoy, G., DeCherney, A. H., Ward, D. C., and Sanyal, M. K. (1990). Pre-embryo biopsy and analysis of blastomeres by in situ hybridisation. *Am. J. Obstet. Gynecol.* **163,** 2013–2019.

Grifo, J. A., Tang, Y. X., Cohen, J., Gilbert, F., Sanyal, M. K., and Rosenwaks, Z. (1992a). Pregnancy after embryo biopsy and coamplification of DNA from X and Y chromosomes. *J. Am. Med. Assoc.* **268,** 727–729.

Grifo, J. A., Boyle, A., Tang, Y. X., and Ward, D. C. (1992b). Preimplantation diagnosis. In situ hybridisation as a tool for analysis. *Arch. Pathol. Lab. Med.* **116,** 393–397.

Handyside, A. (1992a). Prospects for the clinical application of preimplantation diagnosis: the tortoise or the hare? *Hum. Reprod.* **7,** 1481–1483.

Handyside, A. (1992b). Genetic defects in early human embryos. *Dev. Basis Inherited Disord., Br. Soc. Dev. Biol. Meet.* (Abstr.)

Handyside, A. H., and Delhanty, D. A. (1992). Cleavage stage biopsy of human embryos and diagnosis of X-linked recessive disease. *In* "Preimplantation Diagnosis of Human Genetic Disease" (R. G. Edwards, ed.), pp. 239–270. Cambridge Univ. Press, Cambridge, England.

Handyside, A. H., Pattinson, J. K., Penketh, R. J. A., Delhanty, J. D. A., Winston, R. M. L., and Tuddenham, E. G. D. (1989). Biopsy of human pre-embryos and sexing by DNA amplification. *Lancet* **i,** 347–349.

Handyside, A. H., Kontogianni, E. H., Hardy, K., and Winston, R. M. L. (1990). Pregnancies from human preimplantation embryos sexed by Y-specific DNA amplification. *Nature (London)* **344,** 768–770.

Handyside, A. H., Lesko, J., Tarin, J. J., Winston, R. M. L., and Hughes, M. R. (1992). Birth of a normal girl following preimplantation diagnostic testing of cystic fibrosis. *N. Engl. J. Med.* **327,** 905–909.

Hardy, K., and Handyside, A. H. (1992). Biopsy of cleavage stage human embryos and diagnosis of single gene defects by DNA amplification. *Arch. Pathol. Lab. Med.* **116,** 388–392.

Harvey, J. C. (1992). Ethical issues and controversies in assisted reproductive technologies. *Curr. Opinion Obstet. Gynaecol.* **4,** 750–755.

Holding, C., and Monk, M. (1989). Diagnosis of beta-thalassemia by DNA amplification in single blastomeres from mouse preimplantation embryos. *Lancet* **ii,** 532–535.

Holding, C., Bentley, D., Roberts, R., Bobrow, M., and Mathew, C. (1993). Development and validation of laboratory procedures for preimplantation diagnosis of Duchenne muscular dystrophy. *J. Med. Genet.* **30,** 903–909.

Hooper, M., Hardy, K., Handyside, A., Hunter, S., and Monk, M. (1987). HPRT-deficient (Lesch–Nyhan) mouse embryos derived from germline colonization by cultured cells. *Nature (London)* **326,** 292–295.

Hubert, R., Weber, J. L., Schmitt, K., Zhang, L., and Arnheim, N. (1992). A new source of polymorphic DNA markers for sperm typing: analysis of microsatellite repeats in single cells. *Am. J. Hum. Genet.* **51,** 985–991.

Innis, M. A., Gelfand, D. H., Sninsky, J. J., and White, T. J., eds. (1990). "PCR Protocols: A Guide to Methods and Applications" Academic Press, San Diego.

Janny, L., and Menezo, Y. (1993). Cocultures: a new tool for improving results and understanding of assisted reproduction? *Hum. Reprod.* **8,** 69.

Johnson, M. H. (1989). The onset of human identity and its relationship to legislation concerning research on human embryos. *Proc. Int. Conf. Philos. Ethics Reprod. Med. 1st, Dep. Obstet. Gynaecol., Univ. Leeds* pp. 2–7.

Jones, K. W., Singh, L., and Edwards, R. G. (1987). The use of probes for the Y chromosomes in preimplantation embryo cells. *Hum. Reprod.* **2,** 439–445.

Jonveaux, P. (1991). PCR amplification of specific DNA sequences from routinely fixed chromosomal spreads. *Nucleic Acids Res.* **19,** 1946.

Kaufmann, R. A., Morsy, M., Takeuchi, K., and Hodgen, G. D. (1992). Preimplantation genetic analysis. *J. Reprod. Med.* **37,** 428–436.

Kazazian, H. (1989). Use of PCR in the diagnosis of monogenic diseases. *In* "PCR Technology: Principles and Applications for DNA Amplification" (H. Erlich, ed.), pp. 153–169. Stockton Press, New York.

Keohavong, P., and Thilly, W. (1989). Fidelity of DNA polymerases in DNA amplification. *Proc. Natl. Acad. Sci. U.S.A.* **86,** 9253–9257.

Kola, I., and Wilton, L. (1991). Preimplantation embryo biopsy: detection of trisomy in a single cell biopsied from a four cell mouse embryo. *Mol. Reprod. Dev.* **29,** 16–21.

Kwok, S., and Higuchi, R. (1989). Avoiding false positives with PCR. *Nature (London)* **339,** 237–238.

Landegren, U., Kaiser, R., Caskey, C. T., and Hood, L. (1988a). DNA diagnostics—Molecular techniques and automation. *Science* **242,** 229–237.

Landegren, U., Kaiser, R., Saunders, J., and Wood, L. (1988b). A ligase mediated gene-detection technique. *Science.* **24,** 1077–80.

Landegren, U., Kaiser, R., and Hood, L. (1990). Oligonucleotide ligation assay. *In* "PCR Protocols: A Guide to Methods and Applications" (M. A. Innis, D. H. Gelfand, J. J. Sninsky, and T. J. White, eds.), pp. 98–99. Academic Press, San Diego.

Lesch, M., and Nyhan, W. L. (1964). A familial disorder of uric acid metaboism and central nervous system function. *Am. J. Med.* **36,** 561–570.

Li, H., Gyllensten, U. B., Cui, X., Saiki, R. K., Erlich, H. A., and Arnhem, N. (1988). Amplification and analysis of DNA sequences in single human sperm and diploid cells. *Nature (London)* **335,** 414–417.

Li, H., Cui, X., and Arnheim, N. (1990). Direct electrophoretic detection of the allelic state of single DNA molecules in human sperm by using the polymerase chain reaction. *Proc. Natl. Acad. Sci. U.S.A.* **87,** 4580–4584.

Li, H., Xiangfeng, C., and Arnheim, N. (1991). Analysis of DNA sequnce variation in single cells. *Methods: Companion Methods Enzymol.* **2,** 49–59.

Liebaers, I., Sermon, K., Lissens, W., Liu, J., Devroey, P., Tarlatzis, B., and Van-Steirtegheim, A. (1992). Preimplantation diagnosis. *Hum. Reprod.* **7**, 107–110.

Liu, J., Lissens, W., Devroey, P., Van Steirtegheim, A., and Liebaers, I. (1992). Efficiency and accuracy of polymerase chain reaction assay for cystic fibrosis allele DF508. *Lancet* **339**, 1190–1193.

Longo, C. A., Beringer, M. S., and Harley, J. L. (1990). Use of uracil DNA glycosylase to control carry over contamination in polymerase chain reactions. *Gene* **93**, 125–128.

McLaren, A. (1984). Where to draw the line? *Proc. R. Inst. G.B.* **56**, 101–121.

McLaren, A. (1986). Embryo research. *Nature (London)* **320**, 570.

McLaren, A. (1987). Can we diagnose genetic disease in pre-embryos? *New Sci.* Dec. 10th 42–47.

McPherson, M. J., Quirke, P., and Taylor, G. R. (1991). "PCR: A Practical Pproach." IRL Press, Oxford.

Menezo, Y., Guerin, J., and Czyba, J. (1990). Improvements of human early embryo development *in vitro* by co-culture on monolayers of Vero cells. *Biol. Reprod.* **42**, 301–306.

Monk, M., and Holding, C. (1990). Amplification of a b-haemoglobin sequence in individual human oocytes and polar bodies. *Lancet* **335**, 985–988.

Monk, M., Handyside, A., Hardy, K., and Whittingham, D. (1987). Preimplantation diagnosis of deficiency of hypoxanthine phophoribosyl transferase in a mouse model for Lesch–Nyhan syndrome. *Lancet* **ii**, 423–425.

Monk, M., Muggleton-Harris, A. L., Rawlings, E., and Whittingham, D. G. (1988). Preimplantation diagnosis of HPRT-deficient male and carrier female mouse embryos by trophectoderm biopsy. *Hum. Reprod.* **3**, 377–381.

Monk, M., Handyside, A., Muggleton-Harris, A. L., and Whittingham, D. G. (1990). Preimplantation sexing and diagnosis of hypoxanthine phosphoribosyl transferase deficiency in mice by biochemical microassay. *Am. J. Med. Genet.* **35**, 201–205.

Morsy, M., Takeuchi, K., Kaufman, R., Veeck, L., Hodgen, G. D., and Beebe, S. J. (1992). Preclinical models for human pre-embryo biopsy and genetic diagnosis II. Polymerase chain reaction amplification of deoxyribonucleic acid from single lymphoblasts and blastomeres with mutation detection. *Fertil. Steril.* **57**, 437–438.

Muggleton-Harris, A. L. (1990). Proliferation of cells derived from the biopsy of pre-embryos. *In* "Advances in Assisted Reproductive Technologies (Mashiach, S., Ben-Rafael, Z., Laufer, N., and Schenker, J. G., eds), pp. 887–898. Plenum Press, New York.

Muggleton-Harris, A. L. (1992). The preimplantation mammalian embryo, its uses to derive cell lines and diagnose inherited genetic defects. *Treb. Soc. Cat Biol.* **42**, 163–175.

Muggleton-Harris, A. L., and Braude, P. R. (1993). Preimplantation diagnosis of genetic disease. *Curr. Opinion Obstet. and Gynecol.* **5**, 600–605.

Muggleton-Harris, A. L., and Findlay, I. (1991). *In vitro* studies on 'spare' human preimplantation embryos in culture. *Hum. Reprod.* **6**, 85–92.

Muggleton-Harris, A. L., Findlay, I., and Whittingham, D. G. (1990). Improvement of the culture conditions for the development of human preimplantation embryos. *Hum. Reprod.* **5**, 217–220.

Muggleton-Harris, A. L., Glazier, A. M., and Pickering, S. J. (1993). Biopsy of the human blastocyst and polymerase chain reaction (PCR) amplification of the β-globin gene and a dinucleotide repeat motif from 2-6 trophectoderm cells. *Hum. Reprod.* **8** (in press).

Mullis, K. B., and Faloona, F. A. (1987). Specific synthesis of DNA *in vitro* via a polymerase catalysed chain reaction. *In* "Recombinant DNA," Part F (R. Wu, ed.), Methods in Enzymology, Vol. 155, pp. 335–351. Academic press, San Diego.

Nasr-Esfahani, M., Johnson, M. H., and Aitken, J. R. (1990a). The effect of iron and iron chelators on the *in vitro* block to development of the mouse preimplantation embryo: BAT6 a new medium for improved culture of mouse embryos *in vitro*. *Hum. Reprod.* **5**, 997–1003.

Nasr-Esfahani, M. H., Aitken, J. R., and Johnson, M. H. (1990b). Hydrogen peroxide levels in oocytes and early cleavage stage embryos from blocking and non-blocking strains of mice. *Development* **109**, 501–507.

Nave, K. A., and Milner, R. J. (1989). Proteolipid proteins: structure and genetic expression in normal and myelin-deficient mutant mice. *Crit. Rev. Neurobiol.* **5**, 65–91.

Navidi, W., and Arnheim, N. (1991). Using PCR in preimplantation diagnosis. *Hum. Reprod.* **6**, 836–849.

Neufield, E. F. (1991). Lysosomal storage diseases. *Ann. Rev. Biochem.* **60**, 257–280.

Pickering, S. (1992). Preimplantation diagnosis coamplification of a polymorphic repeat sequence may reduce the possibility of misdiagnosis due to contamination. *Dev. Basis Inherited Disord., Br. Soc. Dev. Biol. Meet.* (Abstr.)

Pickering, S. J., Braude, P. R., McConnell, J. C., and Johnson, M. H. (1992). Reliability of detection by PCR of the sickle cell containing region of the beta-globin gene in single human blastomeres. *Hum. Reprod.* **7**, 630–636.

Pickering, S., Braude, P. R., McConnell, M., and Johnson, M. H. (1993). Analysis of human β-globin and CA repeat in isolated human blastomeres. Submitted for publication.

Pinkel, D., Straume, T., and Gray, J. M. (1986). Cytogenetic analysis using quantitative high sensitivity, fluorescent hybridisation. *Proc. Natl. Acad. Sci. U.S.A.* **83**, 2934–2938.

Plachot, M. (1992). Viability of preimplantation embryos. *Baillieres Clin. Obstet. Gynaecol.* **6**, 327–28.

Plachot, M., deGrouchy, J., and Montagut, J. (1987). Multicentric study of chromosome analysis in human oocytes and embryos in an IVF programme. *Hum. Reprod.* **2**, 29. (Abstr.)

Readhead, C., and Hood, L. (1990). The dysmyelinating mouse mutations shiverer (shi) and myelin deficient (shimld). *Behav. Genet.* **20**, 213–234.

Readhead, C., and Muggleton-Harris, A. L. (1991). The shiverer mouse mutation shi/shi: rescue and preimplantation detection. *Hum. Reprod.* **6**, 93–100.

Reid, T., Landes, G., Duckowski, W., Klinger, K., and Ward, D. (1992). Multicolour fluorescence in situ hybridisation for the simultaneous detection of probe sets for chromosomes 13, 18, 21, X and Y uncultured amniotic fluid cells. *Hum. Mol. Genet.* **1**, 307–313.

Roberts, C. J., Lutjen, U. B., Krzyminska, U. B., and O'Neill, C. (1990). Cytogenetic analysis of biopsied preimplantation mouse embryos: implication for prenatal diagnosis. *Hum. Reprod.* **5**, 197–202.

Roberts, R. G., Bentley, D. R., Barby, T. F. M., Manners, E., and Bobrow, M. (1990). Direct diagnosis of carriers of Duchenne and Becker muscular dystrophy by amplification of lymphocyte RNA. *Lancet* **336**, 1523–1526.

Robertson, E. J. (1987). Embryo derived cell lines. *In* "Teratocarcinomas and Embryonic Stem Cells, a Practical Approach" (E. J. Robertson, ed.), pp. 71–112. IRL Press, Oxford.

Robertson, J. A. (1992). Legal and ethical issues arising from the New Genetics. *J. Reprod. Med.* **37**, 521–524.

Saidi, R. K., Scharf, S., Faloona, F., Mullis, K. B., Horn, G. T., Erlich, H., and Arnheim, N. (1985). Enzymatic amplification of b-globin genomic sequences and restriction site analysis for diagnosis of sickle cell anaemia. *Science* **230**, 1350–1354.

Saiki, R. K., Gelfand, D. H., Stoffel, S., Scharf, S. J., Higuchi, R., Horn, G. T., Mullis, K. B., and Erlich, H. A. (1988). Primer-directed amplification of DNA with a thermostable DNA polymerase. *Science* **239**, 487–491.

Sardelli, A. D. (1993). Plateau effect—Understanding PCR limitations. Amplifications. (Perkin Elmer).

Schmiady, H., and Kentenich, H. (1993). Cytological studies of human zygotes exhibiting developmental arrest. *Hum. Reprod.* **8,** 744–751.

Schriver, C. R., Beaudet, A., Sly, W. S., and Volte, D., eds. (1989). "The Metabolic Basis of Inherited Disease," pp. 1623–1839. McGraw-Hill, New York.

Schurs, B. M., Winston, R. M. L., and Handyside, A. H. (1993). Preimplantation diagnosis of aneuploidy using fluorescent in situ hybridisation: evaluation using chromosome 18 specific probe. *Hum. Reprod.* **8,** 296–301.

Sheardown, S., Findlay, I., Turner, A., Greaves, D., Bolton, V., Mitchell, M., Layton, D. M., and Muggleton-Harris, A. L. (1992). Preimplantation diagnosis for an Hbs transgene in biopsied trophectoderm cells and blastomeres of the mouse embryo. *Hum. Reprod.* **7,** 1297–1303.

Simpson, J. L. (1992). Preimplantation genetics and recovery of fetal cells from maternal blood. *Curr. Opinion Obstet. Gynaecol.* **4,** 295–301.

Simpson, J. L., and Carson, S. A. (1992). Preimplantation genetic diagnosis. *N. Engl. J. Med.* **372,** 951–953.

Strom, C. (1992). Reliability of polymerase chain reaction for preimplantation genetic analysis. *Int. Symp. Preimplantation Genet. Assisted Fert., Brussels.* (Abstr.)

Strom, C., Verlinsky, Y., Milayeva, S., Eskov, S., Cieslak, J., Lifchez, A., Valle, J., Morse, J., Ginsberg, N., and Applebaum, M. (1990). Preconception genetic diagnosis of cystic fibrosis. *Lancet* **336,** p. 306.

Summers, P. M., Campbell, J. M., and Miller, M. W. (1988). Normal *in vivo* development of marmoset monkey embryos after trophectoderm biopsy. *Hum. Reprod.* **1,** 89–94.

Takeuchi, K., Sandow, B. A., Morsy, M., Kaufmann, R. A., Beebe, S. J., and Hodgen, G. D. (1992). Preclinical models for human pre-embryo biopsy of genetic diagnosis. 1. Efficiency and normalcy of mouse pre-embryo development after different biopsy techniques. *Fertil. Steril.* **57,** 425–430.

Tarin, J. J., Conaghan, J., Winston, R. M., and Handyside, A. H. (1992). Human embryos biopsy on the 2nd day after insemination for preimplantation diagnosis: removal of a quarter of embryo retards cleavage. *Fertil. Steril.* **58,** 970–976.

Trounson, A. O. (1992). Editorial: Preimplantation genetic diagnosis—counting chickens before they hatch? *Hum. Reprod.* **7,** 583–584.

Varawalla, N. Y., Dokras, A., Old, J. M., Sargent, I. L., and Barlow, D. H. (1991). An approach to preimplantation diagnosis of beta-thalassaemia. *Prenatal Diagn.* **11,** 775–785.

Verlinsky, Y., and Kuliev, A. (1992). Micromanipulation of gametes and embryos in preimplantation genetic diagnosis and assisted fertilisation. *Curr. Opinion Obstet. Gynaecol.* **4,** 720–725.

Verlinsky, Y., Ginsberg, N., Lifchez, A., Valle, J., Moise, J., and Strom, C. I. N. (1990). Analysis of the first polar body: preconception genetic diagnosis. *Hum. Reprod.* **5,** 826–829.

Verlinsky, Y., Cieslak, J., Evsikov, S., Milayeva, S., Strom, C. M., White, M., Valle, J., Moise, J., Ginsberg, N., and Applebaum, M. (1991). Effect of subsequent oocyte and blastomere biopsy on preimplantation development. *Hum. Reprod.* **6,** 136.

Verlinsky, Y., Rechitsky, S., Evsikov, S., White, M., Cieslak, J., Lifchez, A., Valle, J., Moise, J., and Strom, C. M. (1992). Preconception and preimplantation diagnosis for cystic fibrosis. *Prenatal Diagn.* **12,** 103–110.

Warnock, M. (1987). Do human cells have rights? *Bioethics* **1,** 1–14.

Weatherall, D. J. (1991). "The New Genetics and Clinical Practice." Oxford Med. Publ., Oxford.

Weimer, K. E., Hoffman, D. I., Maxson, W. S., Eager, S., Mahlberger, B., Fiore, I., and Cuervo, M. (1993). Embryonic morphology and rate of implatation of human embryos following co-culture on bovine oviductal epithelial cells. *Hum. Reprod.* **8,** 97–101.

West, J. D., Gosden, J. R., Angell, R. R., West, K. M., Glasier, A. F., Thatcher, S. S., and Baird, D. T. (1988). Sexing whole human pre-embryos by in-situ hybridization with a Y-chromosome specific DNA probe. *Hum. Reprod.* **3**, 1010–1019.

Wilton, L. J., and Trounson, A. O. (1989). Biopsy of preimplantation mouse embryos: Development of micromanipulated embryos and proliferation of single blastomeres. *Biol. Reprod.* **40**, 145–152.

Winston, N. J., Johnson, M. H., Pickering, S. J., and Braude, P. R. (1991a). Parthenogenetic activation and development of fresh and aged human oocytes. *Fertil. Steril.* **56**, 904–912.

Winston, N. J., Braude, P. R., Pickering, S. J., George, M. A., Cant, A., Currie, J., and Johnson, M. H. (1991b). The incidence of abnormal morphology and nucleocytoplasmic ratios in 2, 3, and 5 day human pre-embryos. *Hum. Reprod.* **6**, 17–24.

Wu, R., Decorte, R., Swimnen, K., Gorilts, S., Margnen, P., and Cassiman, J. J. (1992). CF mutation detection and sex determination in single cells by PCR in combination with fluorescent in situ hybridisation (FISH). A combined approach for preimplantation diagnosis. *Int. Symp. Preimplantation Genet. Assisted Fert., Brussels.* (Abstr.)

Zhang, L., Cui, X., Schmitt, K., Hubert, R., Navidi, W., and Arnheim, N. (1992). Whole genome amplification from a single cell. Implications for genetic analysis. *Proc. Natl. Acad. Sci. U.S.A.* **89**, 5847–5851.

Structure and Function of the Cyanelle Genome

Wolfgang Löffelhardt* and Hans J. Bohnert†
*Institut für Biochemie und Molekulare Zellbiologie der Universität Wien und Ludwig-Boltzmann-Forschungstelle für Biochemie, A-1030 Vienna, Austria, and †Department of Biochemistry, The University of Arizona, BioSciences West, Tucson, Arizona 85721

I. Introduction

Among the photosynthetic eukaryotes, those that contain chloroplasts— the green algae, mosses, ferns, and higher plants—are but one group of plastid-containing organisms. Different forms of nuclear and plastid genome organization can be found in a number of other algal groups. As more information about the genetic organization of these plastids in different phyla, for example, chromophyta and rhodophyta, becomes known, better-educated guesses can be made about plastid evolution and we can begin to understand the many-faceted metabolic interactions and regulatory connections between the nucleocytoplasmatic and plastic compartments. Comparative molecular biology of plastids of different algae will greatly enhance our understanding of the evolution of photosynthesis. The view of plastid acquisition by an endosymbiotic event, involving an ancestral photosynthetic prokaryote and a heterotrophic eukaryote, is now widely accepted (Margulis, 1981; Gray, 1989). Questions that remain to be answered are: (1) How often did such an endosymbiotic event occur? (2) Were different types of prokaryotic ancestors involved? (3) What was the nature of the evolutionary pressure governing the reduction of the endosymbiont's genome? (4) How did control systems evolve for the genetic and metabolic integration of the former endosymbiont into the cell?

In this chapter we wish to introduce a group of organisms that contain a special type of plastids—cyanelles with cyanobacteria-like morphology and phycobilisomes. What makes these plastids—and the organisms within which they are found—unique is that cyanelles, like cyanobacteria, are surrounded by a peptidoglycan wall. In addition, cyanelles resemble

29

cyanobacteria in their pigment composition and in possessing a carboxysome-like structure. Such endosymbiotic associations between a eukaryote and a blue-green bacterium are sometimes called "endocyanomes." Equally remarkable is the fact that cyanelles have been detected in different eukaryotic cells which are obviously unrelated evolutionarily and systematically. For example, an ameboid "host" containing cyanelles, *Paulinella chromatophora,* has been described (Kies, 1984, 1992), while other cyanelles have been detected in host cells that may represent either red algal, diatomaceous, dinoflagellate, or cryptomonad forms of organization (Kies, 1992).

It is remarkable in a historical sense that cyanelles were at one time considered cyanobacterial endosymbionts of eukaryotic hosts (Pascher, 1929), requiring classification by family and species names (Hall and Claus, 1963). This view became untenable once the limited genome size of the cyanelle found in *Cyanophora paradoxa* was recognized (Herdman and Stanier, 1977). It appears reasonably, although it has never been proven, to assume a genetic interdependence for the other "endocyanomes" also. One exception is *Geosiphon pyriforme,* in which the associated cyanobacteria have been shown to enter the host cell but are also viable outside the host (Mollenhauer, 1992).

The historical term "cyanelle," while it may not be completely satisfactory, has over the past several years come to precisely describe a unique type of plastid genome organization, structure, and metabolic interaction. Maintaining the term "cyanelle" seems preferable since these organelles have preserved more cyanobacterial features than, for example, rhodoplasts. The type specimen is *Cyanophora paradoxa* (Korschikoff, 1924). Two isolates of *Cyanophora* are known which differ in the organization of their cyanelle DNA (Löffelhardt *et al.,* 1983; Breiteneder *et al.,* 1988; see the following discussion). Reflecting the difficulties in systematic, evolutionary, and molecular description of cyanelle-bearing organisms, various other names for these organelles have been proposed to incorporate new views or recent findings. We suggest that the established name be maintained. It is to be expected that in-depth analysis of cyanelle genomes of the less well studied endocyanomes or plastid genomes of other algal classes may uncover even more exotic forms of plastid organization and in a short time additional myopic schemes for distinguishing increasingly different cyanelle types would have to be established.

In this chapter we wish to point out the importance of *Cyanophora* cyanelles for our comprehension of plastid evolution. We summarize recent results in cyanelle biochemistry, wall structure, and genome organization, with special emphasis on the latter. Previous reviews on *C. paradoxa* covering evolution and systematics, growth and cyanelle division, physiology and biochemistry, including molecular biology, are listed for reference (Trench, 1979, 1982a,b; Reisser, 1984; Kies, 1979, 1984; Coleman, 1985; Wasmann et al., 1987; Schenk, 1990; Bohnert and Löffelhardt, 1992; Kies, 1992; Schenk, 1992).

II. *Cyanophora paradoxa* and Its Cyanelles

In recent years investigations of the biochemical functions of *Cyanophora* cyanelles, and of cytosol-organelle interactions, have received less attention than their molecular genetic aspect. We have previously reviewed the work that has been carried out up to approximately 1986 (Wasmann et al., 1987). Since then, several aspects of cyanelle metabolism have been studied by a number of groups (Bayer and Schenk, 1986; Zook et al., 1986; Zook and Schenk, 1986; Schenk et al., 1985, 1987; Burnap, 1987; Burnap and Trench, 1989a–c; Schlichting et al., 1990; Betsche et al., 1992). These aspects include the transport of cytoplasmically synthesized proteins into cyanelles, nitrate assimilation, metabolite transport, lipid composition and biosynthesis, and enzymes of the photorespiratory pathway. We include here a short discussion of the new developments in this area which provide an excellent illustration of differences among cyanelles, chloroplasts, and extant cyanobacteria. While there is a trend to rely more on sequences and sequence comparisons to estimate relationships and evolution, it must be emphasized that metabolism is the ultimate selection principle which determines ecological niches and the success of a species.

Burnap (1987), Burnap and Trench (1989a–c), and Schenk and collaborators, using differential inhibition of the cyanelle and cytosol translation machineries and protein analysis, have provided much useful data on the contribution of the nuclear and cyanelle genetic systems in building up the cyanelle protein complexes. Jehn and Zetsche (1988) studied protein synthesis capacity in isolated cyanelles and after *in vitro* translation of cyanelle RNA. The fatty acid pattern in *Cyanophora paradoxa* was found to be highly unusual, containing predominantly $16:0$, $20:3$, and $20:4$ unsaturated fatty acids, with the $20:4$ predominately found in cyanelle membrane fractions (Kleinig et al., 1986). In a different cyanelle-bearing organism, *Glaucocystis nostochinearum*, Scott (1987) studied the major acyl lipids.

Cyanophora cyanelles, especially in the stationary phase of growth, develop a characteristic protein inclusion body surrounded by thylakoid membranes, which resembles a cyanobacterial carboxysome (Blank, 1985). This structure contains most of ribulose 1,5-bisphosphate carboxylase/oxygenase enzyme (rubisco) of the cyanelle (Mangeney and Gibbs, 1987) whereas phosphoribulokinase is located in the stroma in a soluble form (Mangeney *et al.*, 1987). This means that, as in cyanobacteria, ribulose 1,5-bisphosphate must be shuttled into the structure. This is an interesting aspect since in the past few years cyanobacterial carboxysomes have been recognized as structures with the biochemical function of concentrating CO_2, thereby enhancing the carboxylation reaction of rubisco (Kaplan *et al.*, 1991). From an analysis of cyanobacterial mutants that lack carboxysomes—which are dependent on a high concentration of carbon dioxide for growth—it is apparent that the overall CO_2 concentration is not affected. Rather, the effective CO_2 concentration at the carboxylation site inside the carboxysome appears to be two orders of magnitude lower than in wild-type cyanobacterial cells (Friedberg *et al.*, 1989; Price and Badger, 1989). Gaseous CO_2 appears to be concentrated inside the carboxysome, which is surrounded by a specific proteinaceous membrane. Cyanelles appear to lack this specific membrane (Blank, 1985). It is thus doubtful whether the cyanelle structure functions as a carboxysome.

The lack of this function in cyanelles is in accord with the lack of a carbon dioxide concentration mechanism (A. Goyal and N. E. Tolbert, personal communication) and hence poor assimilation power (unless a different carbon dioxide pump exists which has not been detected as yet). It appears significant that we have not found any of the genes in cyanelles that have been found to be essential for the buildup of a functional carboxysome in cyanobacteria (Schwarz *et al.*, 1992). This also supports the view of the inclusion body as a defective or deficient carboxysome. Pertinent to our view of the cyanelle, Betsche *et al.* (1992) conducted experiments that measured the enzymes of the photorespiratory pathway in *Cyanophora* plastids. The results indicated that a rudimentary pathway existed: several isoforms of hydroxypyruvate reductase and glycolate oxidase were detected, but their activity was only approximately one tenth as high as that of a leaf cell. The authors suggest that the pathway in cyanelles may deviate from the leaf pathway.

One remarkable aspect of cyanelle to cytosol interaction is the inability to maintain physiologically functional cyanelles after isolating them from the cells. Although such isolated organelles appear structurally intact for extended periods of time, certainly much longer than isolated chloroplasts from higher plants, oxygen evolution and other physiological functions decay much more rapidly than similar functions in isolated chloroplasts (Wasmann *et al.*, 1987). It is unclear why this should be the case, considering the protective peptidoglycan wall. We suspect a problem in metabolite transport, such as the inability of cyanelles to retain the metabolites necessary for maintaining an energized state, indicating a type of integration into the cell that is different from that of higher plant plastids. In this context, a recent investigation on the complement of metabolite carriers in the cyanelle envelope indicated remarkable differences to chloroplasts, for example, the lack of a dicarboxylic acid shuttle and an adenosine diphosphate/adenosine triphosphate (ADP/ATP) translocator (Schlichting *et al.*, 1990). Furthermore, it was shown in this study that polyols such as sorbitol penetrate the cyanelle envelope, which is also different than in chloroplasts. This fact clearly necessitates a redesign of the isolation media for cyanelles used to date which are principally media adapted from the work done with chloroplasts. Following this rationale, an isolation medium was generated that relies on glycine betaine as the osmoticum (Schlichting *et al.*, 1990), leading to the isolation of cyanelles that are competent for protein import experiments (C. Neumann-Spallart, personal communication).

A link between biochemistry and molecular genetics has recently been provided by the detection of a nucleus-encoded, cyanelle-located enzyme (Bayer *et al.*, 1990). Ferredoxin-NADP$^+$-oxidoreductase (FNR) was shown to be encoded by a nuclear gene (Bayer *et al.*, 1990; Schenk, 1992), as in higher plants. It is synthesized in the cytosol as a preprotein that must be imported into cyanelles. The gene has recently been detected in an expression-type *Cyanophora* cDNA library (Jakowitsch *et al.*, 1993) with the help of FNR antibodies (Bayer *et al.*, 1990). Import of the pre-FNR protein into the cyanelle appears to be due to a transit peptide sequence (Jakowitsch *et al.*, 1993; see the following discussion), which most likely functions like transit peptides in the import pathway into higher plant chloroplasts (Keegstra *et al.*, 1989). The FNR precursor is imported *in vitro* into isolated cyanelles as well as into isolated pea plastids (J. Jakowitsch and C. Neumann-Spallart, unpublished observations).

The most intriguing feature of cyanelles is that although they reside in the cytoplasm of an eukaryotic host, they are surrounded by a bacterial cell wall. The presence of peptidoglycan has been demonstrated through biochemical analysis in two species which contain cyanelles, *C. paradoxa* (Aitken and Stanier, 1979) and *Glaucocystis nostochinearum* (Scott *et al.*, 1984). All components of an A1γ-type murein, namely *N*-acetylglucosamine, *N*-acetyl muramic acid, L-alanine, D-alanine, D-glutamate, and *m*-diaminopimelic acid, have been detected in the appropriate ratios. This explains the lethal effect of β-lactam antibiotics on these obligatorily autotrophic organisms (Berenguer *et al.*, 1987).

In addition, Kies (1988) reported that growth in the presence of penicillin resulted in the release of *C. paradoxa* cyanelles into the medium, whereas the cyanelles from *G. nostochinearum* and *Gloeochaete wittrockiana* appeared to be degraded within the cells. Kies suggested that the cyanelle outer envelope membrane (OEM) might originate from a host cell vacuole that might turn into a lysosome when the cyanelle is damaged. Only the integrity of the cyanelle wall prevented digestion of the organelle. This, in turn, would have to be interpreted to mean that integration of cyanelles into the *Cyanophora* cell represented a precarious, unstable equilibrium. Such a view as a predator–invader relationship, however, does not easily explain why the majority of cyanelle genes are located in the nuclear DNA, which must mean that the association persisted over long periods of time.

An alternative view is that the cyanelle OEM is derived from the outer membrane of the ancestral cyanobacterial invader and that its observed instability (Giddings *et al.*, 1983) might be due to the lack of lipoprotein (Höltje and Schwarz, 1985) connecting it to the murein layer. The necessary presence of receptors for the import of approximately 800 nucleus-encoded cyanelle polypeptides would be difficult to reconcile with a vacuolar nature of the cyanelle OEM. A final decision between these two possibilities must await analysis of the lipid and protein composition of isolated cyanelle OEM. In addition, understanding the routing of proteins into the cyanelle should provide important data, probably more important than sequence comparisons, on the evolution of *Cyanophora* cyanelles.

Berenguer *et al.* (1987) demonstrated the presence of seven penicillin-binding proteins (PBPs), ranging in size from 110 to 30 kDa, in the cyanelle envelope. Their distribution and binding characteristics for various β-lactam antibiotics suggest that they should be similar to the well-characterized PBPs from *Escherichia coli* (Höltje and Schwarz, 1985), although immunological relatedness might be low. Heterologous hybridizations of PBP gene probes from *E. coli* with cyanelle DNA proved unsuccessful, which might mean that the majority of the PBPs are encoded in the nuclear genome. Several cDNA clones isolated with the aid of antisera directed against affinity-enriched cyanelle PBPs have been obtained and are being analyzed at present (M. Kraus, personal communication). Evidence for the biosynthesis of the soluble precursor uridine-diphosphate-N-acetylmuramyl pentapeptide in the cyanelle stroma through a pathway analogous to that in *E. coli* has been obtained (Plaimauer *et al.*, 1991). In this study, a location in the cyanelle periplasm was suggested for enzymes acting on an already-polymerized peptidoglycan like DD- and LD-carboxypeptidase and endopeptidase. This necessitates the existence of a carrier lipid, undekaprenyl phosphate, in the cyanelle inner envelope membrane (IEM) for the translocation of the soluble precursor to the periplasm, thus creating the membrane-bound activated substrate for the PBPs (Höltje and Schwarz, 1985).

Considering the structure of a peptidoglycan-containing cell envelope, results from *E. coli* showed that the peptidoglycan in the periplasmic space is sandwiched between an outer and an inner membrane, and that the PBPs reside at the outer surface of the inner membrane (Park, 1987). An intriguing question is how the targeting of the PBPs and the other periplasmic enzymes involved in cyanelle wall metabolism is achieved. Are they transported directly across the OEM, like mitochondrial cytochrome c, assuming that the corresponding genes are nuclear? Or is import of cytoplasmic precursors into the cyanelle stroma followed by re-export across the IEM, as in the case of mitochondrial cytochromes c_1 and b_2 (Segui-Real et al., 1992)? Or are some enzymes exported from the cyanelle, meaning that part of the corresponding genes are located in the cyanelle?

Considerable progress has been made in understanding the structure of the cyanelle wall. The 16 major muropeptides obtained after muramidase cleavage of purified peptidoglycan (Höltje and Schwarz, 1985) were isolated by preparative high-performance liquid chromatography (HPLC) and subjected to amino acid analysis and molecular weight determination through plasma-desorption mass spectrometry (Pfanzagl et al., 1993). Five cyanelle muropeptides proved to be identical to monomers, dimers, and a trimer known from E. coli. Eleven muropeptide units showed a constant molecular weight increment of 112 compared with the respective counterparts from E. coli. This substituent was recently identified as N-acetylputrescine, which formed an amide bond with the α-carboxy group of D-glutamate (Pittenauer et al., 1993). The amount of this modification appeared to vary with the growth stage, being most pronounced in the early logarithmic phase of culture growth (Pfanzagl et al., 1993). Such a substitution has been found in prokaryotic cell walls in only a few cases. Cadaverine and putrescine have been reported as substituents at the analogous locus in murein from a small number of strict anaerobic gram-negative cocci that group with gram-positive bacteria upon phylogenetic analysis (Schleifer et al., 1990). It should be noted that cyanobacteria are also gram-negative with some characteristics of gram-positive bacteria. A preliminary search for N-acetylputrescine in the peptidoglycan from the cyanobacterium Synechococcus sp. was negative (U. J. Jürgens, personal communication). Unfortunately, there are no data on the fine structure of cyanobacterial peptidoglycans. A comparison of cell wall architecture in this diverse group with that of the unique organelle wall of C. paradoxa is certainly needed.

III. Chloroplast Genomes from Higher Plants and Green Algae

The structure of chloroplast DNA was reviewed recently (Sugiura, 1992). All our comparisons referring to chloroplast gene structure are based on this compilation of the chlorophyll b plastid lineage. Chloroplast DNA from land plants is very similar in structure and gene content. It encodes all the rRNA and tRNA needed for the organelle-inherent translation system, 21 ribosomal proteins, and 4 subunits of an RNA polymerase. Chloroplasts contribute genetic information for more than half (29) of their thylakoid membrane polypeptides involved in photosynthesis and for 11 subunits of the NADH dehydrogenase complex (ndh) whose function in the organelle remains enigmatic. The majority of the remaining approximately 30 protein genes represent open reading frames (ORF) of still-unknown function. In general, chlorophyte plastomes are characterized by the lack of genes, the products of which function in anabolic pathways. In this context it must be remembered that, in the plant cell, the chloroplast is responsible not only for the biosynthesis of carbohydrates but also for the synthesis of amino acids, fatty acids, isoprenoid lipids, and for nitrogen and sulfur assimilation. Only the gene *rbcL*, encoding the large subunit of rubisco, encodes a protein involved in such an anabolic pathway in higher plant chloroplasts.

Knowledge about chloroplast DNA from chlorophyll *b*-containing algae is less developed, with the exception of *Euglena gracilis* and *Chlamydomonas reinhardtii*. They seem to lack *ndh* genes but harbor a number of genes not encountered on the completely sequenced chloroplast genomes of rice and tobacco, for example, *tufA* and *frxC* (Baldauf and Palmer, 1990; Sugiura, 1992). The sequence of the *E. gracilis* plastid genome has now been completed. A conspicuous feature is the vast number of introns present in protein genes but never in tRNA genes. Some genes are present that are not found in the plastid genomes from other chlorophytes, such as *rpl5*, *rpl12*, and *ccsA*, a gene essential for chlorophyll biosynthesis (Hallick *et al.*, 1993; Orsat *et al.*, 1992). The rDNA spacer resembles that from plastids and cyanelles lacking chlorophyll *b* rather than that from green algae. An explanation for these discrepancies is the presumed origin of euglenophytes from a secondary endosymbiotic event involving an eukaryotic invader, probably a primitive red alga (Reith and Munholland, 1993a). This might mean that features of plastid genome organization are more important phylogenetic traits than, for example, pigment composition.

The plastid genome from *C. reinhardtii* offers an opportunity for combining data from classical and molecular genetics. This plastid genome was the first amenable to transformation by the biolistic approach (Boynton *et al.*, 1988), providing an excellent system for further analysis of plastid functions. The high degree of rearrangements observed in *Chlamydomonas* plastid DNA results in the disruption of many transcription units known from cyanobacterial genomes and the plastid genomes of higher plants (Harris, 1989). Thus the *C. reinhardtii* plastome, though interesting per se, is not very suitable for comparisons of plastid genome organization with phylogenetic implications.

Recently data on the plastid genome organization of charophytic algae have become available. Charophytes are probably the closest relatives to land plants. The details emerging on plastid genome organization in fact support a view for charophytes as progenitors of land plants. In some species *tufA* is a nuclear gene (Baldauf *et al.*, 1990). The tRNA genes in the rDNA spacer contain introns (Manhart and Palmer, 1990) and the *rps12* mRNA is produced through *trans*-splicing (J. R. Manhart, personal communication). All of these characteristics are analogous to the corresponding features of land plants (Sugiura, 1992).

IV. Plastid Genomes from Algae Lacking Chlorophyll *b*

Plastid DNA from red algae, brown algae, and cryptomonads has been studied by several groups (Reith and Munholland, 1993a; Valentin *et al.*, 1992a; Douglas, 1992b). In all but one case one plastid DNA species in the size range of chlorophytic plastomes was characterized. In the brown alga *Pylaiella littoralis* (Loiseaux-de Goër *et al.*, 1988), two circular chromosomes harboring a different set of genes have been obtained from the plastids. Differences in gene complement, relative to the chlorophyll *b* lineage, are found in all these algal phyla (Table I). The small subunit of rubisco is always plastid-encoded; genes for two additional subunits of the ATP synthase and a significantly higher number of ribosomal proteins have been found in several groups. About 60% of the plastid genome from the red alga *Porphyra purpurea* has been sequenced (Reith and Munholland, 1993a). More than 50 genes not encountered on the plastid genomes of higher plants have been reported. Included among these novel genes are a number of genes for enzymes involved in anabolic pathways, for example, the biosynthesis of amino acids and fatty acids (Table I). To date, only one intron has been reported for any plastid gene of an alga lacking chlorophyll *b*: that in *rpeB* (phycocyanin β-subunit) from *Rhodella violacea* (Bernard *et al.*, 1992). In general, intergenic regions are very short; the inverted repeat (IR) is smaller in size than in higher plants or it is completely absent. Together with the very low frequency of introns, these features lead to a higher number of genes without an increase in the size of the plastid genomes.

V. Organization of the Cyanelle Genome

The size of cyanelle DNA from *C. paradoxa* is comparable to that of chloroplast DNA in higher plants, approximately 130 and 140 kbp, respectively, in the two strains known (Löffelhardt *et al.*, 1983). The chromo-

TABLE I

Unusual Genes on Algal Plastid DNAs

Species	Gene	Function/Homology	References
Porphyra purpurea	*psaE*	Component of PS II	Reith (1992)
Cyanidium caldarium	*ompR*	Transcription factor	Kessler *et al.* (1992)
Porphyridium aerugineum			
C. caldarium	*groEL*	Chaperonin	Maid *et al.* (1992)
Odontella sinensis	*atpD*	CF1, δ-subunit	Pancic *et al.* (1991)
O. sinensis	*atpG*	CF0, subunit II	Pancic *et al.* (1992)
C. caldarium			Valentin *et al.* (1992a)
Antithamnion sp.			Valentin *et al.* (1992a)
Cryptomonas Φ	*hlpA*	Histone-like protein	Wang and Liu (1991)
Pavlova lutherii	*hsp70*	Chaperonin	Scaramuzzi *et al.* (1992a)
P. purpurea			Reith and Munholland (1991)
Cryptomas Φ			Wang and Liu (1991)
C. caldarium	*cpcL*	Phycobilisome linker	Valentin *et al.* (1992b)
C. caldarium	*apcL*	Phycobilisome linker	Valentin *et al.* (1992b)
Cylindrotheca sp.	*apcA*	Acyl carrier protein	Hwang and Tabita (1991)
Cryptomonas Φ			Wang and Liu (1991)
P. purpurea	*ilvB*	Acetolactate synthase	Reith and Munholland (1993b)
P. purpurea	*argB*	*N*-acetylglutamate kinase	Reith and Munholland (1993b)
Euglena gracilis	*ccsA*	Chlorophyll biosynthesis	Orsat *et al.* (1992)

some contains an IR, as it is present in many other plastid DNAs. The size of the cyanelle IR is approximately 10.5 kb. The two segments of this repeat separate the small single-copy (SSC) region (approximately 17 kb) from the large single-copy (LSC) DNA region (approximately 90 and 100 kb, respectively, for the two strains). The strain used in most of the molecular genetic studies described here is that originally isolated by Pringsheim and deposited in culture collections of algae around the world (e.g., LB 555 UTEX; strain 2980 of the Algal Culture Collection, Göttingen). The second strain (no. 1555) was isolated by Kies (University of Hamburg). In spite of high overall sequence homology, the restriction pattern of cyanelle DNA from the Kies strain is completely different from that of the Pringsheim strain. However, 16 protein genes mapped to analogous positions on the cyanelle genome of both strains. Thus, differences in restriction fragment patterns can be ascribed to numerous small insertions and/or deletions and inversions in intergenic regions (Breiteneder *et al.*, 1988). The focus on cyanelle gene organization, which will be reviewed below, should eventually be supplemented by including studies on the genetic system of the host. Using fluorescence-assisted sorting of mith-ramycin- or DAPI-stained nuclei (De Rocher *et al.*, 1990) we measured DNA amounts per nucleus (Genome sizes) of the two strains. Both have indistinguishable DNA amounts, equivalent to 0.27 +/- 0.04 pg DNA/nucleus. This value might be the C or $2C$, or an even a higher C value, since nothing is known about this aspect of the life of *Cyanophora*. To put this DNA amount into perspective: a C value of approximately 0.2 pg/nucleus would be nearly identical to the corresponding value for *Arabidopsis thaliana*. The nuclear rDNA units from both isolates have been cloned and compared. The intergenic spacer of the Kies isolate is larger (7.4 kb versus 4.3 kb) than that of the Pringsheim isolate and also shows a different restriction pattern. The rRNA coding regions and the transcribed spacers from both isolates are indistinguishable with respect to these criteria (D. Aryee, unpublished observations). Such a diversity in plastome restriction pattern and in structure of nuclear rDNA units exceeds that observed between two strains of the same species and might justify a renaming of the two isolates of *C. paradoxa*.

A. rRNA Genes

Cyanelle DNA from both strains known contain the rRNA genes within the 10.5 kb IR (Breiteneder *et al.*, 1988). Each repeat includes one operon of rRNA genes, which are arranged in the following order: 5'–16S–*trn*A–*trn*I–23S–5S–3', as in chloroplasts (Sugiura, 1992). The rDNA unit occupies approximately half of the IR, 16S rDNA being located at the border facing the SSC region (Fig. 1), a feature which is only paralleled in the IR of *Chlamydomonas reinhardtii*.

Unlike higher plant chloroplast tRNA genes in this operon, the cyanelle counterparts lack introns (Janssen *et al.*, 1987). The rRNA genes are transcribed as one large primary transcript. We have sequenced the central part of this region mainly to learn about the 3'-terminal end of the 16S rRNA, which has a functional role in the recognition of start sites of translation (Bonham-Smith and Bourque, 1989). The cyanelle 16S rRNA 3' end shows a sequence complementary to ribosome binding sites (rbs) in bacterial mRNA. Since sequences similar to such rbs, which may base-pair with the 16S rRNA, are found at the 5' end of most cyanelle transcripts close to either ATG (methionine) or GTG (valine) start codons (see later discussion), it is safe to assume that these sites are actually used, although

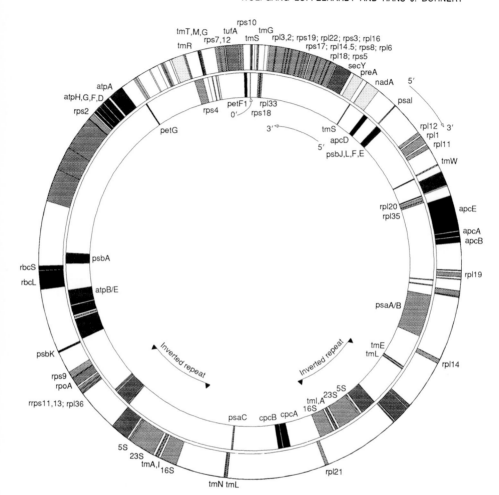

FIG. 1   A gene map of the chromosome of the cyanelle from *Cyanophora paradoxa* 55UTEX, approximately 127,000 bp, is shown. In general only published genes are identified by gene names, with the exception of the genes for ribosomal proteins and ATP synthase subunits, where all genes known at present are shown. Unpublished genes in (partially) sequenced portions of the chromosome are indicated by shaded regions without denomination. The direction of transcription is indicated throughout. Genes encoding subunit proteins in a pathway are marked by identical shading.

the final proof has not been provided. In fact, cyanelle rbs are recognized as binding sites in *E. coli* (Michalowski *et al.*, 1991a; R. Flachmann, unpublished observations).

About 1 kb of the 16S rRNA gene has been sequenced by Giovannoni *et al.* (1988) who, upon comparing the corresponding sequences from other organisms, arrived at the conclusion that cyanelles are equidistant from extant cyanobacteria and chloroplasts. Analogous results were obtained when the 5'-terminal 512 bp of the cyanelle 23S gene were compared with the corresponding sequences from cyanobacteria and plastids (Janssen *et al.*, 1987). As was also observed for cyanobacteria, cyanelle 23S rRNA shows hidden nicking (Trench, 1982a) which causes the appearance of several substoichiometric, degraded RNA species on denaturing gels. These nicks appear to be generated *in vivo* upon the assembly of the ribosome (Marsh, 1979). Sequences for 5S RNA were obtained at the RNA level and revealed distinct cyanobacterial signatures in secondary structures (Maxwell *et al.*, 1986).

B. tRNA Genes

The 4S RNA fraction obtained from isolated, sucrose gradient-purified cyanelles has been resolved by two-dimensional polyacrylamide acid (PAA) gel electrophoresis, yielding about 40 RNA species. Of these, 29 RNAs were identified as cyanelle tRNAs by aminoacylation using *E. coli* aminoacyl-tRNA synthetases (Kuntz *et al.*, 1984). Single tRNA species were found for 7 amino acids, 2 isoacceptors for 6 amino acids, 3 for two amino acids, and 4 for leucine. The tRNAs for cysteine, glutamic acid, glutamine, and glycine could not be identified by that method. Using the individual tRNAs as well as heterologous chloroplast probes in hybridizations with cyanelle DNA, we measured the tRNA gene complement (26 genes) in cyanelles. Twenty-one tRNAs mapped to the LSC region, 3 to the SSC region, and 2 to the IR (Kuntz *et al.*, 1984).

Since then, 14 tRNA genes have been characterized by sequencing, 2 of these being located in the IR. The duplicated genes $trnI^{GAU}$ and $trnA^{UGC}$ were the first to be analyzed during sequencing of the rDNA spacer (Janssen *et al.*, 1987). The tRNAs can be folded into the common cloverleaf structure and require post-transcriptional addition of the 3'-terminal -CCA-OH. The small size of the rDNA spacer (287 bp) and the very short distance (3 bp) between the spacer tRNAs are in excellent agreement with data obtained for plastids from chromophytes (Markowicz *et al.*, 1988; Delaney and Cattolico, 1989), rhodophytes (Maid and Zetsche, 1991), and *Cryptomonas* Φ (Douglas and Durnford, 1990).

Two out of the three tRNA genes predicted for the SSC region were recently sequenced: $trnN^{GUU}$ and $trnL^{UAG}$. They are clustered upstream of *psaC*, but are not cotranscribed with this gene (Rhiel *et al.*, 1992). In the LSC region, several of the mapped loci did reveal tRNA genes upon sequencing. In addition, new genes were identified, for example, for glutamic acid and glycine. In the vicinity of one IR adjacent to $trnE^{UUC}$, Evrard *et al.* (1988) observed a gene for a leucine tRNA with the anticodon UAA, which is split by a 232-bp class I intron—the only intron found to date in any cyanelle gene. Thus far it has not been possible to demonstrate a self-splicing activity for this intron.

It is interesting that this leucine tRNA gene also contains an intron in cyanobacteria (Xu *et al.*, 1990; Kuhsel *et al.*, 1990). The presence of this cyanobacterial intron in the corresponding plastic genes suggests that at least some introns represent old features predating the prokaryote to eukaryote transition. In the central part of the LSC region, *trnS*^{GGA} and *trnG*^{GCC} were localized between *petF1* and *rps18/rpl33* (Kuntz *et al.*, 1988). The two tRNA genes are transcribed independently since the ORF65 that separates them yields a monocistronic RNA of 400 b (Evrard *et al.*, 1990c). Michalowski *et al.* (1991b) identified a *trnS*^{UGA} gene yielding a transcript of approximately 100 b between the genes for a putative prenyltransferase and *nadA* (Fig. 1). Recently, *rps4* was found flanked by four tRNA genes: *trnG, trnM, trnT,* and *trnR;* and *trnW* was detected in the vicinity of the L11 and L10 operons (C. B. Michalowski, unpublished observations).

C. Genes for Proteins of the Translation Apparatus

Using heterologous probes from the large ribosomal protein gene cluster of higher plant chloroplasts (Zhou *et al.*, 1989; Markmann-Mulisch and Subramanian, 1988), several homologous genes were found in one region of the cyanelle chromosome (Löffelhardt *et al.*, 1990). Evrard *et al.* (1990a,b) and Michalowski *et al.* (1990b) determined the structure of this portion of the cyanelle genome, which contained a larger number of genes than higher plant plastids, equivalent to the eubacterial S10 and *spc* operons (Lindahl and Zengel, 1986). For one of these novel plastid genes, *rpl3*, the product could be detected in cyanelles and in transgenic *E. coli* through the use of heterologous antibodies directed against ribosomal protein L3 from *E. coli* (Evrard *et al.*, 1990b). Some of these genes have also been reported by Bryant and Stirewalt (1990).

In cyanelles, the S10 and *spc* operons are fused and give rise to a primary transcript of 7500 nucleotides that is successively cleaved into smaller RNAs, although the primary transcript is relatively stable. The *str* ribosomal protein gene operon is located close to this fused S10-*spc* operon, separated by approximately 2.7 kb (Kraus *et al.*, 1990). It is interesting that the gene *rps10* (located at the 5'-terminal of the bacterial S10 operon) is not contiguous with the cyanelle S10-*spc* operon (Fig. 1). It has been translocated to the 3' end of the *str* operon and appears to be cotranscribed with *tufA* (Neumann-Spallart *et al.*, 1991). Formally, the location of the genes *petFI*, two tRNA genes, and the *rpl33* and *rps18* genes, which form a transcription unit as on higher plant plastid DNAs (Evrard *et al.*, 1990a,b), may be considered the result of a transposition event that separated *rps10* from the rest of the genes of the S10-*spc* operon in cyanelles (Fig. 1).

The *rpl23* gene—a pseudogene in spinach plastid DNA (Thomas *et al.*, 1988), *infA*—a pseudogene in tobacco plastid DNA (Sugiura, 1992), *rpl36*, and two genes derived from the prokaryotic α-operon (*rps11, rpoA;* Lindahl and Zengel, 1986), which are all found in the large ribosomal protein gene cluster of higher plant plastid DNAs, are absent from the cyanelle gene cluster. Results from heterologous hybridizations (Löffelhardt *et al.*, 1991) and sequence data indicate that some of these genes reside on a different locus on cyanelle DNA, in the order 5'-*rpl36-rps13-rps11-rpoA-rps9*-3' (V. L. Stirewalt *et al.*, personal communication). Of the additional genes, *rps13* occupies the same position that it does in the *E. coli* α-operon, whereas *rps9*—also a nuclear gene in higher plants—forms a bicistronic operon with *rpl13* in *E. coli* (Lindahl and Zengel, 1986).

As in chloroplasts, *rps4*, a component of the α-operon in *E. coli* but not in *Bacillus subtilis* (Boylan *et al.*, 1989), occupies a solitary position in a different part of the cyanelle genome (C. B. Michalowski, unpublished observations). The same applies for *rps14* (V. L. Stirewalt *et al.*, personal communication), which in *E. coli* is part of the *spc* operon. In the vicinity of the *petB/petD* region, a cluster was recently detected containing three out of four genes of the *E. coli* L11 and L10 operons—*rpl11, rpl1*, and *rpl12* (C. B. Michalowski, unpublished observations). These are nuclear genes in higher plants. The cyanelle ribosomal protein gene complement includes in addition the clustered genes *rpl20* and *rpl35* (Bryant and Stirewalt, 1990) and *rps2, rpl19*, and *rpl21*, which are not linked (V. L. Stirewalt *et al.*, personal communication). Table II lists the 14 genes that are absent from higher plant plastid genomes as well as the 4 genes that have been identified on higher plant plastomes but not on cyanelle DNA. Three of these 4 ribosomal protein genes are also missing from the plastome of *Euglena gracilis* (Hallick *et al.*, 1993).

Higher plant chloroplasts encode 21 ribosomal proteins. This number is much higher in cyanelles where so far 32 ribosomal protein genes have been identified (Fig. 1). Additional genes might still be detected, with *rpl24* as one candidate based on hybridization results using heterologous probes (Löffelhardt *et al.*, 1991). Thus cyanelles encode more than half of the protein genes for the organelle ribosome. By comparison, the plastome of *E. gracilis* contains two additional genes, *rpl5* and *rpl12*, which are absent from higher plant plastid DNA (Hallick *et al.*, 1993).

The number of ribosomal proteins encoded by plastids from algae lacking chlorophyll *b* is similar to that of cyanelles. The *str* operon of *Cryptomonas* Φ contains *rps10* at the 3' end as in cyanelles and in addition *rpoA*,

TABLE II

Ribosomal Protein Gene Complement of Cyanelle DNA Compared with Plastid Genomes in Higher Plants

Genes found in cyanelles but not chloroplasts		Chloroplast genome location of genes missing in cyanelles		
Novel gene	Location[a]	Missing gene	Location[a]	Species
rpl1	L11 and L10	rps16	Separate	Tobacco
rpl3	S10-spc	rpl32	Separate	All
rpl5	S10-spc	rps15	Separate	All
rpl6	S10-spc	rpl23	S10-spc	Spinach
rpl11	L11 and L10			
rpl12	L11 and L10			
rpl18	S10-spc			
rpl19	Separate			
rpl35	rps20-rpl35			
rps5	S10-spc			
rps9	Alpha			
rps10	str			
rps13	alpha			
rps17	S10-spc			

[a] The nomenclature of the corresponding operons of *E. coli* was used. "Separate" indicates that the gene is not found in a transcription unit with other ribosomal protein genes.

rpl13, and *rps9* upstream of *rps12* (Douglas, 1991). In the plastome of *Cryptomonas* Φ, the operon *rpl13-rps9*, which is bicistronic in *E. coli*, has been incorporated together with the α-operon gene *rpoA* into the *str*-operon. The gene *rpl13* has not been reported before for any plastid genome.

A high number of novel ribosomal protein genes have recently been detected on the genome of the red alga *Porphyra purpurea* (Reith and Munholland, 1993a). A portion of the large cluster of genes ranging from *rpl2* to *rpl6* has been sequenced. It contained all the genes found in cyanelles and, in addition, *rpl29* (*E. coli* S10 operon) and *rpl24* (*E. coli spc* operon). Also, genes *rps1* and *rps6,* which have never been detected on plastid genomes before, are present on the rhodoplast genome.

Upon analysis of the organization of ribosomal protein genes in prokaryotes and in plastids, it becomes apparent that the gene transfer to the nuclear genome was accompanied by shuffling of the remaining organelle cistrons, thus creating new transcription units. A comparison of such operons among different plastid types in our opinion demonstrates the phylogenetic relations involved.

With the exception of a partial sequence consisting of two genes (*rpl15* and *secY;* Nakai *et al.*, 1992), data are missing on the organization of cyanobacterial S10, *spc,* and α-operons that might deviate from what is found in *E. coli.* Most cyanelle ribosomal protein genes give better identity scores if they are compared with *B. subtilis* than with *E. coli* (Michalowski *et al.,* 1990b). The *spc* operon of *B. subtilis* contains the gene for initiation factor I, *infA,* which is a solitary gene in *E. coli* but is part of the large ribosomal protein gene cluster of higher plant chloroplasts (Sugiura, 1992). The cyanobacterial origin of cyanelles is clearly shown in a comparison of the *str*-operon sequences from plastids and cyanobacteria (Kraus *et al.,* 1990). Elongation factor Tu is generally observed to be plastid-encoded in all groups of algae with the exception of some charophytes (Baldauf *et al.,* 1990), whereas EF-G is the product of a nuclear gene (Löffelhardt *et al.,* 1990).

D. Genes for Components of the Photosynthetic Apparatus

Cyanelle gene maps based on heterologous hybridizations have been published (Bohnert *et al.,* 1985; Lambert *et al.,* 1985). All chloroplast gene probes used yielded positive results. In many cases the corresponding genes have been sequenced and some novel genes that are nucleus-encoded in higher plants have been identified.

1. Photosystem I

Higher plants contain 17 subunits in the PS I-LHC I complex, 5 of which are chloroplast-encoded (Sugiura, 1992). Homologs to some chloroplast *psa* genes were detected on the cyanelle genome. Genes *psaA/B* encoding the reaction center heterodimer have been sequenced (V. L. Stirewalt *et al.,* personal communication) and give rise to a 6-kb mRNA (Löffelhardt *et al.,* 1987). The *psaC* gene encoding the 9-kDa protein harboring the two [4Fe-4S] centers F_A and F_B was recently identified in the SSC region of cyanelle DNA (Rhiel *et al.,* 1992). Identity scores for this highly conserved protein are in the range of 90–95% compared with plastid and cyanobacterial counterparts.

2. Photosystem II

From the 12 known chloroplast genes for subunits of PS II, 6 have been characterized on the cyanelle genome thus far. Janssen *et al.* (1989) reported the sequence of the cyanelle *psbA* gene encoding the D1 protein of the photosystem II reaction center. Cyanelle D1 shows higher overall homology with chloroplast than with cyanobacterial counterparts.

However, in contrast to plastids from higher plants and chlorophyll b-containing algae, the cyanelle gene contains a specific region at its carboxy-terminal end, an insertion of 7 codons, which is also found in cyanobacteria and in the plastids from algae lacking chlorophyll b, like the Rhodophyceae (Maid and Zetsche, 1990), Phaeophyceae (Winhauer et al., 1991), and Xanthophyceae (Scherer et al., 1991). Furthermore, psbK has been identified (Stirewalt and Bryant, 1989a) as closely resembling the homolog from Synechococcus sp. PCC 6301 (Fukuda et al., 1989). Cytochrome b_{559} is encoded in chloroplasts and cyanobacteria by the adjacent genes psbE/F. In cyanelles this operon contains in addition the genes psbI and psbJ, which are assumed to specify smaller intrinsic membrane proteins of PS II (Cantrell and Bryant, 1987). This arrangement is similar to that in cyanobacteria and Euglena plasmids (Cushman et al., 1988).

3. Cytochrome b_6/f Complex

As in chloroplasts, four components are encoded by the cyanelle genome, and the Rieske-iron-sulfur-protein appears to be the product of a nuclear gene (J. Jakowitsch, unpublished observations). Partial sequence information indicating its location near rpl12 is available for petD encoding subunit iV and petB encoding cytochrome b_6 (C. B. Michalowski, unpublished observations). These genes are cotranscribed in 1400-b mRNA. PetA, encoding cytochrome f, is located betwen the atpB/E and psbD/C transcription units. It appears to be transcribed as a bicistronic mRNA of 1500 b together with a downstream ORF (V. L. Stirewalt et al., personal communication; Löffelhardt et al., 1987). PetG, encoding subunit V of the intrinsic membrane complex, was characterized by sequencing and identified by homology (Stirewalt and Bryant, 1989b).

4. Other Components of the Electron Transport Chain

Neumann-Spallart et al. (1990) recently reported the nucleotide sequence of gene petFI, encoding ferredoxin I. Its 450-b mRNA represents one of the few monocistronic transcripts in cyanelles. Here, the N-terminal amino acid sequence of the purified protein (Stefanovic et al., 1990) could be shown to correspond to the gene sequence. Sequence identity in this case was most pronounced toward the counterparts from cyanobacteria followed by algae lacking chlorophyll b, green algae, and higher plants (Table III). Evrard et al. (1990c) described a reading frame for a 6.5-kDa hydrophobic protein—presumably an intrinsic protein of thylakoid membranes—in cyanelles that is also encoded by higher plant plastomes.

TABLE III

Identity Scores of Cyanelle Ferredoxin I Compared with Its Homologs from Cyanobacteria and Plastids

Organism	%Identity[a]
Synechococcus sp. PCC 7418	79.6
Synechococcus sp. PCC 7942	77.8
Fischerella sp.	77.6
Anabaena sp. PCC 7937	75.6
Cyanidium caldarium	76.5
Porphyra purpurea	71.9
Bumilleriopsis filiformis	69.8
Chlamydomonas reinhardtii	75.0
Scendesmus quadricauda	72.4
Silene pratensis	70.4
Spinacea oleracea	66.0
Marchantia polymorpha	65.3

[a] Identity scores were calculated from alignments in which between 95 and 98 amino acids of 98 residues overlapped (Neumann-Spallart *et al.*, 1990).

5. Phycobilisomes

Cyanelles and red algal plastids appear to encode the chromophoric polypeptide components of the phycobilisomes (PBS), whereas the linker polypeptides are products of nuclear genes (Egelhoff and Grossman, 1983; Burnap and Trench, 1989b). However, gene probes for two rod linkers from *Calothrix* sp. PCC 7601 (Tandeau de Marsac *et al.*, 1988) gave distinct signals with cyanelle DNA, but not with nuclear DNA from *C. paradoxa* (C. Neumann-Spallart, unpublished observations). This indicates that some linker genes might also reside on plastid DNAs of PBS-containing algae (Valentin *et al.*, 1992b). Biliprotein genes have been identified on the cyanelle genome prior to the detection of their cyanobacterial counterparts (Lemaux and Grossman, 1984). Transcription units composed of the genes for the α- and β-subunits of phycocyanin, *cpcA/B*, in the center of the SSC region, and of allophycocyanin, *apcA/B*, have been sequenced (Bryant *et al.*, 1985; Lemaux and Grossman, 1985). The 97-kDa "anchor" phycobiliprotein gene, *apcE*, is located adjacent to *apcA/B* on the same strand (Bryant, 1988) but is transcribed separately, which exactly matches the organization in cyanobacteria (Capuano *et al.*, 1991). This large protein functions in the excitation energy transfer from the PBS to PS II. It contains one biliprotein domain and three conserved domains of 150 amino acids each that show homology to rod linker polypeptides. Based on these structural domains, an interaction of the 97-kDa polypeptide with three cylindrical blocks of allophycocyanin subunits in the PBS core has been suggested (Bryant, 1988). A sixth gene, *apcD*, which encodes an additional allophycocyanin subunit, has been found in cyanelle DNA and is unlinked to the other biliprotein genes (Michalowski *et al.*, 1990a).

Biliprotein genes have recently been detected on the plastomes of several rhodophytic algae and cryptomonads (Shivji *et al.*, 1992; Valentin *et al.*, 1992b; Douglas, 1992b; Reith and Munholland, 1993a). The α-subunit of phycoerythrin appears to be nucleus- or nucleomorph-encoded in a cryptomonad, *Chroomonas* sp. This alga contains a light-harvesting system consisting of soluble biliproteins located inside the thylakoid lumen (Jenkins *et al.*, 1990).

6. ATP Synthase

The bipartite set of genes observed in cyanobacteria (Cozens and Walker, 1987) and chloroplasts is also encountered on the cyanelle genome. The bicistronic transcription unit *atpB/E* is located adjacent to *rbcL* on the opposite strand (Wasmann, 1985; Löffelhardt *et al.*, 1987), a feature typical for higher plant chloroplasts, but different in cyanobacteria. The second gene cluster is remarkable in being adjacent to the upstream genes 5'-*rpoB-rpoC1-rpoC2-rps2*-3', a trait in genome organization that is shared only with chloroplasts. However, there is a difference in gene composition. While the gene order found in the cyanobacterial operon is preserved, two additional genes, *atpD* and *atpG*, are present and gene *atpI* which is present in chloroplasts is absent from the cyanelle gene cluster (V. L. Stirewalt *et al.*, personal communication).

In the plastomes of the diatom *Odontella sinensis* (Pancic *et al.*, 1992) and of red algae (Valentin *et al.*, 1992a), *atpD*, *atpG*, and *atpI* are present in the respective operons, which lack only *atpC* compared with the cyanobacterial gene cluster. Interestingly, the cluster in *Porphyra purpurea* (Reith and Munholland, 1993a) apears to be the result of a fusion of the prokaryotic *rpoB* operon, the S2 operon (containing the *tsf* gene encoding elongation factor Ts), and the cyanobacterial *atp* gene cluster in the correct order: 5'-*rpoB*–*rpoC1*–*rpoC2*–*rps2*–*tsf*–*atpI*–*atpH*–*atpG*–*atpF*–*atpD*–*atpA*-3', without any missing gene. The *Porphyra* gene order and number could represent the gene arrangement established after a primary endosymbiotic event.

7. Rubisco

The genes for both subunits of rubisco on cyanelle DNA have been characterized (Heinhorst and Shively, 1983; Wasmann, 1985; Valentin and Zetsche, 1990a). Cyanelle *rbcS* was the first reported plastid-encoded *rbcS* gene whereas SSU from higher plants, green algae, and euglenoids are the products of nuclear genes. The two genes are cotranscribed, as in cyanobacteria (Starnes *et al.,* 1985). An identical arrangement of the two subunit genes has been detected in the plastomes of a number of algae lacking chlorophyll *b*, for example, in three brown algae (Boczar *et al.,* 1989; Valentin and Zetsche, 1990b; Assali *et al.,* 1991), two red algae (Valentin and Zetsche, 1990b,c), a cryptophycean alga (Douglas and Durnford, 1989), and a diatom (Hwang and Tabita, 1989).

E. Genes for Proteins Involved in Biosynthetic Pathways

1. NAD Biosynthesis

An open reading frame located close to the S10-*spc* ribosomal protein gene cluster (Fig. 1) has been characterized (Michalowski *et al.,* 1991a). Comparisons with sequences in the data banks suggested that this ORF329 had homology with bacterial genes encoding *nadA,* quinolinate synthetase (E.C.4.6.1.3). The enzyme catalyzes the condensation of iminoaspartic acid with dihydroxyacetone phosphate to generate quinolinic acid. Cyanelle *nadA* has 33.7% and 38.9% identity with the functionally identified genes from *E. coli* and *Salmonella typhimurium,* respectively. If conservative exchanges are permitted, homology is in the range of 65 to 74%. No homology for this coding region is found in the completely sequenced chloroplast genomes.

NAD biosynthesis is an essential function of all organisms, most of which possess a salvage pathway for this compound (for review see Foster and Moat, 1980). A biosynthetic pathway for NAD starting from low-molecular-weight compounds of basic metabolism has up to now been studied biochemically and genetically only in microorganisms (Flachmann *et al.,* 1988; Foster *et al.,* 1990), although it is to be expected that plants possess this pathway. The detection of *nadA* in *Cyanophora paradoxa* cyanelles is the first genetic indication that this pathway is present in eukaryotic cells.

2. Isoprenoid Pathway

Cyanelle DNA contains an open reading frame, ORF323 (Michalowski *et al.,* 1991b), with homology to gene *crtE,* originally reported to encode prephytoene pyrophosphate dehydrogenase, from *Rhodobacter capsulatus* (Armstrong *et al.,* 1989). This ORF is transcribed in cyanelles and is, at least *in vitro,* translated into a peptide of the size expected for this sequence. Reinvestigation of the function of the *crtE* gene product, now also identified in higher plants where it is nucleus-encoded (Kuntz *et al.*, 1992), indicated an earlier role in the pathway, that is, the activity of a geranyl-geranyl pyrophosphate synthase. In a test system based on genes for carotenogenic enzymes from *Erwinia uredovora* (Misawa *et al.*, 1990) and their expression in transgenic *E. coli*, the cyanelle "*crtE*" gene product did not show geranyl-geranyl pyrophosphate synthase activity (G. Sandmann, personal communication). Sequence comparisons (Löffelhardt *et al.*, 1993) point to the function of a higher prenyl transferase. The best identity score was obtained with a hexaprenyl pyrophosphate synthase from yeast mitochondria (Ashby and Edwards, 1990). Thus we assume that the cyanelle protein encoded by ORF323 catalyzes consecutive additions either *in cis* or *in trans* of isopentenyl pyrophosphate to farnesyl pyrophosphate (Sherman *et al.*, 1989) leading to undekaprenyl pyrophosphate, or leading to the precursor of the nonaprenyl side chain of plastoquinone, respectively. In the latter case this enzyme activity should also be present in chloroplasts whereas in the former case it might be confined to the peptidoglycan-containing cyanelles.

In addition to the "extra" genes absent from higher plant chloroplast genomes which have been discussed before, cyanelles share a number of novel plastid genes with other algal groups. Cyanelles and *Porphyra* rhodoplasts have a number of genes in common that encode enzymes involved in the biosynthesis of amino acids, carotenoids, chlorophyll, and fatty acids (Reith and Munholland, 1993a; V. L. Stirewalt *et al.*, personal communication; C. B. Michalowski, unpublished observations). To date, the number of markers on cyanelle DNA—either completely or partially sequenced—exceeds 140 genes. Approximately 85% of the genome is sequenced (V. L. Stirewalt *et al.*, personal communication; C. B. Michalowski, unpublished observations). Owing to the absence of introns in protein genes, to small intergenic distances, and to the small size of the IR segment, a gene number exceeding that of chloroplasts can be accommodated, although the cyanelle genome is slightly smaller than the typical chloroplast genome.

F. Characteristics of Cyanelle Genes

Generalizations about cyanelle gene structure become approachable with the availability of an increased number of sequences compiled in the references. In many cases, heterologous hybridizations using probes from either cyanobacterial or plastid genes are successful in finding the corresponding cyanelle gene (Löffelhardt *et al.*, 1985). However, the lack of a hybridization result does not necessarily mean that the gene is missing.

Genes on cyanelle DNA have a characteristic bias for either A or T in the third codon position whenever this is possible. Genes are densely packed and intergenic regions are extremely A- and T-rich. Putative control regions of gene expression show the following three features.

First, sequence elements which closely resemble the well-known "-35" and "-10" promoter boxes of bacterial genes can always be found. In several genes, such elements have been identified in regions that may form a stem-loop structure. Stem-loop structures, the second conspicuous feature, are very often found close to the termination codons of identified genes. In cases where two genes are separated by only a few nucleotides, we have found instances where a stem-loop structure might be formed that involves the protein initiation codon for the following gene. Folding of stems up to and exceeding approximately 40 base pairs ($\Delta G = \sim -40$ kcal) with four to five nucleotide loops has been possible. Since such structures have been found ubiquitously at positions which separate operons (identified by transcripts analysis; Michalowski *et al.*, 1990b), we consider them functional features. Further in-depth analysis should provide the data. The third obvious feature is the presence of "ribosome-binding sites" with complementarity to the cyanelle 16S rRNA 3' end at virtually every position that contains an open reading frame with homology to functionally identified bacterial genes. In the promoter region of one gene, *nadA*, a sequence with close similarity to bacterial cAMP receptor protein binding sites has been detected (Michalowski *et al.*, 1991a).

VI. Protein Transport into and within Cyanelles

A. Import of Proteins

Like plastids, cyanelles have to import the majority of their proteins, even when we consider that the dense spacing of genes and the near total lack of introns (as far as sequences are known) allow more functional genes to be encoded on the cyanelle DNA than on higher plant chloroplast DNA. The first example for such an imported protein is ferredoxin-NADP$^+$ oxidoreductase because it is synthesized as a preprotein on cytosolic 80S ribosomes (Bayer *et al.*, 1990). Recently, the sequence of the corresponding cDNA was determined (Jakowitsch *et al.*, 1993). The mature protein, which is overall highly conserved to other FNR, lacks the N-terminal extension thought to be responsible for attachment to phycobilisomes in the enzyme from the cyanobacterium *Synechococcus sp.* (Schluchter and Bryant, 1992) and behaves in this respect like the higher plant enzymes.

The amino-terminal transit peptide sequence shows little, if any, homology with transit peptides of higher plant pre-FNR enzymes or with other N-terminal extensions in general. This is not surprising since plastid presequences are notoriously variable. However, overall characteristics of the presequence, such as a lack of charge at the extreme amino-terminal and the presence of hydroxyl amino acids, suggest a function for the transit peptide in stroma targeting. Functionally, the protein import apparatus for cyanelles appears similar to that of chloroplasts.

B. Routing within Cyanelles

Chloroplast proteins that are destined to the thylakoid lumen are subject to an additional independent targeting or routing mechanism that depends on the presence of a N-terminal leader sequence. This may constitute either the N-terminal more hydrophobic part of composite transit sequences in the case of nuclear gene products or the targeting signal of chloroplast genes (Smeekens et al., 1990). This second organellar protein translocation machinery, which is also invoked for the functionally equivalent targeting to the mitochondrial intermembrane space, is assumed to originate from the ancestral prokaryotic endosymbiont. The retention of prokaryotic preprotein translocases in plastid and mitochondrial membranes is postulated by the "conservative sorting" hypothesis (Hartl and Neupert, 1990).

In Fig. 1 we have included an open reading frame (ORF492; Flachmann et al., 1993) that is homologous to the E. coli secY gene (Akiyama and Ito, 1987) whose product is essential for growth and is considered a part of the protein export complex (for review see Wickner et al., 1991; Bieker and Silhavy, 1990). The cyanelle ORF492 is located at the 3'-end of the spc operon (Michalowski et al., 1990b). E. coli secY occupies the same position, although it is cotranscribed with the ribosomal protein genes of the spc operon. It appears that this ORF492—which is transcribed in the organelles as a monocistronic mRNA separate from the spc operon—encodes an Sec Y-like protein in cyanelles. This protein shows a sequence identity of 28.1% with the E. coli SecY counterpart and is functional in E. coli as shown by a complementation assay. Cyanelle SecY restored growth at 42°C in transformed, thermosensitive secY mutants of E. coli (Shiba et al., 1984; Flachmann et al., 1993).

Very recently, secY genes have also been identified on the plastid genomes of Cryptomonas Φ (Douglas, 1992a) and the chromophytic alga Pavlova lutherii (Scaramuzzi et al., 1992b) by sequence similarity. In these cases the identity scores toward the cyanelle protein are in the range of 35 to 38%. Cyanelles and other plastids without chlorophyll b might well encode additional subunits of the preprotein translocase, as exemplified by the recent detection of *secA* on the plastid genome of a chromophyte (Scaramuzzi *et al.*, 1992c) and a rhodophyte (Valentin, 1993).

However, the nature of the cyanelle membrane(s) harboring SecY and the other components of the translocase remains to be demonstrated through the use of specific antibodies. Our working hypothesis assumes that cyanelles and most likely all plastid types possess a thylakoid-bound, SecY-dependent protein transport system for the translocation of lumenal polypeptides. In the case of higher plant chloroplasts, the respective genes must reside on the nuclear genome. In addition, we think that in cyanelles the IEM may also contain SecY protein and the transport machinery. In contrast to the intermembrane space of the plastid envelope, the periplasmic space of cyanelles is a defined compartment containing a number of identified proteins, including enzymes involved in wall synthesis (Plaimauer *et al.*, 1991). The most likely candidates for proteins leaving the organelle would be the preproteins of enzymes involved in the biosynthesis of the peptidoglycan wall and OEM proteins.

VII. The Cyanelle and Its Position in Plastid Evolution

As a bridge organism, *C. paradoxa* has often been included in phylogenetic trees constructed from different traits using different computing programs. Owing to the problems still inherent in phylogenetic algorithms and the possible substitional and constitutional bias encountered in the sequences to be analyzed (Lockhart *et al.*, 1992), the results should be considered not only with interest, but also with caution. Initially, cyanelle rRNA genes have been used for comparisons. Phylogenies derived from 16S rRNA show cyanelles and all types of plastids to place well within the cyanobacterial radiation (Giovannoni *et al.*, 1988; Urbach *et al.*, 1992), with a pronounced relation of cyanelles to plastids from *Cryptomonas* Φ and red algae (Douglas, 1992b). When 5S rRNA is taken as trait, all plastids group together, with cyanelles coming closest to *Porphyra* and *Euglena* plastids (Wolters *et al.*, 1990). Based on D1 protein sequences, *Cyanophora* groups with the rhodophyte *C. caldarium* and the xanthophycean alga *Bumilleriopsis filiformis* between the cyanobacteria and prochlorophytes on one hand and the branch containing green algae, *Euglena*, and higher plants on the other hand (Scherer *et al.*, 1991).

Ribosomal protein genes *rpl33, rps18, rpl2, rps19,* and *rpl22* constituted the data set for trees that indicate a shorter evolutionary distance between cyanelles and higher plant plastids than between the latter and *Euglena* plastids (Evrard *et al.*, 1990a). Trees constructed from the nucleotide sequences of the three *str* operon genes (Kraus *et al.*, 1990) were most conclusive for *rps7*, which is the least conserved one: here cyanelles occupy an intermediate position between chloroplasts and cyanobacteria, coming somewhat closer to the latter. With *rps12* and especially *tufA*, the order of branching of the algal chloroplasts becomes uncertain. A wealth of sequence information is available concerning ferredoxin I: in two trees obtained with different analysis programs, cyanelles group with cyanobacteria, the plastids from rhodophytic algae and Xanthophyceae being the next closest relatives (M. Kraus, unpublished observations; Lüttke, 1991).

A recently published tree using sequences of the *rpoC1* gene products (Palenik and Haselkorn, 1992) confirms the results of 16S rRNA data, that is, that the cyanelle is the closest known relative to the ancestor of chloroplasts. When the large subunit (LSU) of rubisco was used as a phylogenetic marker, cyanelles appeared to be separated from plastids of rhodophytes, chromophytes, and *Cryptomonas* Φ, grouping with the chlorophyll *b*-type plastids (Valentin *et al.*, 1992a). SSU-derived trees (Assali *et al.*, 1991; Valentin *et al.*, 1992a) show similar features: *Cyanophora* cyanelle *rbcS* groups with cyanobacteria which also lack an amino-terminal insertion in the SSU protein. Cyanelle *rbcS* is separated from other plastic types, especially those from rhodophytes and *Cryptomonas*. In contrast, rhodophytes and chromophytes include a 31-amino acid insertion and chlorophytes contain an additional 12 amino acids. This grouping is in contrast to phylogenetic trees comparing numerous other traits, indicating that rubisco, which has been postulated to originate from a lateral gene transfer from β- or γ-purple bacteria (Assali *et al.*, 1991), might not be a suitable marker for evolutionary relationships. This may be due to the extensive interactions between LSU and SSU in the holoenzyme and the numerous ligand-to-protein interactions, resulting in considerable constraints on the evolution of the subunits and their genes.

Upon analysis of their trees, several authors claim a monophyletic origin for plastids whereas others favor a polyphyletic origin, with the possibility that cyanelles constitute a separate event. In our opinion neither the data sets that are available nor the present state of phylogenetic analysis allows a definitive answer to this question. Nuclear gene sequences from *C. paradoxa* have become available only recently. The trees deduced from FNR sequences support an intermediate position for cyanelles between plant chloroplasts and—slightly closer—cyanobacteria (Jakowitsch *et al.*, 1993). Very recently a trait pertinent to the eukaryotic host became available: 18S rRNA data show a close relationship between *C. paradoxa*, *Glaucocystis nostochinearum*, and the cryptophycean algae *Cryptomonas* Φ and *Pyrenomonas salina*. They are sister groups, that is, they share a common evolutionary history (D. Bhattacharya, personal communication). This is also supported by some of the plastid gene-derived trees mentioned above. Interestingly, due to its mitotic apparatus, *C. paradoxa* was once classified as a cryptomonad (Pickett-Heaps, 1972).

VIII. Concluding Remarks

In summarizing the partly controversial results of phylogenetic analyses, we are inclined to put more emphasis on comparing features of genome, operon, and gene organization than merely isolated sequences. Gene clusters like the one consisting of *rpo* genes, *rps2*, and *atp* genes, which are shared among cyanelles, rhodoplasts, and chloroplasts, but not found in this composition on cyanobacterial genomes, are a strong indication of a common origin for all plastid types from a single primary endosymbiotic event. As the only semiautonomous organelles that did not lose the prokaryotic wall, cyanelles constitute an early branch which appears to have led to evolutionary confinement. The other branch, possibly best characterized as primitive rhodoplasts, became the ancestors of all other plastid types (Reith and Munholland, 1993a). In the case of the plastids from chromophytes and euglenoids, (multiple) secondary endosymbiotic events in which both partners were eukaryotes are envisaged (Douglas, 1992b). This view of an ancient origin for the *C. paradoxa* "endocyanom" is in accord with the observed parallels in the protein import apparatus of cyanelles and chloroplasts.

Acknowledgments

We wish to thank all our colleagues and co-workers who have contributed to our knowledge of cyanelle genome and gene structure over the past 14 years. During that time work has been supported by the Deutsche Forschungsgemeinschaft, EMBL (Heidelberg), the Austrian Research Council (projects S 29/06 and S6008-BIO to W. Löffelhardt), National Science Foundation, and the Arizona Agricultural Experiment Station, Tucson, Arizona. We thank Christine Michalowski and Marty Wojciechowski for help with phylopenetic trees and Christoph Neumann-Spallart for drawing the map of the cyanelle genome. The cyanelle genome sequencing project was supported by a grant from the U.S. Department of Agriculture-National Research Institute (Plant Genome) to H. J. Bohnert and Don A. Bryant (Pennsylvania State University).

References

Aitken, A., and Stanier, R. Y. (1979). Characterization of peptidoglycan from the cyanelles of *Cyanophora paradoxa. J. Gen. Microbiol.* **212**, 218–223.

Akiyama, Y., and Ito, K. (1987). The SecY membrane component of the bacterial protein export machinery: analysis by new electrophoretic methods for integral membrane proteins. *EMBO J.* **6**, 3465–3470.

Armstrong, G. A., Alberti, M., Leach, F., and Hearst, J. E. (1989). Nucleotide sequence, organization, and nature of the protein products of the carotenoid biosynthesis gene cluster of *Rhodobacter capsulatus*. *Mol. Gen. Genet.* **216**, 254–268.

Ashby, M. N., and Edwards, P. A. (1990). Elucidation of the deficiency in two yeast coenzyme Q mutants. Characterization of the structural gene encoding hexaprenyl pyrophosphate synthetase. *J. Biol. Chem.* **265**, 13157–13164.

Assali, N. E., Martin, W. F., Sommerville, C. C., and Loiseaux-de Goër, S. (1991). Evolution of the Rubisco operon from prokaryotes to algae: structure and analysis of the *rbcS* gene of the brown alga *Pylaiella littoralis*. *Plant Mol. Biol.* **17**, 853–863.

Baldauf, S. L., and Palmer, J. D. (1990). Evolutionary transfer of the chloroplast *tufA* gene to the nucleus. *Nature (London)* **344**, 262–265.

Baldauf, S. L., Manhart, J. R., and Palmer, J. D. (1990). Different fates of the chloroplast *tufA* gene following its transfer to the nucleus in green algae. *Proc. Natl. Acad. Sci. U.S.A.* **87**, 5317–5321.

Bayer, M. G., and Schenk, H. E. A. (1986). Biosynthesis of proteins in *Cyanophora paradoxa*: I. Protein import into the endocyanelle analyzed by micro two-dimensional gel electrophoresis. *Endocytobiosis Cell Res.* **3**, 197–202.

Bayer, M. G., Maier, T. L., Gebhart, U. B., and Schenk, H. E. A. (1990). Cyanellar ferredoxin-NADP⁺-oxidoreductase of *Cyanophora paradoxa* is encoded by the nuclear genome and synthesized on cytoplasmatic 80S ribosomes. *Curr. Genet.* **17**, 265–267.

Berenguer, J., Rojo, F., dePedro, M. A., Pfanzagl, B. and Löffelhardt, W. (1987). Penicillin-binding proteins in the cyanelles of *Cyanophora paradoxa*, a eukaryotic photoautotroph sensitive to b-lactam antibiotics. *FEBS Lett.* **224**, 401–405.

Bernard, C., Thomas, J. C., Mazel, D., Mousseau, A., Castets, A. M., Tandeau de Marsac, N., and Dubacq, J. P. (1992). Characterization of the genes encoding phycoerythrin in the red alga *Rhodella violacea*: Evidence for a splitting of the *rpeB* gene by an intron. *Proc. Natl. Acad. Sci. U.S.A.* **89**, 9564–9568.

Betsche, T., Schaller, D., and Melkonian, M. (1992). Identification and characterization of glycolate oxidase and related enzymes from the endocyanotic alga *Cyanophora paradoxa* and from pea leaves. *Plant Physiol.* **98**, 887–893.

Bieker, K. L., and Silhavy, T. J. (1990). The genetics of protein secretion in *E. coli*. *Trends Genet.* **6**, 329–334.

Blank, R. J. (1985). Is the central inclusion body of the cyanelle of *Cyanophora paradoxa* a carboxysome? *Endocytobiosis Cell Res.* **2**, 113–118.

Boczar, B. A., Delaney, T. P., and Cattolico, R. A. (1989). Gene for the ribulose-1,5-bisphosphate carboxylase small subunit protein of the marine chromophyte *Olisthodiscus luteus* is similar to that of a chemoautotrophic bacterium. *Proc. Natl. Acad. Sci. U.S.A.* **86**, 4996–4999.

Bohnert, H. J., and Löffelhardt, W. (1992). Molecular genetics of cyanelles from *Cyanophora paradoxa*. In "Algae and Symbiosis" (W. Reisser, ed.), pp. 379–397. Biopress, Bristol.

Bohnert, H. J., Michalowski, C. B., Bevacqua, S., Mucke, H., and Löffelhardt, W. (1985). Cyanelle DNA from *Cyanophora paradoxa*. Physical mapping and location of protein coding regions. *Mol. Gen. Genet.* **201**, 565–574.

Bonham-Smith, P. C., and Bourque, D. P. (1989). Translation of chloroplast-encoded mRNA: potential initiation and termination signals. *Nucleic Acids Res.* **17**, 2057–2080.

Boylan, S. A., Suh, J.-W., Thomas, S. M., and Price, C. W. (1989). Gene encoding the alpha core subunit of *Bacillus subtilis* RNA polymerase is cotranscribed with the genes for initiation factor 1 and ribosomal proteins B, S11, and L17. *J. Bacteriol.* **171**, 2553–2562.

Boynton, J. E., Gillham, N. W., Harris, E. H., Hosler, J. P., Johnson, A. M., Jones, A. R., Randolph-Anderson, B. L., Robertson, D., Klein, T. M., Shark, K. B., and Sanford, J. C. (1988). Chloroplast transformation in *Chlamydomonas* with high velocity microprojectiles. *Science* **240**, 1534–1538.

Breiteneder, H., Seiser, C., Löffelhardt, W., Michalowski, C., and Bohnert, H. J. (1988). Physical map and protein gene map of cyanelle DNA from the second known isolate of *Cyanophora paradoxa* (Kies-strain). *Curr. Genet.* **13**, 199–206.

Bryant, D. A. (1988). Genetic analysis of phycobilisome biosynthesis, assembly, structure and function in the cyanobacterium *Synechococcus* sp. PCC7002. *In* "Light-Energy Transduction in Photosynthesis: Higher Plants and Bacterial Models" (S. E. Stevens and D. A. Bryant, eds.), pp. 62-90. Amer. Soc. Plant Physiol., Rockville, Maryland.

Bryant, D. A., and Stirewalt, V. L. (1990). The cyanelle genome of *Cyanophora paradoxa* encodes ribosomal proteins not encoded by the chloroplast genomes of higher plants. *FEBS Lett.* **259**, 273–280.

Bryant, D. A., deLorimier, R., Lambert, D. H., Dubbs, J. M., Stirewalt, V. L., Stevens, S. E., Porter, R. D., Tam, J., and Jay, E. (1985). Molecular cloning and nucleotide sequence of the a and b subunits of allophycocyanin from the cyanelle genome of *Cyanophora paradoxa*. *Proc. Natl. Acad. Sci. U.S.A.* **82**, 3242–3246.

Burnap, R. L. (1987). Biogenesis of the cyanelles of *Cyanophora paradoxa:* An example of inter-genomic cooperation. PhD Thesis, Univ. of California, Santa Barbara.

Burnap, R. L., and Trench, R. K. (1989a). The biogenesis of the cyanellae of *Cyanophora paradoxa*. I. Polypeptide composition of the cyanellae. *Proc. R. Soc. London, Ser. B* **238**, 53–72.

Burnap, R. L., and Trench, R. K. (1989b). The biogenesis of the cyanellae of *Cyanophora paradoxa*. II. Pulse-labelling of cyanellar polypeptides in the presence of transcriptional and translational inhibitors. *Proc. R. Soc. London, Ser. B* **238**, 73–87.

Burnap, R. L., and Trench, R. K. (1989c). The biogenesis of the cyanellae of *Cyanophora paradoxa*. III. *In vitro* synthesis of cyanellar polypeptides using separated cytoplasmic and cyanellar RNA. *Proc. R. Soc. London, Ser. B* **238**, 89–102.

Cantrell, A., and Bryant, D. A. (1987). Nucleotide sequence of the genes encoding cytochrome b-559 from the cyanelle genome of *Cyanophora paradoxa*. *Photosynth. Res.* **16**, 65–81.

Capuano, V., Braux, A.-S., Tandeau de Marsac, N., and Houmard, J. (1991). The "anchor polypeptide" of cyanobacterial phycobilisomes. Molecular characterization of the *Synechococcus* sp. PCC 6301 *apcE* gene. *J. Biol. Chem.* **266**, 7239–7247.

Coleman, A. W. (1985). Cyanophyte and cyanelle DNA: A search for the origins of plastids. *J. Phycol.* **21**, 371–379.

Cozens, A. L., and Walker, J. E. (1987). The organization and sequence of the genes for ATP synthase subunits in the cyanobacterium *Synechococcus* 6301. Support for an endosymbiotic origin of chloroplasts. *J. Mol. Biol.* **194**, 359–383.

Cushman, J. C., Christopher, D. A., Little, M. C., Hallick, R. B., and Price, C. A. (1988). Organization and expression of the PSBE, PSBF, ORF38 and ORF42 gene loci on the *Euglena gracilis* chloroplast genome. *Curr. Genet.* **13**, 173–180.

Delaney, T. P., and Cattolico, R. A. (1989). Chloroplast ribosomal DNA organization in the chromophytic alga *Olisthodiscus luteus*. *Curr. Genet.* **15**, 221–229.

De Rocher, E. J., Harkins, K., Galbraith, D. W., and Bohnert, H. J. (1990). Developmentally regulated systemic endopolyploidy in succulents with small genomes. *Science* **250**, 99–101.

Douglas, S. E. (1991). Unusual organization of a ribosomal protein operon in the plastid genome of *Cryptomonas F:* evolutionary considerations. *Curr. Genet.* **19**, 289–294.

Douglas, S. E. (1992a). A *secY* homologue is found in the plastid genome of *Cryptomonas* Φ *FEBS Lett.* **298**, 93–96.

Douglas, S. E. (1992b). Probable evolutionary history of cryptomonad algae. *In* "Origins of Plastids" (R. Lewin, ed.), pp. 265–290. Chapman & Hall, London.

Douglas, S. E., and Durnford, D. G. (1989). The small subunit of ribulose-1,5-bisphosphate carboxylase is plastid-encoded in the chlorophyll c-containing alga *Cryptomonas* Φ *Plant. Mol. Biol.* **13**, 13–20.

Douglas, S. E., and Durnford, D. B. (1990). Sequence analysis of the plastid rDNA spacer region of the chlorophyll c-containing alga *Cryptomonas* Φ *DNA Sequence* **1**, 55–62.

Egelhoff, T., and Grossman, A. R. (1983). Cytoplasmic and chloroplast synthesis of phycobilisome polypeptides. *Proc. Natl. Acad. Sci. U.S.A.* **80**, 3339–3343.

Evrard, J. L., Kuntz, M., Straus, N. A., and Weil, J. H. (1988). A class I intron in a cyanelle gene from *Cyanophora paradoxa*: phylogenetic relationship to plant chloroplast genes. *Gene* **71**, 115–122.

Evrard, J. L., Kuntz, M., and Weil, J. H. (1990a). The nucleotide sequence of five ribosomal protein genes from the cyanelles of *Cyanophora paradoxa*: implications concerning the phylogenetic relationship betwen cyanelles and chloroplasts. *J. Mol. Evol.* **30**, 16–25.

Evrard, J. L., Johnson, C., Janssen, I., Löffelhardt, W., Weil, J. H., and Kuntz, M. (1990b). The cyanelle genome of *Cyanophora paradoxa*, unlike chloroplast genomes, codes for the ribosomal L3 protein. *Nucleic Acids Res.* **18**, 1115–1119.

Evrard, J. L., Weil, J. H., and Kuntz, M. (1990c). An ORF potentially encoding a 6.5 kDa hydrophobic protein in chloroplasts is also present in the cyanellar genome of *Cyanophora paradoxa*. *Plant Mol. Biol.* **15**, 779–781.

Flachmann, R., Kunz, N., Seifert, J., Gütlich, M., Wientjes, F. J., Läufer, A., and Gassen, H. G. (1988). Molecular biology of pyridine nucleotide biosynthesis in *E. coli*. Cloning and characterization of quinolinate synthesis genes *nadA* and *nadB*. *Eur. J. Biochem.* **175**, 221–228.

Flachmann, R., Michalowski, C. B., Löffelhardt, W., and Bohnert, H. J. (1993). SecY, an integral subunit of the bacterial preprotein translocase is encoded by a plastid genome. *J. Biol. Chem.* **268**, 7514–7519.

Foster, J. W., and Moat, A. G. (1980). Nicotinamide adenine dinucleotide biosynthesis and pyridine nucleotide cycle metabolism in microbial systems. *Microbiol. Rev.* **44**, 83–105.

Foster, J. W., Park, Y. K., Penfound, T., and Spector, M. P. (1990). Regulation of NAD metabolism in *Salmonella typhimurium*: molecular sequence analysis of the bifunctional *nadR* regulator and the *nadA-pnuC* operon. *J. Bacteriol.* **172**, 4187–4196.

Friedberg, D., Kaplan, A., Ariel, R., Kessel, M., and Seiffers, J. (1989). The 5'-flanking region of the gene encoding the large subunit of ribulose-1,5-bisphosphate carboxylase/ oxygenase is crucial for growth of the cyanobacterium *Synechococcus* sp. strain PCC7942 at the level of CO_2 in air. *J. Bacteriol.* **171**, 6069–6076.

Fukuda, M., Meng, B. Y., Hayashida, N., and Sugiura, M. (1989). Nucleotide sequence of the *psbK* gene of the cyanobacterium, *Anacystis nidulans* 6301. *Nucleic Acids Res.* **17**, 7521.

Giddings, T. H., Wasmann, C., and Staehelin, L. A. (1983). Structure of the thylakoids and envelope membranes of the cyanelles of *Cyanophora paradoxa*. *Plant Physiol.* **71**, 409–419.

Giovannoni, S. J., Turner, S., Olsen, G. J., Barns, S., Lane, D. J., and Pace, R. R. (1988). Evolutionary relationships among cyanobacteria and green chloroplasts. *J. Bacteriol.* **170** 3584–3592.

Gray, M. W. (1989). The evolutionary origins of organelles. *Trends Genet.* **5**, 294–299.

Hall, W. T., and Claus, G. (1963). Ultrastructural studies on the blue-green algal symbiont in *Cyanophora paradoxa* Korschikoff. *J. Cell Biol.* **19**, 551–563.

Hallick, R. B., Hong, L., Drager, R. G., Favreau, M. R., Montfort, A., Orsat, B., Spielmann, A., and Stutz, B. (1993). Complete sequence of *Euglena gracilis* chloroplast DNA. *Nucleic Acids Res.* **21**, 3537–3544.

Harris, E. H. (1989). "The *Chlamydomonas* Source Book." Academic Press, San Diego.

Hartl, F.-U,., and Neupert, W. (1990). Protein sorting to mitochondria: evolutionary conservation of folding and assembly. *Science* **247**, 930–938.

Heinhorst, S., and Shively, J. M. (1983). Encoding of both subunits of ribulose-1,5-bisphosphate carboxylase by organelle genome of *Cyanophora paradoxa*. *Nature (London)* **304**, 373–374.

Herdman, M., and Stanier, R. Y. (1977). The cyanelle: Chloroplast or endosymbiotic prokaryote? *FEMS Microbiol. Lett.* **1**, 7–12.

Höltje, J.-V., and Schwarz, U. (1985). In "Molecular Cytology of *Escherichia Coli*" (N. Nanninga, ed.), pp. 77–119. Academic Press, London.

Hwang, S.-R., and Tabita, F. R. (1989). Cloning and expression of the chloroplast-encoded *rbcL* and *rbcS* genes from the marine diatom *Cylindrotheca* sp. strain N1. *Plant Mol. Biol.* **13**, 69–79.

Hwang, S.-R., and Tabita, F. R. (1991). Acyl carrier protein-derived sequence encoded by the chloroplast genome in the marine diatom *Cylindrotheca* sp. strain N1. *J. Biol. Chem.* **266**, 13492–13494.

Jakowitsch, J., Bayer, M. G., Maier, T. L., Lüttge, A., Gebhart, U. B., Brandtner, M., Hamilton, B., Neumann-Spallart, C., Michalowski, C. B., Bohnert, H. J., Schenk, H. E. A., and Löffelhardt, W. (1993). Sequence analysis of pre-ferredoxin-NADP$^+$-reductase cDNA from *Cyanophora paradoxa* specifying a precursor for a nucleus-encoded cyanelle polypeptide. *Plant Mol. Biol.* **21**, 1023–1033.

Janssen, I., Mucke, H., Löffelhardt, W., and Bohnert, H. J. (1987). The central part of the cyanelle rDNA unit of *Cyanophora paradoxa*: Sequence comparison with chloroplasts and cyanobacteria. *Plant Mol. Biol.* **9**, 479–484.

Janssen, I., Jakowitsch, J., Michalowski, C. B., Bohnert, H. J., and Löffelhardt, W. (1989). Evolutionary relationship of *psbA* genes from cyanobacteria, cyanelles and chloroplasts. *Curr. Genet.* **15**, 335–340.

Jehn, U., and Zetsche, K. (1988). *In vitro* synthesis of the cyanelle proteins of *Cyanophora paradoxa* by isolated cyanelles and cyanelle RNA. *Planta* **173**, 58–60.

Jenkins, J., Hiller, R. G., Speirs, J., and Godovac-Zimmermann, J. (1990). A genomic clone encoding a cryptophyte phycoerythrin a-subunit. Evidence for three a-subunits and an N-terminal transit sequence. *FEBS Lett.* **273**, 191–194.

Kaplan, A., Schwarz, R., Liemann-Hurwitz, J., and Reinhold, L. (1991). Physiological and molecular aspects of the inorganic carbon-concentrating mechanism in cyanobacteria. *Plant Physiol.* **97**, 851–855.

Keegstra, K., Olsen, L. J., and Theg, S. M. (1989). Chloroplastic precursors and their transport across the envelope membranes. *Annu. Rev. Plant Physiol.* **40**, 471–501.

Kessler, U., Maid, U., and Zetsche, K. (1992). An equivalent to bacterial ompR genes is encoded on the plastid genome of red algae. *Plant Mol. Biol.* **18**, 777–780.

Kies, L. (1979). Zur systematischen Einordnung von *Cyanophora paradoxa*, *Gloeochaete wittrockiana* and *Glaucocystis nostochinearum*. *Ber. Dtsch. Bot. Ges.* **92**, 445–454.

Kies, L. (1984). Cytological aspects of blue-green algal endosymbiosis. In "Endocytobiology, Endosymbiosis and Cell Biology" (W. Reisser, D. Robinson, and C. Starr, eds.), pp. 7–19. DeGruyter, Berlin.

Kies, L. (1988). The effect of penicillin on the morphology and ultrastructure of *Cyanophora*, *Gloeochaete* and *Glaucocystis* (*Glaucocystophyceae*) and their cyanelles. *Endocytobiosis Cell Res.* **5**, 361–372.

Kies, L. (1992). Glaucocystophyceae and other protists harbouring prokaryotic endocytobionts. In "Algae and Symbiosis" (W. Reisser, ed.), pp. 353–377. Biopress, Bristol, England.

Kleinig, H., Beyer, P., Liedvogel, B., and Lütke–Brinkhaus, F. (1986). *Cyanophora paradoxa*: Fatty acids and fatty acid synthesis *in vitro*. *Z. Naturforsch.* **41c**, 169–171.

Korschikoff, A. A. (1924). Protistologische Beobachtungen I. *Cyanophora paradoxa*. *Russ. Arch. Protistol.* **3**, 57–74.

Kraus, M., Götz, M., and Löffelhardt, W. (1990). The cyanelle *str* operon from *Cyanophora paradoxa*: sequence analysis and phylogenetic implications. *Plant Mol. Biol.* **15**, 561–573.

Kuhsel, M. G., Strickland, R., and Palmer, J. D. (1990). An ancient group I intron shared by eubacteria and chloroplasts. *Science* **250**, 1570–1573.

Kuntz, M., Crouse, E. J., Mubumbila, M., Burkard, G., Weil, J. H., Bohnert, H. J., Mucke, H., and Löffelhardt, W. (1984). Transfer RNA gene mapping studies on cyanelle DNA from *Cyanophora paradoxa*. *Mol. Gen. Genet.* **194**, 508–512.

Kuntz, M., Evrard, J. L., and Weil, J. H. (1988). Nucleotide sequence of the tRNA-Ser(GGA) and tRNA-Gly(GCC) genes from cyanelles of *Cyanophora paradoxa*. *Nucleic Acids Res.* **16**, 8733.

Kuntz, M., Römer, S., Suire, C., Hugueney, P., Weil, J. H., Schantz, R., and Camara, B. (1992). Identification of a cDNA for the plastid-located geranylgeranyl pyrophosphate synthase from *Capsicum annuum*: correlative increase in enzyme activity and transcript level during fruit ripening. *Plant J.* **2**, 25–34.

Lambert, D. H., Bryant, D. A., Stirewalt, V. L., Dubbs, J. M., Stevens, S. E., and Porter, R. D. (1985). Gene map for the *Cyanophora paradoxa* cyanelle genome. *J. Bacteriol.* **164**, 659–664.

Lemaux, P. G., and Grossman, A. R. (1984). Isolation and characterization of a gene for a major light-harvesting polypeptide from *Cyanophora paradoxa*. *Proc. Natl. Acad. Sci. U.S.A.* **81**, 4100–4104.

Lemaux, P. G., and Grossman, A. R. (1985). Major light-harvesting polypeptides encoded in polycistronic transcripts in a eukaryotic alga. *EMBO J.* **4**, 1911–1919.

Lindahl, L., and Zengel, J. M. (1986). Ribosomal genes in *Escherichia coli*. *Annu. Rev. Genet.* **20**, 186–193.

Lockhart, P. J., Howe, C. J., Bryant, D. A., Beanland, T. J., and Larkum, A. W. D. (1992). Substitutional bias confounds inference on cyanelle origins from sequence data. *J. Mol. Evol.* **34**, 153–162.

Löffelhardt, W., Mucke, H., Crouse, E. J., and Bohnert, H. J. (1983). Comparison of the cyanelle DNA from two different strains of *Cyanophora paradoxa*. *Curr. Genet.* **7**, 139–144.

Löffelhardt, W., Michalowski, C. B., and Bohnert, H. J. (1985). Sequence homologies between cyanobacterial and chloroplast genes. *Endocytobiosis Cell Res.* **2**, 213–231.

Löffelhardt, W., Mucke, H., Breiteneder, H., Aryee, D. N. T., Seiser, C., Michalowski, C., Kaling, M., and Bohnert, H. J. (1987). The cyanelle genome of *Cyanophora paradoxa*. Chloroplast and cyanobacterial features. *Ann. N.Y. Acad. Sci.* **503**, 550–553.

Löffelhardt, W., Kraus, M., Pfanzagl, B., Götz, M., Brandtner, M., Markmann-Mulisch, U., Subramanian, A. R., Michalowski, C. B., and Bohnert, H. J. (1990). Cyanelle genes for components of the translation apparatus. *In* "Endocytobiology IV" (P. Nardon, ed.), pp. 561–564. INRA, Paris.

Löffelhardt, W., Michalowski, C. B., Kraus, M., Pfanzagl, B., Neumann-Spallart, C., Jakowitsch, J., Brandtner, M., and Bohnert, H. J. (1991). Rps10 and six other ribosomal protein genes from the S10/*spc* operon not encountered on higher plant plastid DNA are located on the cyanelle genome of *Cyanophora paradoxa*. *In* "The Translational Apparatus of Photosynthetic Organelles" (R. Mache, E. Stutz, and A. R. Subramanian, eds.), pp. 155–165. Springer-Verlag, Berlin.

Löffelhardt, W., Michalowski, C. B., Kraus, M., Neumann-Spallart, C., Jakowitsch, J., Flachmann, R., and Bohnert, H. J. (1993). Towards the complete structure of the *Cyanophora paradoxa* cyanelle genome. *In* "Endocytobiology V" (H. Ishikawa, ed.), pp. 189–194. Tübingen Univ. Press, Tübingen, Germany.

Loiseaux-de Goër, S., Markowicz, Y., Dalmon, J., and Audren, H. (1988). Physical maps of the two circular plastid DNA molecules of the primitive brown alga *Pylaiella littoralis*. *Curr. Genet.* **14**, 155–162.

Lüttke, A. (1991). On the origin of chloroplasts and rhodoplasts: protein sequence comparison. *Endocytobiosis Cell Res.* **8**, 75–92.

Maid, U., and Zetsche, K. (1990). The psbA gene from a red alga resembles those from cyanobacteria and cyanelles. *Curr. Genet.* **17**, 255–259.

Maid, U., and Zetsche, K. (1991). Structural features of the plastid ribosomal RNA operons of two red algae: *Antithamnion* sp. and *Cyanidium caldarium*. *Plant Mol. Biol.* **16**, 537–546.

Maid, U., Steinmüller, R., and Zetsche, K. (1992). Structure and expression of a plastid-encoded *groEL* homologous heat shock gene in a thermophilic unicellular red alga. *Curr. Genet.* **21**, 521–525.

Mangeney, E., and Gibbs, S. P. (1987). Immunocytochemical localization of ribulose-1,5-bisphosphate carboxylase/oxygenase in the cyanelles of *Cyanophora paradoxa* and *Glaucocystis nostochinearum*. *Eur. J. Cell Biol.* **43**, 65–70.

Mangeney, E., Hawthornthwaite, A. M., Codd, G. A., and Gibbs, S. P. (1987). Immunocytochemical localization of phosphoribulose kinase in the cyanelles of *Cyanophora paradoxa* and *Glaucocystis nostochinearum*. *Plant Physiol.* **84**, 1028–1032.

Manhart, J. R., and Palmer, J. D. (1990). The gain of two chloroplast tRNA introns marks the green algal ancestors of land plants. *Nature (London)* **345**, 268–270.

Margulis, L. (1981). "Symbiosis in Cell Evolution." Freeman, San Francisco.

Markmann-Mulisch, U., and Subramanian, A. R. (1988). Nucleotide sequence and linkage map position of the genes for ribosomal proteins L14 and S8 in the maize chloroplast genome. *Eur. J. Biochem.* **170**, 507–514.

Markowicz, Y., Mache, R., and Loiseaux-de Goër, S. (1988). Sequence of the plastid rDNA spacer region of the brown alga *Pylaiella littoralis* (L.) Kjellm. Evolutionary significance. *Plant Mol. Biol.* **10**, 465–469.

Marsh, L. E. (1979). Electrophoretic analysis of the ribosomal RNAs of some primitive eukaryotic algae. Ph.D. Thesis, Univ. of Delaware, Newark.

Maxwell, E. S., Liu, J., and Shively, J. M. (1986). Nucleotide sequence of *Cyanophora paradoxa* cellular and cyanelle-associated 5S ribosomal RNAs: the cyanelle as a possible intermediate in plastid evolution. *J. Mol. Evol.* **23**, 300–304..

Michalowski, C. B., Schmitt, J. M., and Bohnert, H. J. (1989). Expression during salt stress and nucleotide sequence of cDNA for ferredoxin-NADP+-reductase from *Mesembryanthemum crystallinum*. *Plant Physiol.* **89**, 817–822.

Michalowski, C. B., Bohnert, H. J., and Löffelhardt, W. (1990a). A novel allophycocyanin gene (*apcD*) from *Cyanophora paradoxa* cyanelles. *Nucleic Acids Res.* **18**, 2186.

Michalowski, C. B., Pfanzagl, B., Löffelhardt, W., and Bohnert, H. J. (1990b). The cyanelle *S10-spc* ribosomal protein gene operon from *Cyanophora paradoxa*. *Mol. Gen. Genet.* **224**, 222–231.

Michalowski, C. B., Flachmann, R., Löffelhardt, W., and Bohnert, H. J. (1991a). Gene *nadA*, encoding quinolinate synthetase, is located on the cyanelle DNA from *Cyanophora paradoxa*. *Plant Physiol.* **95**, 329–330.

Michalowski, C. B., Bohnert, H. J., and Löffelhardt, W. (1991b). ORF323 with homology to *crtE*, specifying prephytoéne pyrophosphate dehydrogenase, is encoded by cyanelle DNA in the eukaryotic alga *Cyanophora paradoxa*. *J. Biol. Chem.* **266**, 11866–11870.

Misawa, N., Nakagawa, M., Kobayashi, K., Yamano, S., Izawa, Y., Nakamura, K., and Harashima, K. (1990). Elucidation of the *Erwinia uredovora* carotenoid biosynthetic pathway by functional analysis of the gene products expressed in *Escherichia coli*. *J. Bacteriol.* **172**, 6704–6712.

Mollenhauer, D. (1992). *Geosiphon pyriforme. In* "Algae and Symbiosis" (W. Reisser, ed.), pp. 339–351. Biopress, Bristol, England.

Nakai, M., Tanaka, A., Omata, T., and Endo, T. (1992). Cloning and characterization of the *secY* gene from the cyanobacterium *Synechococcus* sp. PCC 7942. *Biochim. Biophys. Acta* **1171**, 113–116.

Neumann-Spallart, C., Brandtner, M., Kraus, M., Jakowitsch, J., Bayer, M. G., Maier, T. L., Schenk, H. E. A., and Löffelhardt, W. (1990). The *petFI* gene encoding ferredoxin I is located close to the *str* operon on the cyanelle genome of *Cyanophora paradoxa*. *FEBS Lett.* **268**, 55–58.

Neumann-Spallart, C., Jakowitsch, J., Kraus, M., Brandtner, M., Bohnert, H. J., and Löffelhardt, W. (1991). *Rps10*, unreported for plastic DNAs, is located on the cyanelle genome and is cotranscribed with the *str* operon. *Curr. Genet.* **19**, 313–315.

Orsat, B., Montfort, A., Chatellard, P., and Stutz, E. (1992). Mapping and sequencing of an actively transcribed *Euglena gracilis* chloroplast gene (ccsA) homologous to the Arabidopsis thaliana nuclear gene cs(ch-42). *FEBS Lett.* **303**, 181–184.

Palenik, B., and Haselkorn, R. (1992). Multiple evolutionary origins of prochlorophytes, the chlorophyll b-containing prokaryotes. *Nature (London)* **355**, 265–267.

Pancic, P. G., Strotmann, H., and Kowallik, K. V. (1991). The d subunit of the chloroplast ATPase is plastid-encoded in the diatom *Odontella sinensis*. *FEBS Lett.* **280**, 387–392.

Pancic, P. G., Strotmann, H., and Kowallik, K. V. (1992). Chloroplast ATPase genes in the diatom *Odontella sinensis* reflect cyanobacterial characters in structure and arrangement. *J. Mol. Biol.* **224**, 529–536.

Park, J. T. (1987). The murein sacculus. *In* "*Escherichia coli* and *Salmonella typhimurium*" (F. C. Neidhardt, ed.), Vol. 1, pp. 23–30. Am. Soc. Microbiol., Washington, D.C.

Pascher, A. (1929). Studien über Symbiosen. I. Über einige Endosymbiosen von Blaualgen in Einzellern. *Jahrb. Wiss. Bot.* **71**, 386–462.

Pfanzagl, B., Pittenauer, E., Allmaier, G., Schmid, E., Martinez, J., de Pedro, M. A., and Löffelhardt, W. (1993). Structure and biosynthesis of cyanelle peptidoglycan from *Cyanophora paradoxa. In* "Bacterial Growth and Lysis" (M. A. de Pedro, J.-V. Höltje, and W. Löffelhardt, eds.), pp. 47–55. Plenum, New York.

Pickett-Heaps, J. (1972). Cell Division in *Cyanophora paradoxa. New Phytol.* **71**, 561–567.

Pittenauer, E., Allmaier, G., Schmid, E. R., Pfanzagl, B., Löffelhardt, W., Quintela Fernandez, C., de Pedro, M. A., and Stanek, W. (1993). Structural characterization of the cyanelle peptidoglycan of *Cyanophora paradoxa* by ^{252}Cf-plasma desorption mass spectrometry and fast atom bombardment/tandem mass spectrometry. *Biol. Mass Spectrom.* **22**, 524–536.

Plaimauer, B., Pfanzagl, B., Berenguer, J., de Pedro, M. A., and Löffelhardt, W. (1991). Subcellular distribution of enzymes involved in the biosynthesis of cyanelle murein in the protist *Cyanophora paradoxa*. *FEBS Lett.* **284**, 168–172.

Price, G. D., and Badger, M. R. (1989). Expression of human carbonic anhydrase in the cyanobacterium *Synechococcus* PCC7942 creates a high CO_2-requiring phenotype. *Plant Physiol.* **91**, 505–513.

Reisser, W. (1984). Endosymbiotic cyanobacteria and cyanellae. *In* "Encyclopedia Plant Physiology, New Series" (H. F. Linskens and J. Heslop-Harrison, eds.), Vol. 17, pp. 91–112. Springer-Verlag, Berlin.

Reith, M. (1992). *psaE* and *trnS*(GCA) are encoded on the plastid genome of the red alga *Porphyra umbilicalis*. *Plant Mol. Biol.* **18**, 773–775.

Reith, M., and Munholland, J. (1991). An *hsp70* homolog is encoded on the plastid genome of the red alga, *Porphyra umbilicalis*. *FEBS Lett.* **294**, 116–120.

Reith, M., and Munholland, J. (1993a). A high-resolution gene map of the chloroplast genome of the red alga *Porphyra purpurea*. *Plant Cell* **5**, 465–475.

Reith, M., and Munholland, J. (1993b). Two amino-acid biosynthetic genes are encoded on the plastid genome of the red alga *Porphyra umbilicalis*. *Curr. Genet.* **23**, 59–65.

Rhiel, E., Stirewalt, V. L., Gasparich, G. E., and Bryant, D. A. (1992). The *psaC* genes of *Synechococcus* sp. PCC7002 and *Cyanophora paradoxa*: cloning and sequence analysis. *Gene* **112**, 123–128.

Scaramuzzi, C. D., Stokes, H., and Hiller, R. G. (1992a). Characterization of a chloroplast-encoded *secY* homologue and *atpH* from a chromophytic alga. *FEBS Lett.* **304**, 119–123.

Scaramuzzi, C. D., Hiller, R. G., and Stokes, H. W. (1992b). Identification of a chloroplast-encoded *secA* gene homologue in a chromophytic alga: possible role in chloroplast protein translocation. *Curr Genet.* **22**, 421–427.

Scaramuzzi, C. D., Stokes, H., and Hiller, R. G. (1992c). Heat shock Hsp70 protein is chloroplast-encoded in the chromophytic alga *Pavlova lutherii*. *Plant Mol. Biol.* **18**, 467–476.

Schenk, H. E. A. (1990). *Cyanophora paradoxa*: a short survey. *In* "Endocytobiology IV" (P. Nardon, ed.), pp. 199–209. INRA, Paris.

Schenk, H. E. A. (1992). Cyanobacterial symbioses. *In* "The Prokaryotes" (A. Ballows, H. G. Truper, M. Dworkin, W. Harder, and K. H. Schleifer, eds.), Vol. 4, pp. 3819–3854. Springer-Verlag, New York.

Schenk, H. E. A., Poralla, K., Härtner, T., Deimel, R., and Thiel, D. (1985). Lipids in *Cyanophora paradoxa*. I. Unusual fatty acid pattern of *Cyanocyta korschikoffiana*. *Endocytobiosis Cell Res.* **2**, 233–238.

Schenk, H. E. A., Bayer, M. G., and Maier, T. (1987). Nitrate assimilation and regulation of biosynthesis and disintegration of phycobiliproteids by *Cyanophora paradoxa*. Indications for a nitrate store function of the phycobiliproteids. *Endocytobiosis Cell Res.* **4**, 167–176.

Scherer, S., Herrmann, G., Hirschberg, J., and Boeger, P. (1991). Evidence for multiple xenogenous origins of plastids: comparison of *psbA* genes with xanthophyte sequence. *Curr. Genet.* **19**, 503–507.

Schleifer, K. H., Leuteritz, M., Weiss, N., Ludwig, W., Kirchhof, G., and Seidel-Rufer, H. (1990). Taxonomic study of anaerobic, gram-negative, rod-shaped bacteria from breweries: emended description of *Pectinatus cerevisiiphilus* and description of *Pectinatus frisingensis* sp. nov., *Selenomonas lecticifex* sp. nov., *Zymophilus raffinosivorans* gen. nov., sp. nov., and *Zymophilus paucivorans* sp. nov. *Int. J. Syst. Bacteriol.* **40**, 19–28.

Schlichting, R., Zimmer, W., and Bothe, H. (1990). Exchange of metabolites in *Cyanophora paradoxa* and its cyanelles. *Bot. Acta* **103**, 392–398.

Schluchter, W. M., and Bryant, D. A. (1992). Molecular characterization of ferredoxin-NADP$^+$oxidoreductase in cyanobacteria: cloning and sequence of the *petH* gene of *Synechococcus* sp. PCC 7002 and studies on the gene product. *Biochemistry* **31**, 3092–3102.

Schwarz, R., Lieman-Hurwitz, J., Hassidim, M., and Kaplan, A. (1992). Phenotypic complementation of high CO_2-requiring mutants of the cyanobacterium *Synechococcus* sp. strain PCC 7942 by inosine 5'-monophosphate. *Plant Physiol.* **100**, 1987–1993.

Scott, O. T. (1987). Major acyl lipids of cyanelles from *Glaucocystis nostochinearum*. *Ann. N.Y. Acad. Sci.* **503**, 555–558.

Scott, O. T., Castenholz, R. W., and Bonnett, H. T. (1984). Evidence for a peptidoglycan envelope in the cyanelles of *Glaucocystis nostochinearum* Itzigsohn. *Arch. Microbiol.* **139**, 130–138.

Segui-Real, B., Stuart, R. A., and Neupert, W. (1992). Transport of proteins into the various subcompartments of mitochondria. *FEBS Lett.* **313**, 2–7.

Sherman, M. M., Petersen, L. A., and Poulter, C. D. (1989). Isolation and characterization of isoprene mutants of *Escherichia coli*. *J. Bacteriol.* **171**, 3619–3628.

Shiba, K., Ito, K., Yura, T., and Ceretti, D. P. (1984). A defined mutation in the protein export gene within the *spc* ribosomal protein operon of *Escherichia coli:* isolation and characterization of a new temperature-sensitive *secY* mutant. *EMBO J.* **3,** 631–636.

Shivji, M. S., Li, N., and Cattolico, R. A. (1992). Structure and organization of rhodophyte and chromophyte plastid genomes: implications for the ancestry of plastids. *Mol. Gen. Genet.* **232,** 65–73.

Smeekens, S., Weisbeek, P., and Robinson, C. (1990). Protein transport into and within chloroplasts. *Trends Biochem. Sci.* **15,** 73–76.

Starnes, S. M., Lambert, D. H., Maxwell, E. S., Stevens, S. E., Porter, R. D., and Shively, J. M. (1985). Cotranscription of the large and small subunit genes of ribulose-1,5-bisphosphate carboxylase/oxygenase in *Cyanophora paradoxa. FEMS Microbiol. Lett.* **28** 165–169.

Stefanovic, S., Bayer, M. G., Tröger, W., and Schenk, H. E. A. (1990). *Cyanophora paradoxa* Korsch.: Ferredoxin partial amino-terminal amino acid sequence, phylogenetic/taxonomic evidence. *Endocytobiosis Cell Res.* **6,** 219–226.

Stirewalt, V. L., and Bryant, D. A. (1989a). Nucleotide sequence of the *psbK* gene of the cyanelle of *Cyanophora paradoxa. Nucleic Acids Res.* **17,** 10096.

Stirewalt, V. L., and Bryant, D. A. (1989b). Molecular cloning and nucleotide sequence of the *petG* gene of the cyanelle genome of *Cyanophora paradoxa. Nucleic Acids Res.* **17,** 10095.

Sugiura, M. (1992). The chloroplast genome. *Plant Mol. Biol.* **19,** 149–168.

Tandeau de Marsac, N., Mazel, D., Damerval, T., Guglielmi, G., Capuano, V., and Houmard, J. (1988). Photoregulation of gene expression in the filamentous cyanobacterium *Calothrix* sp. PCC 7601: light harvesting complexes and cell differentiation. *Photosynth. Res.* **18,** 99–132.

Thomas, F., Massenet, O., Dorne, A. M., Briat, J. F., and Mache, R. (1988). Expression of the *rpl23, rpl2* and *rps19* genes in spinach chloroplasts. *Nucleic Acids Res.* **16,** 203–209.

Trench, R. K. (1979). The cell biology of plant–animal symbiosis. *Annu. Rev. Plant Physiol.* **30,** 485–531.

Trench, R. K. (1982a). Physiology, biochemistry, and ultrastructure of cyanellae. *In* "Progress in Phycological Research" (F. E. Round and D. J. Chapman, eds.), Vol. 1, pp. 257–288. Elsevier, Amsterdam.

Trench, R. K. (1982b). Cyanelles. *In* "On the Origin of Chloroplasts" (J. A. Schiff, ed.), pp. 55–76. Elsevier/North-Holland, New York.

Urbach, E., Robertson, D. L., and Chisholm, S. W. (1992). Multiple evolutionary origins of prochlorophytes within the cyanobacterial radiation. *Nature (London)* **355,** 267–270.

Valentin, K. (1993). SecA on the plastid genome of a red alga. *Mol. Gen. Genet.* **236,** 245–250.

Valentin, K., and Zetsche, K. (1990a). Nucleotide sequence of the gene for the large subunit of Rubisco from *Cyanophora paradoxa*—phylogenetic implications. *Curr. Genet.* **18,** 199–202.

Valentin, K., and Zetsche, K. (1990b). Rubisco genes indicate a close phylogenetic relation between the plastids of chromophyta and rhodophyta. *Plant Mol. Biol.* **15,** 575–584.

Valentin, K., and Zetsche, K. (1990c). Structure of the Rubisco operon from the unicellular red alga *Cyanidium caldarium:* evidence of a polyphyletic origin of the plastids. *Mol. Gen. Genet.* **222,** 425–430.

Valentin, K., Cattolico, R. A., and Zetsche, K. (1992a). Phylogenetic origin of the plastids. *In* "Origins of Plastids" (R. Lewin, ed.), pp. 195–222. Chapman & Hall, London.

Valentin, K., Maid, U., Emich, A., and Zetsche, K. (1992b). Organization and expression of a phycobiliprotein gene cluster from the unicellular red alga *Cyanidium caldarium. Plant Mol. Biol.* **20,** 267–276.

Wang, S., and Liu, X. Q. (1991). The plastid genome of *Cryptomonas* Φ encodes an hsp70-like protein, a histone-like protein, and an acyl carrier protein. *Proc. Natl. Acad. Sci. U.S.A.* **88**, 10783–10787.

Wasmann, C. C. (1985). The cyanelle and the cyanelle genome of *Cyanophora paradoxa*. Ph.D. Thesis, Michigan State Univ., East Lansing.

Wasmann, C. C., Löffelhardt, W., and Bohnert, H. J. (1987). Cyanelles: organization and molecular biology. *In* "The Cyanobacteria" (P. Fay and C. Van Baalen, eds.), pp. 303–324. Elsevier, Amsterdam.

Wickner, W., Driessen, A. J. M., and Hartl, F.-U. (1991). The enzymology of protein translocation across the *Escherichia coli* plasma membrane. *Annu. Rev. Biochem.* **60**, ·101–124.

Winhauer, T., Jaeger, S., Valentin, K., and Zetsche, K. (1991). Structural similarities between *psbA* genes from red algae and brown algae. *Curr. Genet.* **20**, 177–180.

Wolters, J., Erdmann, V. A., and Stackebrandt, E. (1990). Current status of the molecular phylogeny of plastids. *In* "Endocytobiology IV" (P. Nardin, ed.), pp. 545–552. INRA, Paris.

Xu, M. Q., Kathe, S. D., Goodrich-Blair, H., Nierzwicki-Bauer, S. A., and Shub, D. A. (1990). Bacterial origin of a chloroplast intron: conserved self-splicing group I introns in cyanobacteria. *Science* **250**, 1566–1570.

Zhou, D. X., Quigley, F., Massenet, O., and Mache, R. (1989). Cotranscription of the S10- and spc-like operons in spinach chloroplasts and identification of their gene products. *Mol. Gen. Genet.* **216**, 439–445.

Zook, D., and Schenk, H. E. A. (1986). Lipids in *Cyanophora paradoxa*. III. Lipids in cell compartments. *Endocytobiosis Cell Res.* **3**, 203–211.

Zook, D., Schenk, H. E. A., Thiel, D., Poralla, K., and Härtling, T. (1986). Lipids in *Cyanophora paradoxa:* II. Arachidonic acid in the lipid fractions of the endocyanelle. *Endocytobiosis Cell Res.* **3**, 99–104.

Diverse Distribution and Function of Fibrous Microtubule-Associated Proteins in the Nervous System

Thomas A. Schoenfeld* and Robert A. Obar†
*Departments of Psychology and Biology and the Neuroscience Program, Clark University, Worcester, Massachusetts 01610, and †Alkermes, Inc., Cambridge, Massachusetts 02139

I. Introduction

Microtubule-associated proteins (MAPs) are a diverse set of molecules associated with cytoplasmic microtubules. They are believed to play important roles in assembly of tubulin heterodimers into microtubules and in regulation of the dynamic instability of tubulin polymers. These proteins are also observed to form lateral extensions between adjacent microtubules, between microtubules and the other constituents of the cytoskeleton (intermediate filaments, microfilaments), and between microtubules and membranous organelles when they are either anchored to or transported along the cytoskeleton. These extensions or cross-bridges may contribute significantly to the density and stability of microtubule arrays. MAPs may be important intermediaries as well between microtubules and cellular events such as calcium influx or stimulation by neurotrophic factors. The response of MAPs to these events, through the action of a variety of protein kinases, may cause alterations in microtubule assembly and stability, and in the interactions between microtubules and other cellular constituents. MAPs are particularly prominent in the nervous system, where their diverse distribution in different classes and compartments of cells at different stages of development has led to the view that they play diverse roles in neuronal function (Matus *et al.*, 1983; Vallee and Bloom, 1984; Vallee *et al.*, 1984; Olmsted, 1986; Matus, 1988a,b; Meininger and Binet, 1989; Wiche, 1989; Tucker, 1990; Burgoyne, 1991; Goedert *et al.*, 1991; Hirokawa, 1991; Wiche *et al.*, 1991).

A recent review of MAPs in this series focused on their molecular biology and biochemistry as observed in a variety of different tissues

(Wiche *et al.*, 1991). The focus of this chapter is on the cellular and subcellular distribution in the vertebrate nervous system of fibrous MAPs, those MAPs observed to extend as side arms from microtubules to other subcellular elements. The topics are organized to give particular emphasis to the functional roles with which the various fibrous MAPs have been associated and which they are presumed to play in nervous tissues: the establishment and maintenance of cellular form, the differentiation of particular types of neuronal and supporting cells in diverse regions of the nervous system, and the generation and regulation of the outgrowth, plasticity, and stabilization of neuronal processes. The substantial literature addressing the role of MAPs in the etiology of neurodegenerative disorders such as Alzheimer's disease has been amply reviewed elsewhere (Selkoe, 1986; Davies, 1988; Klunk and Abraham, 1988; Grundke-Iqbal and Iqbal, 1989; Selkoe *et al.*, 1990; Kosik, 1989, 1991, 1992; Lien *et al.*, 1991) and so will not be discussed here.

II. Definition of Microtubule-Associated Proteins

MAPs are nontubulin proteins which copurify to constant stoichiometry with tubulin either during cycles of temperature-dependent microtubule assembly and disassembly (Murphy and Borisy, 1975; Sloboda *et al.*, 1975) or during taxol-stimulated microtubule assembly (Vallee, 1982). Although MAPs are thought to provide important links between microtubules and other cellular elements, in some instances bona fide MAPs may be associated with other cytoskeletal constituents even in the absence of microtubules (e.g., MAP 2 may be associated with actin in dendritic spines: see Section III, B, 2). For fibrous MAPs (see the later discussion), microscopic localization on ("decoration" of) microtubules *in vitro* and *in vivo* has been an additional criterion (Dentler *et al.*, 1975; Murphy and Borisy, 1975). Interestingly, promoting microtubule assembly has not been an essential criterion for distinguishing MAPs from other cytoskeletal proteins, even though all fibrous MAPs and many other MAPs purportedly show this property (see Section V).

MAPs have traditionally been categorized in terms of molecular mass, thermostability, and structural and functional attributes. For purposes of this chapter, we will refer to three molecular mass classes—high (HMM), intermediate (IMM), and low (LMM)—which reflect a number of recent advances in our understanding of the diversity of these proteins, particularly the increasing number of recognized IMM proteins (see Table I). Cutting across these classes, the demonstration of thermal stability (e.g., for MAP 2, tau protein, and all of the IMM MAPs to date)

TABLE I

Major Categories of Nervous Tissue MAPs According to the Molecular Mass of the Principal Subunits

Category	Species[a]
High molecular mass (HMM) (± 300 kDa)	MAP 1A
	MAP 1B
	Cytoplasmic dynein
	MAPs 2a/2b
Intermediate molecular mass (IMM) (± 200 kDa)	MAP 4
	Bovine MAP 4 (190 kDa MAP, MAP U)
	205–210-kDa MAPs
	MAP 3
Low molecular mass (LMM) (± 100 kDa)	Kinesin
	Dynamin
	STOPs
	MAP 2c
	Chartins
	Tau

[a] These names reflect the most common current usage in the field (see text for citations). Alternative nomenclature based on biochemical and immunochemical homologies, especially for MAPs 1A, 1B, and 2c, and the IMM MAPs has been discussed by Tucker (1990), Wiche *et al.* (1991), and Aizawa *et al.* (1991).

or lability (e.g., for MAPs 1A and 1B and the chartins) reflects molecular differences that are not yet fully understood (Vallee, 1985; Olmsted, 1986) but nevertheless underscore increasingly apparent familial relationships, for example, among the IMM proteins (Aizawa *et al.*, 1990) or between MAPs 1A and 1B (Schoenfeld *et al.*, 1989; Hammarback *et al.*, 1991; Langkopf *et al.*, 1992). Interestingly, but circumstantially, the effect of phosphorylation on the affinity of a particular MAP for microtubules is correlated with its thermostability: phosphorylation tends to suppress this affinity in thermostable MAPs and enhance it in thermolabile MAPs (see Section V).

As for structural and functional attributes, three categories are recognized. The fibrous MAPs, which are the focus of this chapter, are structural proteins that are believed to act as cross-bridges in the cytomatrix. This group currently includes MAPs 1A and 1B, the MAP 2 polypeptides, IMM MAPs, and tau. The lateral extensions or cross-bridges formed by fibrous MAPs may have diverse consequences and functions within cells: regulating shape and size, outgrowth, plasticity and stability, even cellular differentiation. Note that MAP 2 and other fibrous MAPs may also regulate activity of molecular motor MAPs (see next section). The molecular motor MAPs are enzymes that drive intracellular transport. These include cytoplasmic dynein, the kinesin family, and possibly dynamin. They have been been the focus of several recent reviews (Vallee and Shpetner, 1990; Vallee and Bloom, 1991) and will not be discussed here. The assembly regulatory proteins are a less explicitly designated group of putative MAPs that have not been shown to be either fibrous proteins or mechanoenzymes but do appear in various ways to regulate the assembly of tubulin into microtubules. These include the stable tubule-only peptides, or STOPs (Job *et al.*, 1985), and the chartins (Magendantz and Solomon, 1985) (see Section V,B,1).

III. Neuronal Form

A role for MAPs in the establishment and regulation of neuronal form is inferred from their varied but specific associations with the three principal structural compartments or domains of neurons—axonal, somatic, and dendritic. Antibodies to some MAPs (MAP 2, tau) are now frequently used as compartment-specific markers (Bruckenstein and Higgins, 1988a,b), although there is evidence that all major MAPs are expressed in all three compartments under some circumstances. However, it is also becoming clear that MAPs are not likely to play a determining role in the actual differentiation of neuronal processes into the two regions—axonal and dendritic—that produce this compartmentalization. Instead, MAPs may be critical in supporting and modulating these differences once they are already established, to invest dendrites and axons with certain properties not derivable from microtubules alone, such as relative size and degree of branching.

A. Somatodendritic and Axonal Compartments

1. MAP 2

The distribution of MAP 2 was the first among the MAPs to be extensively studied and characterized. Immunocytochemical studies have almost universally demonstrated a preferential subcellular expression of MAP 2 in dendrites and somata but not axons in most regions of the adult central nervous system (CNS) *in vivo*. Matus *et al.* (1981) first reported that a polyclonal antibody to unfractionated HMM brain MAPs stained only dendrites and somata in neurons. This antibody was later reported to react with MAP 2 but not MAP 1 in immunoblots of differentiated material (Matus *et al.*, 1983).

Miller *et al.* (1982), using a polyclonal antibody prepared against highly purified MAP 2, provided more direct evidence that MAP 2 is restricted to the somatodendritic compartment. Moreover, they showed that an antibody to the regulatory subunit of cyclic adenosine monophosphate (cAMP)-dependent type II protein kinase (RII) stains in an identical pattern—only dendrites and somata. This subunit had been shown previously to bind to MAP 2 (Vallee *et al.*, 1981), and somatodendritic staining could be blocked by preadsorption of the RII antibody with a MAP 2-enriched microtubule fraction.

Subsequent studies using antibodies directed against isolated MAP 2 demonstrated that in stained sections of the adult rat brain, MAP 2 is indeed largely confined to the somatodendritic compartment of the neurons examined (Bernhardt and Matus, 1984; Caceres *et al.*, 1984b; DeCamilli *et al.*, 1984; Huber and Matus, 1984b). The largely somatodendritic localization of MAP 2 has been confirmed numerous times in a variety of tissues, from mature cell cultures (Caceres *et al.*, 1984a, 1986; Kosik and Finch, 1987) to postmortem human brain (Sims *et al.*, 1988; Trojanowski *et al.*, 1989b).

Nevertheless, there is also evidence that MAP 2 is occasionally associated with axons. In biochemical studies, although MAP 2 is found enriched in microtubule-rich fractions taken from gray matter, it is also present in low but detectable amounts in axon-enriched white matter (Vallee, 1982; Binder *et al.*, 1986). When MAP 2 immunoreactivity has been found in axon bundles, it has been expressed in a limited number of specific axon classes of both the CNS and peripheral nervous system (PNS) (Caceres *et al.*, 1984b; Papasozomenos *et al.*, 1985; Yee *et al.*, 1988) (see Section IV). It has also been restricted to certain anti-MAP 2 monoclonal antibodies (Caceres *et al.*, 1984b), or found in axon-associated astrocytes rather than axons (Papasozomenos and Binder, 1986)(see Section IV). There is also evidence for transient expression of MAP 2, and particularly the low-molecular mass isotype MAP 2c, in axons at early stages of development both *in vitro* (Caceres *et al.*, 1986; Kosik and Finch, 1987; Hernandez *et al.*, 1989) (see Section III,A,5) and *in vivo* (Tucker and Matus, 1987; Tucker *et al.*, 1988a,b)(see the later discussion and Section III,A,5) or following direct injection of MAP 2 into the cell bodies of mature neurons in culture (Okabe and Hirokawa, 1989)(see the following discussion).

The mechanism for sorting MAP 2 into the somatodendritic compartment of neurons is not known. MAP 2 mRNA is found in dendrites as well as perikarya (Garner *et al.*, 1988; Tucker *et al.*, 1989; Bruckenstein *et al.*, 1990; Kleiman *et al.*, 1990). However, an mRNA sequence specific for the lower molecular mass MAP 2c is found only in perikarya (Papandrikopolou *et al.*, 1989). Localization of synthesis in the somatodendritic compartment does not by itself account for the regionalization of MAP 2 there, since the mRNA for axonal proteins such as tau and MAP 1 is also localized in that compartment, rather than in the axon (Fig. 1), as are the ribosomal sites of message translation (Kosik *et al.*, 1989a)(see Sections III,A,2 and III,A,3). Moreover, the roughly comparable distribution of both MAP 2 and tau mRNA in both neuronal dendrites and somata suggests that the location of MAP synthesis may not be a necessary factor in the targeting of these proteins for dendrites and somata versus axons. Nevertheless, the unique presence of MAP 2 mRNA in distal dendritic segments may be a factor in the distal distribution of MAP 2. Interestingly, MAP 2 injected into the cell body of a neuron *in vitro* migrates to the axon as well as to dendrites, although it is ultimately sequestered in the somatodendritic compartment (Okabe and Hirokawa, 1989). Perhaps some of the reported nondendritic localization of MAP 2 antigen is the result of leakiness in the transport process (i.e., failing to discriminate between tau protein and MAP 2) during high rates of protein synthesis. Thus, compartmentalization may be regulated more by the relative affinity of a protein for a subcellular locale than by the site of its synthesis

MAP 2 might form only weak associations with the cytoskeleton of the typical axon or may be competitively displaced by tau or another MAP in many axons (Okabe and Hirokawa, 1989). Regarding the latter hypothesis, there is evidence (Kim *et al.*, 1986) that MAP 2 displaces tau from whole-brain microtubules *in vitro* more successfully than tau displaces

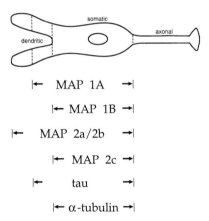

FIG. 1   Distribution of mRNA for various MAP isotypes and α-tubulin within the dendritic and somatic compartments of neurons (see text for citations). The drawing makes a distinction between mRNA restricted to the cell soma or extending as well into distal or proximal dendritic segments.

MAP2. However, these data may be more characteristic of dendritic microtubules or depend on the phosphorylation state of tau, which is normally phosphorylated in dendrites and dephosphorylated in axons (Papasozomenos and Binder, 1987)(see Section III,A,2). In this regard, it would be useful to focus studies of MAP 2–tau competitive binding on tissue extracts enriched in axons or somata-dendrites and to specifically manipulate the phosphorylation state.

There is also evidence that MAP 2 staining of developing or peripheral nerve axons is associated with expression of MAP 2c (Tucker and Matus, 1987; Tucker et al., 1988a,b; Hernandez et al., 1989; see also Yee et al., 1988, for reaction of their MAP 2 antibody to an LMM moiety on immunoblots of soleus nerve extract). Thus, the association of MAP 2c in the MAP 2 molecular complex may increase the affinity of the complex for axonal binding sites. The regulation of MAP 2c expression may thus hold the key to the subcellular compartmentalization of MAP 2 in some neurons. On the other hand, antibodies that are monospecific for the HMM MAP 2 doublet, and do not recognize MAP 2c, stain peripherally directed axons of spinal motor neurons and pseudounipolar sensory neurons (Papasozomenos et al., 1985), indicating that the expression of MAP 2c is not likely to be essential to the appearance of MAP 2 in axons generally. Moreover, a 70-kDa MAP corresponding to MAP 2c is found in calf brain gray matter but not in white matter (Vallee, 1982).

To date, only one of the anti-MAP 2 antibodies in use is known to recognize a phosphorylation site (AP18; Wille et al., 1992). Moreover, only one of the antibodies recognizes an individual MAP 2 isotype (c; Papandrikopoulou et al., 1989), and that has not been used to study isotype distribution in the nervous system. However, certain inferences about distribution can be made where only one or two of the subunits is biochemically prominent or where anti-MAP 2 antibodies react with only two of the subunits. For example, of the HMM isoforms, MAP 2a is only detected in preparations of adult brain whereas 2b is detected in both developing and mature tissues (Binder et al., 1984; Burgoyne and Cumming, 1984; Riederer and Matus, 1985), leading to the inference that immunocytochemical staining patterns in developing tissue with anti-MAP 2 antibodies (e.g., staining of Purkinje cell somata; Burgoyne and Cumming, 1984) represents the expression of the 2b but not the 2a isoform (see Section V). The development of antibodies that recognize either all three isoforms (antibody AP14 of Binder et al., 1986; polyclonal antibody of Fischer et al., 1987) or only the HMM doublet (antibody C of Huber and Matus, 1984b; monoclonal antibody I2 of Fischer et al., 1987) has permitted inferential assessment of the distribution of MAP 2c by subtractive comparisons of the staining patterns from the two antibodies (Tucker et al., 1988a,b; see review in Tucker, 1990). In adult tissues, the two classes of antibody yield nearly identical staining patterns, corresponding to the loss of expression of MAP 2c in adult tissue. However, in embryonic and neonatal tissue, the anti-HMM antibody stains only a subset of the pattern associated with the antibody that also reacts with MAP 2c. For example, the latter stains axons in spinal cord white matter and spinal nerves of the embryonic quail that are not stained with the anti-HMM antibody or seen in the adult with either antibody (Tucker *et al.*, 1988a).

Despite its occasional association with axons, MAP 2 immunoreactivity holds particular promise as a dendrite-specific marker in neuroanatomical studies. For example, it offers certain advantages over the Golgi stain in studies of dendritic arborization. In addition to the greater ease of immunocytochemistry compared with the Golgi method, anti-MAP 2 antibodies can be expected to mark the dendrites of every neuron of a given class in a tissue section, whereas the Golgi stain, for unknown reasons, reacts with only about 1–10% of the cells in a given section (Scheibel and Scheibel, 1978). Thus, the spatial relationships of the dendrites of neighboring neurons can be more effectively and comprehensively visualized when they are marked with MAP 2 immunoreactivity (e.g., the vertical and tangential arrangements of neocortical pyramidal cell apical dendrites; Escobar *et al.*, 1986).

2. Tau protein

In contrast to MAP 2, tau protein has been found to be enriched in the axonal compartment of neurons. Tytell *et al.* (1984) first reported that tau, but not HMM MAPs, can be detected biochemically in the optic nerve and tract. Binder *et al.* (1985) then reported that tau may be associated exclusively with axons. Their biochemical analyses of bovine brain showed that axon-enriched white matter contains about three times the amount of tau (relative to tubulin) that gray matter does. Moreover, immunocytochemical analysis of rat brain sections through the cerebellum, medulla, and midbrain revealed that tau immunoreactivity (using monoclonal antibody Tau-1) is localized in axons, particularly white matter tracts and axonal bundles coursing through gray matter, but not in neuronal somata or dendrites. Other antibodies to tau present a similar pattern of predominantly axonal immunoreactivity in rat brain (Kowall and Kosik, 1987; Brion *et al.*, 1988). Antibodies to tau also stain only axons in normal human brain (Wood *et al.*, 1986; Kowall and Kosik, 1987; Trojanowski *et al.*, 1989a). Moreover, *in vitro* studies of tau immunoreactivity in neurons of mature cultures have generally confirmed these findings as well (Peng *et al.*, 1986; Kosik and Finch, 1987).

However, a variety of studies have shown that tau is not restricted to axons in all neurons and in all circumstances. Vallee (1982) demonstrated biochemically that tau protein is present in both gray and white matter from calf brain at equivalent ratios to tubulin, although he noted that the relative amounts of the individual isotypes are different between gray and white matter. Papasozomenos and Binder (1987) discovered that antibody Tau-1 reacts with neuronal dendrites and perikarya as well as axons throughout the CNS when the tissue is pretreated with alkaline phosphatase. This corresponds to a 50–65% increase in the binding of Tau-1 to the protein on immunoblots. Moreover, similar staining of all three neuronal compartments was achieved with a second monoclonal antibody (Tau-2; Papasozomenos and Binder, 1987) that binds to tau even without dephosphorylation, although immunoreactivity was strong in monkey brain sections but not in rat brain sections.

These data indicate that the subcellular distribution of tau may vary only by the degree of phosphorylation of the protein, with axonal tau being dephosphorylated and somatodendritic tau being phosphorylated at or near the Tau-1 antigenic site. However, this may not hold true for all neurons. For example, Purkinje cell somata react only poorly with anti-tau antibodies with or without phosphatase pretreatment (Papasozomenos and Binder, 1987). Moreover, the somata of some olfactory bulb mitral, tufted, and periglomerular cells (Papasozomenos and Binder, 1987)(see Fig. 2) and cerebellar granule cells, among a few others (Migheli *et al.*, 1988), are stained by anti-tau antibodies without prior phosphatase treatment. Cultured hippocampal neurons are also reported to show tau immunoreactivity in all neuronal compartments without dephosphorylation, even at a time of sufficient *in vitro* differentiation to demonstrate a somatodendritic restriction of MAP 2 (Dotti *et al.*, 1987). On the other hand, the axons of cultured sympathetic ganglia neurons reportedly express phosphorylated tau (Peng *et al.*, 1986).

It is of interest that antibodies to tau stain the somata and dendrites of cerebral cortical pyramidal cells in Alzheimer's disease postmortem tissue (Tau-1; Wood *et al.*, 1986; 5E2; Kowall and Kosik, 1987). This is in fact where the neurofibrillary tangles (NFTs) diagnostic of this disease are located. However, such antibodies are largely reactive only with axons in normal human brain. Moreover, both antibodies stain NFTs without prior dephosphorylation, although Tau-1 stains many more NFT-bearing neurons after such treatment (Wood *et al.*, 1986). A prevalent interpretation of these data is that tau becomes hyperphosphorylated or at least hypodephosphorylated during the course of the disease (reviews in Grundke-Iqbal and Iqbal, 1989; Kosik, 1991) and consequently accumulates in the somatodendritic compartment (Kowall and Kosik, 1987),

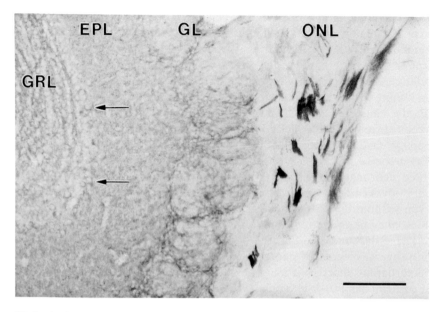

FIG. 2   Anti-tau protein immunoreactivity (antibody Tau-1) in restricted subsets of olfactory receptor neuron axons where they course through the olfactory nerve layer (ONL) of the ventral olfactory bulb of the rat. Mitral cell somata are lightly labeled (arrows). GRL, granule cell layer; EPL, external plexiform layer; GL, glomerular layer. Bar = 200 μm.

perhaps because phosphorylation is a somatodendritic or antiaxonal targeting signal (Kosik *et al.*, 1989a).

The demonstration of tau in the somatodendritic compartment of most neurons under some circumstances helps to resolve a paradox, namely, how tau can only rarely be found in the somatodendritic region when that is precisely where it must be synthesized, since axons do not have ribosomes and hence are expected to lack a mechanism for tau synthesis (Binder *et al.*, 1985). Recently, Kosik *et al.* (1989a) have affirmed that tau mRNA is coextensive with neuronal ribosomes and is thereby restricted to the somatodendritic compartment of neurons (Fig. 1). This suggests that tau must be transported to the axon. Nevertheless, despite the presence of tau in the somata and dendrites of neurons, its prominent enrichment in axons raises a number of questions regarding its fate following synthesis that have yet to be addressed. What form does tau take following message translation? Is it immediately phosphorylated, forming a somatodendritic pool from which the dephosphorylated axonal form of tau is drawn? Is some tau in the soma or dendrite dephosphorylated, or does some remain unphosphorylated? If so, does dephosphorylation/unphosphorylation become a targeting signal for transport to the axon, and why can't we detect more of this form of tau in the soma and dendrites? Or does the axon itself have phosphatases for the dephosphorylation of tau? In either case, what regulates the distribution of tau among the three compartments? The injection technique used by Okabe and Hirokawa (1989) to study the distribution of MAP 2 within neurons would be extremely useful in addressing a similar question about tau.

3. MAPs 1A and 1B

In contrast to heat-stable MAP 2 and tau, the MAP 1 family of thermolabile proteins has been found to be prominent in axons as well as in somata and dendrites. Vallee (1982) first reported that MAP 1 is equally enriched biochemically in gray matter and white matter preparations of bovine brain, which encouraged the expectation that it would be expressed comparably in both the axonal and somatodendritic compartments of neurons. This has been borne out in immunocytochemical studies. Matus and colleagues (Matus *et al.*, 1983; Huber and Matus, 1984a,b), Wiche *et al.* (1983) and Vallee and co-workers (Bloom *et al.*, 1984, 1985) all reported that antibodies to MAP 1 but not to MAP 2 stain axons in addition to somata and dendrites in a variety of regions of the CNS. Using monospecific antibodies to the differentiated MAP 1 polypeptides, 1A and 1B, Vallee and colleagues (Bloom et al. 1984, 1985; Hirokawa *et al.*, 1985) further demonstrated that this inclusive axonal–somatodendritic subcellular staining pattern is characteristic of both MAP 1A and MAP 1B. This pattern was also observed in subsequent studies of these two fibrous MAP 1 species (Bernhardt *et al.*, 1985; Riederer and Matus, 1985; Matus and Riederer, 1986; Riederer *et al.*, 1986; Cambray-Deakin *et al.*, 1987; Shiomura and Hirokawa, 1987a,b; Benjamin *et al.*, 1988; Tucker and Matus, 1988; McKerracher *et al.*, 1989; Okabe *et al.*, 1989; Schoenfeld *et al.*, 1989; Sato-Yoshitake *et al.*, 1989; Viereck *et al.*, 1989).

In vitro studies of MAP 1A and 1B distribution have also found that immunoreactivity is found in all compartments of cultured neurons, in contrast to the more selective staining by antibodies to MAP 2 and tau protein (Bloom *et al.*, 1984, 1985; Peng *et al.*, 1986). Although quantitative variations in the staining of dendrites and axons by anti-MAP 1A and anti-MAP 1B antibodies have been observed, that is, anti-MAP 1A antibodies stain dendrites more prominently than axons whereas anti-MAP 1B antibodies stain axons more prominently (Bloom *et al.*, 1984, 1985), qualitative subcellular variations in the expression of these two proteins have been noted only in early development (Schoenfeld *et al.*, 1989)(see Section V,A,1).

Evidence is accumulating that phosphorylated epitopes of MAP 1B are not expressed in dendrites and that phosphorylated MAP 1B is expressed only in axons and especially in developing axons (see also Sections V,A,1 and V,A,3,b). Luca *et al.* (1986) reported that a monoclonal antibody (called "anti-MAP 1B-3") to a phosphorylated epitope of MAP 1B also cross-reacts with phosphoepitopes of MAP 1A and the high- and middle-molecular mass neurofilament polypeptides. This was one of the first reports that isolated MAPs 1A and 1B are phosphoproteins and the first suggestion that they might be related in an evolutionary as well as a structural sense.

We have recently found (Schoenfeld *et al.*, 1989) that anti-MAP 1B-3 immunoreactivity in the developing and mature cerebellum corresponds to two other staining patterns. The staining of developing parallel fiber axons with this antibody is also seen with monospecific antibodies to MAP 1B (e.g., our anti-MAP 1B-4), whereas the staining of basket cell axons in the mature cerebellum with this antibody is also seen with antibodies to phosphorylated neurofilament polypeptides (Sternberger and Sternberger, 1983). The absence of any staining of Purkinje cell somata or dendrites at any age with this antibody, in contrast to the appearance of such staining at later ages with monospecific antibodies to MAP 1B, is evidence that MAP 1B is not phosphorylated in the somatodendritic compartment.

Curiously, although anti-MAP 1B-3 cross-reacts with MAP 1A as well as with MAP 1B and neurofilament polypeptides, immunostaining of rat cerebellum does not produce a pattern which is uniquely characteristic of MAP 1A immunoreactivity as seen with monospecific anti-MAP 1A antibodies. Immunoreactivity patterns are characteristic only of MAP 1B or neurofilament protein expression. Thus, we cannot at present characterize the subcellular or even cellular distribution of phosphorylated MAP 1A. Such characterization will undoubtedly require the use of an antibody specific to phosphoepitopes on MAP 1A (deMey *et al.*, 1987).

Comparable findings regarding phospho-MAP 1B subcellular expression were also presented recently by Sato-Yoshitake *et al.* (1989), who raised two monoclonal antibodies to MAP 1B, one recognizing nonphosphorylated epitopes (antibody 5E6) and the other reactive only with a phosphorylated epitope (antibody 1B6). The antibodies did not cross-react with MAP 1A. The authors do not report whether either antibody showed any cross-reactivity with phosphorylated or nonphosphorylated neurofilament polypeptides, although immunostaining did not produce any patterns uniquely characteristic of antineurofilament immunoreactivity (e.g., there was no staining of basket cell axons in the cerebellum). They found that 5E6 produced staining of both somata-dendrites and axons (e.g., both Purkinje cells and parallel fibers, both gray and white matter of the spinal cord), whereas 1B6 stained only axons. This was true in adulthood as well as in infancy. The staining pattern for 5E6 corresponds precisely to that for our MAP 1B-4 antibody. However, the staining pattern for 1B6 differs from that for our anti-MAP 1B-3 antibody, in that anti-MAP 1B-3 does not stain classes of axons in adulthood which are also immunoreactive with anti-MAP 1B-4 and other antibodies monospecific to MAP 1B (e.g., cerebellar parallel fibers).

Given the less restricted expression of the MAP 1 polypeptides throughout the neuron in comparison with the distributions of MAP 2 and tau protein, it is of interest that the synthesis of MAP 1A and MAP 1B is at least as restricted subcellularly as that reported for the other major MAPs. Using cDNA probes to MAP 5, a homolog of MAP 1B, and MAP 1, a homolog of MAP 1A, Tucker *et al.* (1989) found that MAP 5 mRNA is restricted to neuronal cell somata whereas MAP 1 mRNA is found in the proximal region of dendrites as well as in cell somata, but does not extend into the entire dendritic domain (Fig. 1). Recall that tau mRNA, like MAP 1A mRNA, also extends from the cell soma into proximal dendrites, whereas MAP 2 mRNA can be found throughout the somatodendritic domain (see Sections III,A,1 and III,A,2). These data affirm our earlier conclusion that the site of synthesis does not by itself represent a general targeting signal for the intracellular distribution of MAPs.

4. 200-kDa MAPs

Proteins in this class of MAPs have received far less attention regarding their distribution in the nervous system than the other so-called major MAPs. In most cases, they have been minor or unresolvable bands on sodium dodecyl sulfate (SDS) gels of nervous system tissue and have been identified only by immunoblotting (Parysek *et al.*, 1984; Huber *et al.*, 1985; Shiomura and Hirokawa, 1989) or by Northern blot hybridization with cDNA probes (Aizawa *et al.*, 1990). Recently, researchers have begun to think of the different molecular species as part of a distinct and homologous family (Kotani *et al.*, 1988; Huber and Matus, 1990). Aizawa *et al.* (1990) proposed the term "MAP-U" to designate this entire class of MAPs because their distribution among different tissues of the body is so ubiquitous. However, despite comparable electrophoretic mobility and thermostability, the anatomical distribution of these MAPs within the nervous system is rather heterogeneous and not really ubiquitous.

MAPs 3 (180 kDa) and 4 (220–240 kDa) have been, until recently, the most extensively studied species in this class. MAP 3 was identified as a distinct MAP along with some of the earliest reports of the distribution of MAPs 1 and 2 (Matus *et al.*, 1983; Riederer and Matus, 1985; Bernhardt *et al.*, 1985; Matus and Riederer, 1986). MAP 3 antibodies produce a very distinctive immunoreactive pattern in adult rat brain: they stain only neurofilament-rich axons and glial cells. In neonatal brain, on the other hand, MAP 3 antibodies stain neuronal somata and dendrites, too, as well as developing axons, such as cerebellar parallel fibers, that are not immunoreactive in adults. In contrast to MAPs 1A, 1B, 2, and tau (see earlier sections), the intracellular distribution of MAP 3 immunoreactivity is different in neuronal cell cultures than in brain sections. It is associated with cell bodies and dendrites as well as axons in cultured neurons (Huber *et al.*, 1985; Peng *et al.*, 1985). Like MAP 3, MAP 4 is also found in glial cells, but exclusively so, because it is not found in neurons (Parysek *et al.*, 1985) (see Section IV).

The other MAP species in this class are identified only by molecular mass. A 190-kDa MAP is found in both cell bodies and axons of enteric neurons in the small intestine but at far lower levels than is true of MAPs 1, 2, and tau. This MAP is reported to be barely detectable in the CNS (Murofushi *et al.*, 1989), although studies of MAP-U, for which the bovine 190-kDa MAP is a prototype, show that bovine cerebrum (but not cerebellum) contains quantities of MAP-U comparable to most other organs (Aizawa *et al.*, 1990).

A 205-kDa MAP, on the other hand, is distributed widely in the CNS, with antibodies staining all compartments of neurons and staining glial cells in both brain and spinal cord (Shiomura and Hirokawa, 1989). Interestingly, antibodies to this 205-kDa MAP also stain developing parallel fiber axons in the neonatal cerebellum. The overall pattern of immunoreactivity is remarkably like that seen with MAP 1B antibodies (Schoenfeld *et al.*, 1989; Sato-Yoshitake *et al.*, 1989), although immunoblots revealed no recognition of HMM MAPs. Moreover, this species differs from MAP 1B in being heat stable.

Finally, a 210-kDa MAP was detected biochemically in cultured neurons of sympathetic ganglia, where it was found associated only with neuritic axons rather than the mass of cell bodies and dendrites. This pattern is identical to that found for tau protein but is more restricted than what has been found for MAP 3 and MAPs 1A and 1B (Peng *et al.*, 1986; see earlier discussion).

We should note that a consensus has begun to arise (Aizawa *et al.*, 1991; Chapin *et al.*, 1991b) that MAP U, MAP 4, and MAP 3 are homologous protein species. However, as noted above, the observed patterns of immunoreactivity in nervous tissue for antibodies to these species are somewhat divergent, leading us to surmise that cellular and regional variations in epitope expression may be significant.

5. Determination of Somatodendritic and Axonal Compartments

Is differential MAP expression a determinant of the polarization of neurons into distinct somatodendritic and axonal compartments, or is it more a manifestation or outcome of such polarity? A variety of evidence supports a compromise answer: MAPs are not determinants of polarity as such but are essential to maintaining it once it is established.

Dendrites and axons are intrinsically different in the orientation of their microtubules. In axons, the (+) ends of microtubules are uniformly directed distally, whereas in typical mature dendrites, about half of the microtubules direct their (+) ends proximally rather than distally (Burton and Paige, 1981; Burton, 1988; Baas et al., 1988; Black and Baas, 1989). Since the transport of membranous organelles from the soma into both axons and dendrites is determined by the orientation of microtubules, this difference in microtubule orientation may directly regulate the compartmentalization of organelles which, in turn, promotes a number of the fundamental differences between dendrites and axons, such as shape, metabolic activity, and growth capacity (Black and Baas, 1989).

Interestingly, however, the differentiation of microtubule orientation does not immediately emerge at the earliest stage of neurite differentiation, when the axon differentiates from the other neurites. As observed in neurons developing in culture, initially all outgrowing neurites are comparable in length, shape, and growth rate (stage 2; Dotti et al., 1988; Ferreira and Caceres, 1989) (see Fig. 3). Moreover, initially they all show axon-typic microtubule orientation, that is, (+) ends directed distally (Baas et al., 1989). However, eventually one neurite begins to elongate more rapidly than the others, without taper, once it extends to a threshold length about 10 μm longer than the other neurites (stage 3; Dotti et al., 1988; Ferreira and Caceres, 1989; Goslin and Banker, 1989). This neurite always becomes the neuron's sole axon, and the other neurites become dendrites. Stage 3 provides the first sign of polarization: one longer neurite destined to be the axon vs. many shorter neurites destined to be dendrites. However, a "dendritic" neurite can become an axon if the "axonal" neurite is lost at this stage (Dotti and Banker, 1987; Goslin and Banker, 1989). This may derive from the axon-typic microtubule orientation that still persists in all neurites at this stage. Ultimately, elongation and tapering follows in the remaining neurites as they differentiate into dendrites (stage 4; Dotti et al., 1988; Ferreira and Caceres, 1989). It is here that a mixed microtubule orientation first appears in these processes (Baas et al., 1989), further accentuating the polarization of the neuron.

The differential expression of MAPs is correlated only with later

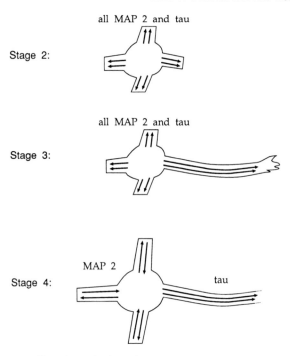

FIG. 3   Development of neurite and microtubule polarity and the expression of MAP 2 and tau protein as seen in cultured neurons. Arrows designate the orientation of the (+) ends of microtubules. The longer neurite in stage 3 eventually becomes the neuron's sole axon while the others become dendrites (stage 4). Only in bona fide dendrites (stage 4) do microtubules show mixed orientation. By this stage, MAP 2 becomes restricted to dendrites, whereas tau protein becomes restricted to axons in most cases. (After Baas *et al.*, 1989.)

stages of neurite differentiation (Fig. 3). Initially, MAP 2 and tau protein are each found distributed in all neurites at stage 2, before an axon differentiates. Moreover, when an elongating axon establishes polarity (stage 3), MAP 2 and tau protein are still uniformly distributed throughout the neuron at this stage (Alaimo-Beuret and Matus, 1985; Caceres *et al.*, 1986; Matus *et al.*, 1986; Dotti *et al.*, 1987; Kosik and Finch, 1987). However, once the remaining neurites begin to differentiate as bona fide dendrites, showing more extensive elongation, tapering, and mixed microtubule orientation, MAP 2 begins to segregate to these processes over the axon (Caceres *et al.*, 1984a, 1991; Alaimo-Beuret and Matus, 1985; Matus *et al.*, 1986; Peng *et al.*, 1986; Dotti *et al.*, 1987; Dotti and Banker, 1987; Kosik and Finch, 1987; Bruckenstein and Higgins, 1988b), as it does in the adult *in vivo*. Likewise, tau ultimately becomes restricted to axons in several kinds of neuronal cultures (cerebral cortex; Kosik and

Finch, 1987; sympathetic ganglia; Peng et al., 1986; Bruckenstein and Higgins, 1988a; cerebellum; Caceres and Kosik, 1990), as it does in vivo. Curiously, in hippocampal cultures, tau protein remains in dendrites after these differentiative steps and does not become restricted to axons, although only the juvenile isotype is expressed (Dotti et al., 1987).

Thus, the first step in neuronal polarization—the differentiation of an axon from the other neurites—is not determined by the restricted expression of tau to the axon, since all stage 3 neurites express tau protein. Nevertheless, the essential differentiative event—the elongation of the axonal neurite—cannot occur or be sustained if tau is not expressed. The addition of tau antisense nucleotides to cultures of cerebellar macroneurons, which inhibits the expression of tau protein, prevents the formation of an axon if the nucleotides are added before stage 3 (within 24 hr after plating; Caceres and Kosik, 1990); it results in the retraction of the axon if they are added at stage 4 (72 hr after plating; Caceres et al., 1991). Thus, while tau does not signal or drive a particular pattern of outgrowth, its association with the microtubules of the outgrowing axon is in some way required for that outgrowth to occur, perhaps through the stabilization of its enormously elongated microtubular array (see Section V,A,2).

On the other hand, the second step in neuronal polarization—the differentiation of the other neurites into dendrites—is at least correlated with the restricted expression of MAP 2 to dendritic neurites and can be sustained in the absence of tau (Caceres and Kosik, 1990; Caceres et al., 1991). Recent work demonstrates that inhibition of MAP 2 expression in cultured differentiated neurons, by transfection of MAP 2 antisense nucleotides, also inhibits the outgrowth and branching of neurites (Dinsmore and Solomon, 1991). Although this study did not attempt to determine whether the transfection affects dendritic differentiation specifically, inasmuch as it did not extend to a time when MAP 2 expression was restricted in control cells to dendritic neurites, the effect on neurites appears quite different from that observed with tau inhibition (Caceres and Kosik, 1990). The role of other MAPs, particularly MAPs 1A and 1B, in the maintenance of either axons or dendrites has not been studied.

B. Shape and Size of Dendrites and Axons

One of the defining features of neurons compared with other cells in the body is that they vary markedly in shape and size (Ramon y Cajal, 1895; Hillman, 1979). This is manifested in a number of ways, including the branching patterns of neuronal processes, particularly dendrites; the occurrence of the special branchlets known as dendritic spines; and the diameter or caliber of dendrites and axons. The cytoskeleton undoubtedly provides the intracellular support for variations in neuronal shape and size, and MAPs may figure ubiquitously in maintaining these variations in different neurons.

1. Dendritic Branching

A variety of evidence points to a role for MAP 2 in dendritic branching. First, within the somatodendritic domain of neurons, MAP 2 is expressed more prominently in dendrites than somata, with the distribution in distal branches at least comparable to if not exceeding the distribution in the proximal trunks (Bernhardt and Matus, 1984; Caceres *et al.*, 1984b; DeCamilli *et al.*, 1984; Niinobe *et al.*, 1988). On the other hand, the distribution of MAPs 1A and 1B does not tend to show such variation between somata and dendrites; if anything, these MAPs may not be distributed to the distal tips of dendritic branches or even in all dendrites (Bloom *et al.*, 1984; Huber and Matus, 1984a,b; Schoenfeld *et al.*, 1989). In developing dendrites, MAP 2 is most prominently localized in distal dendrites at branch points, though not in growth cones (Bernhardt and Matus, 1982; Matus, 1988b). MAP 1A, on the other hand, is found more proximally in the soma and dendritic trunks, and MAP 1B is not even found in this domain until late in development (Sato-Yoshitake *et al.*, 1989; Schoenfeld *et al.*, 1989).

In cultured neurons, MAP 2 expression in primary neurites is correlated with the degree of branching; unbranched neurons do not express MAP 2 (Chamak *et al.*, 1987; Ferreira *et al.*, 1990). A similar pattern is found *in vivo*. Bipolar neurons having an unbranched dendrite in addition to an axon do not express MAP 2 but do express MAP 1B and tau protein (retinal bipolar cell; DeCamilli *et al.*, 1984; Tucker and Matus, 1988; olfactory receptor neuron; Viereck *et al.*, 1989). On the other hand, amacrine-type cells (lacking an axon) having either principal dendritic branches or spine-like branches express MAP 2 more strongly than MAP 1A/1B (retinal amacrine cell; Tucker and Matus, 1988; olfactory bulb granule cell; Viereck *et al.*, 1989). A pathological shift in branch points toward the more distal segments of dendrites, as seen in Purkinje cells following postnatal X-irradiation (Matus *et al.*, 1990), is associated with a comparable shift in MAP 2 distribution. On the other hand, MAP 1A and tubulin remain more or less evenly distributed throughout the somatodendritic domain following this treatment.

Since branching requires depolymerized tubulin (Matus *et al.*, 1986), the role of MAPs and particularly MAP 2 may be to stabilize the branch points already established rather than to drive the branching or determine its configuration (Friedrich and Aszódi, 1991). Indeed, a variety of evidence indicates that the supporting extraneuronal environment may be the strongest determinant of the branching characteristics of neurons (Hillman, 1979; Chamak *et al.*, 1987; Bruckenstein and Higgins, 1988b; Bruckenstein *et al.*, 1989).

2. Dendritic Spines

Microtubules do not extend from the dendritic trunk into spines (Peters *et al.*, 1976). Likewise, antitubulin immunoreactivity is not localized in spines (Caceres *et al.*, 1983, 1984b; Binder *et al.*, 1986). Thus, one might expect that MAPs would not be found in spines, in the absence of tubulin and microtubules. This has been the determination of several laboratories which have found that MAPs in dendrites (MAPs 1A, 1B, 2) are confined to the trunk (Bernhardt and Matus, 1984; Bloom *et al.*, 1984; DeCamilli *et al.*, 1984; Huber and Matus, 1984a,b; Escobar *et al.*, 1986; Riederer *et al.*, 1986).

Nevertheless, several other laboratories have reported finding anti-MAP 2 immunoreactivity in dendritic spines, including in association with postsynaptic densities (Caceres *et al.*, 1983, 1984b; Binder *et al.*, 1986; Morales and Fifkova, 1989). MAP 2 would certainly be the best candidate for a spinous MAP, since it is the most extensively distributed in dendrites, particularly distal branches, of any MAP (see earlier discussion). Moreover, there is a growing expectation that MAP 2, as a major dendritic phosphoprotein, will be shown to play a critical role in synaptic plasticity (Coss and Perkel, 1985; Morales and Fifkova, 1989) (see Section V,B). Furthermore, there is evidence that MAP 2 can be found in growing dendrites in the absence of tubulin and microtubules (Bernhardt and Matus, 1982; but see Caceres *et al.*, 1986), providing another case in which MAP 2 may not be strictly microtubule associated.

How do we account for the discrepant reports? Both groups have used combinations of paraformaldehyde and glutaraldehyde as well as other fixatives and chemicals such as picric acid (Binder *et al.*, 1986) and periodate and lysine (Escobar *et al.*, 1986). Both groups have tended to use vibrating microtome-generated brain sections processed immunocytochemically with peroxidase-antiperoxidase and diaminobenzidine (DAB), but other protocols have been employed as well (e.g., secondary antibodies conjugated to colloidal gold; Morales and Fifkova, 1989; to rhodamine; DeCamilli *et al.*, 1984; to peroxidase; Escobar *et al.*, 1986). The original demonstration of MAP 2 in spines (Caceres *et al.*, 1983) was questioned because of its use of DAB, which is known to produce false positives because apparently it can diffuse from the original site of antigen–antibody binding and peroxidase-catalyzed oxidation (Novikoff *et al.*, 1972). Bernhardt and Matus (1984) seemed to confirm this problem by observing MAP 2 immunostaining in spines but only when "over-staining" occurred, using a protocol that had previously demonstrated the presence of actin in spines without overstaining (Matus *et al.*, 1982). Unfortunately, no details or documentation of the overstaining were provided.

On the other hand, MAP 2 immunolabeling in spines has been contrasted in the same experiments involving peroxidase–DAB with the absence of tubulin immunolabeling extending into spines from the trunk (Caceres *et al.*, 1983, 1984b; Binder *et al.*, 1986). DeCamilli *et al.* (1984) provided for a different kind of control for artifactual DAB diffusion by using secondary antibodies conjugated to something other than peroxidase (rhodamine), thereby not requiring DAB for visualization, and found extensive staining of dendritic trunks but not spines at the light microscope level.

However, Morales and Fifkova (1989) used this same strategy to *find* MAP 2 in spines, colocalized with actin. There are a number of reasons why this report is compelling. The authors used colloidal gold–conjugated secondary antibody, which is known to produce precise labeling of cellular structures (DeCamilli *et al.*, 1983; Hirokawa *et al.*, 1985; Riederer et al., 1986). They also employed a postembedding protocol for electron microscopic (EM) immunocytochemistry, in which the plastic embedding medium might tend to restrict the potential migration of cellular constituents like MAP 2 prior to immunocytochemical processing while still permitting adequate antibody penetration. Moreover, Morales and Fifkova demonstrated that MAP 2 is precisely associated with microfilaments, the postsynaptic density and other structures within spines also being immunoreactive to actin antibody, and yet, as expected, MAP 2 is associated with dendritic but not axonal microtubules. Finally, the colocalization of MAP 2 and actin is consistent with the significant actin-binding properties of MAP 2 (Sattilaro *et al.*, 1981; Selden and Pollard, 1986) and the proposed roles of both proteins in synaptic plasticity (see Section V,B). Since the presence of tubulin inhibits the interaction of MAP 2 and actin (Sattilaro, 1986), the absence of microtubules from dendritic spines may be essential to the function of MAP 2 and actin in spines.

Does this call into question other reports of a failure to localize MAP 2 in spines? Not necessarily. As discussed above, many of the other studies employed significant controls of their own to support the reliability of their respective observations, positive as well as negative. The most likely factor underlying the demonstration of or failure to demonstrate MAP 2 in spines is that the MAP 2 in spines contains epitopes not expressed elsewhere in dendrites and not recognized by a number of antibodies. There does not appear to be an obvious difference in binding to

MAP 2 isotypes, since every antibody discussed here is reported to recognize both MAPs 2a and 2b (with MAP 2c recognition usually not determined).

Nevertheless, there is at least one additional variation in MAP 2 distribution that is correlated with the demonstration of spine staining by some of these antibodies: the localization of anti-MAP 2 immunoreactivity in the dendrite-like peripheral branch processes of pseudounipolar sensory neurons. Two antibodies that demonstrate anti-MAP 2 immunoreactivity in spines also stain these peripheral processes, as well as the axons of spinal motor neurons (AP9, AP13; Papasozomenos et al., 1985). A MAP 2 antibody that fails to demonstrate MAP 2 in spines also fails to demonstrate MAP 2 in these peripheral processes (DeCamilli et al., 1984; see also Hernandez et al., 1989, for a MAP 2 antibody that also fails to stain the peripheral process of pseudounipolar neurons). In all of these cases, the cell body and proximal stem process, but not the central branch process, is stained. The functional importance of this correspondence remains to be clarified, but at least it supports the contention that different antibodies may reveal subtle but significant variations in the subcellular distribution of MAP 2 antigenic sites.

3. Dendritic and Axonal Caliber

The size or caliber of neuronal processes varies with the sum of two factors: the volume of the cytoskeleton and the volume of the membranous organelles (Sasaki-Sherrington et al., 1984). The volume of the cytoskeleton in turn is determined by both the number of elements (microtubules and neurofilaments) and their spacing or density (Hillman, 1979). In axons where neurofilaments are far more numerous than microtubules, neurofilament number is highly correlated with and thought to be a determinant of caliber (Friede and Samorajski, 1970; Hoffman et al., 1984; Szaro et al., 1990). However, where microtubules are more prominent components of the cytoskeleton, as in dendrites and small-caliber, unmyelinated axons (Wuerker and Kirkpatrick, 1972; Peters et al., 1976; Hillman, 1979), their number and density may contribute significantly to the determination of caliber (Hillman, 1979).

MAPs in turn are in a position to play a role in the determination of microtubule density and hence dendritic and axonal caliber. All of the major fibrous MAPs (MAPs 1A, 1B, 2, and tau protein) have been shown to be components of lateral extensions or cross-bridges between microtubules and other microtubules, neurofilaments, or membranous organelles (Hirokawa et al., 1985, 1988a,b; Shiomura and Hirokawa, 1987a,b; Sato-Yoshitake et al., 1989; see review in Hirokawa, 1991). Each fibrous MAP has a microtubule-binding domain at one end of the molecule (or major polypeptide) and a projection domain at the other end (see Fig. 4 for MAP 2 domains). The projection domain is believed to constitute the fibrous sidearm that extends from the microtubule to other cytoplasmic elements (Vallee, 1980; Voter and Erickson, 1982; Vallee and Davis, 1983), since removal of the projection arm by proteolysis produces microtubules devoid of fibrous side arms (Vallee and Borisy, 1977). Since shortening or removal of the projection arm also produces microtubules that are more closely spaced than normal (Friden *et al.*, 1988; Lewis *et*

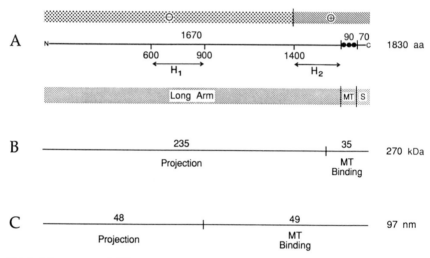

FIG. 4   Linear maps of different domains along the length of a fully extended HMM MAP 2 molecule. (A). Using the number of amino acid residues (aa), this shows the relative size of the N-terminal long arm (1670 aa), the microtubule-binding domain (MT: 90 aa), and the C-terminal short arm (S: 70 aa). Two segments lying between residues 600-900 (H_1) and 1400-1670 (H_2) are rich in proline and/or glycine (Kindler *et al.*, 1990) and thus may function as hinge domains. Residue 1400 marks an approximate boundary between a largely negatively charged N-terminal domain and a mostly positively charged C-terminal domain (Aizawa *et al.*, 1990). (B). This shows the proportional mass of the projection and microtubule-binding fragments yielded by proteolytic digestion of MAP 2 (Vallee and Borisy, 1977). Placed in register with the linear display in A, this map accords well with observations that (1) the microtubule-binding fragment is positively charged whereas the projection fragment is negatively charged (Vallee and Borisy, 1977); (2) limited proteolytic digestion prevents the appearance of side arms but not the assembly and bundling of microtubules (Vallee and Borisy, 1977); and (3) the site for proteolytic removal of the projection fragment corresponds to the middle of the H_2 domain (Dingus *et al.*, 1991; Wille *et al.*, 1992). (C). The lengths of intact MAP 2 and the proteolytically separated projection and microtubule-binding fragments are shown after glycerol spraying and rotary shadowing (Wille *et al.*, 1992). The drawing shows that the measurable physical lengths of specific molecular domains may be disproportionate to their mass (electrophoretic mobility) or sequence length.

al., 1989) (Fig. 5), regulating microtubule density or spacing may be an important function of fibrous MAPs.

In principle, MAPs of different length could produce greater or lesser spacing between microtubules and other cytoskeletal elements (Aizawa

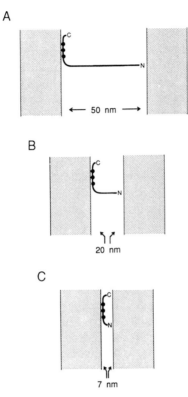

FIG. 5   Variations in intertubular spacing attributable to variations in the length of the long projection arm of MAP 2. A nominal length of about 50 nm in an unaltered long arm (A) is based on recent measurements by Wille *et al.* (1992) (see also Fig. 4). Removing internal segments of the N-terminal long projection arm (B) produces tighter than normal bundling of microtubules [e.g., 20 nm following transfection of the FS construct, (Lewis *et al.*, 1989)]. Proteolytic removal of nearly the entire N-terminal projection arm (C) still results in the aggregation of microtubules but with negligible spacing (about 7 nm; Vallee and Borisy, 1977; Friden *et al.*, 1988). Microtubules are represented as shaded rectangles. The microtubule binding domain of each MAP 2 molecule is represented by three solid, interconnected circles, from which extend a C-terminal arm and an N-terminal arm. The right-angle bend between the N-terminal arm and the microtubule-binding domain represents the approximate position of hinge domain H_2. Drawings are roughly to scale in representing linear dimensions, with the exception of the C-terminal microtubule-binding fragment, including the short arm, which is underrepresented according to recent measurements (Wille *et al.*, 1992) (see Fig. 4).

et al., 1990). Thus, LMM tau protein, which in unfixed tissue is about 50 nm when fully elongated (Hirokawa *et al.*, 1988b), should promote closer spacing and greater density than IMM 190-kDa MAP, which has a contour length of about 100 nm (Murofushi *et al.*, 1986), or the HMM MAPs, which in unfixed tissue are 100–200 nm when elongated (Voter and Erickson, 1982; Hirokawa *et al.*, 1985, 1988a,b; Shiomura and Hirokawa, 1987a,b; Sato-Yoshitake *et al.*, 1989; Wille *et al.*, 1992). This in fact has been demonstrated in a variety of ways. For example, the preferential association of MAP 2 with dendrites and tau protein with axons (see above) is correlated with the tendency of dendrites to be larger in caliber and have less densely packed microtubules than unmyelinated axons, as seen both *in vivo* (Wuerker and Kirkpatrick, 1972; Chen *et al.*, 1992) and *in vitro* (Bartlett and Banker, 1984; Chen *et al.*, 1992). Moreover, lengthy treatment (18–24 hr) of cultured neurons with taxol, which causes microtubules to become densely packed (Letourneau and Ressler, 1984), also causes a concomitant loss of MAP 2 and chartins but not tau protein or 210-kDa MAP, suggesting that MAP 2 (and chartins?) may have something to do with the normally greater spacing between microtubules (Black, 1987a). Indeed, when MAP 2 is added to microtubules polymerized *in vitro* from pure tubulin, the specific volume of the microtubule pellet increases and the packing density of the microtubules examined with the electron microscope declines, whereas the addition of tau protein has an undetectable effect on either volume or packing density (Black, 1987b).

The higher resolution quick-freeze deep-etch technique reveals that LMM tau induces measurable spacing of about 10–30 nm (surface-to-surface) between adjacent microtubules *in vitro* (Hirokawa *et al.*, 1988b), whereas the IMM 190-kDa MAP induces average spacing of 30 nm (surface-to-surface) between adjacent microtubules (Murofushi *et al.*, 1986), and the HMM MAPs are associated with measurable spacing of 40–100 nm (surface-to-surface) between adjacent microtubules or between microtubules and neurofilaments (Hirokawa *et al.*, 1985, 1988a,b; Shiomura and Hirokawa, 1987a; Sato-Yoshitake *et al.*, 1989). Similarly, in the processes of insect ovarian Sf9 cells transfected with expression vectors for either HMM MAP 2, LMM MAP 2c, or LMM tau (Chen *et al.*, 1992), microtubule spacing (surface-to-surface) in MAP 2-expressing cells is three times (62 nm) that measured on average in MAP 2c- or tau-expressing cells (20–21 nm).

Several theoretical models have been proposed that consider the molecular interactions involved in a microtubule spacing function by MAPs. One model (Sasaki *et al.*, 1983; Brown and Berlin, 1985; Friden *et al.*, 1988) proposes that the projection arms of fibrous MAPs, extending out in all directions from the microtubule surface, surround the microtubule with a coat, creating what one group calls a "MAP-tube" (Sasaki *et al.*, 1983). Since the projection domain is largely negatively charged (Fig. 4), in contrast to the positively charged microtubule domain (Sattilaro, 1986; Aizawa *et al.*, 1990; Wiche *et al.*, 1991), the coat surrounding the microtubule is thus negative as well. Therefore, in this model, microtubules decorated with MAPs interact largely by electrostatic forces (repulsion; Brown and Berlin, 1985; Friden *et al.*, 1988; weak attraction; Sasaki *et al.*, 1983) and spacing is regulated by the thickness of the coat (i.e., the length of the projection arms; see Fig. 6A) and by conditions that might modulate electrostatic forces, such as pH (Brown and Berlin, 1985). A

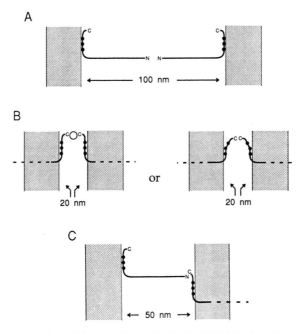

FIG. 6   End-to-end models of intertubule spacing by MAP 2. The drawings illustrate three different modes of interaction between the long and/or short projection arms of MAP 2 molecules shown as extending away from the microtubule binding domain at the microtubule surface. (A). Long arm–long arm (after Sasaki *et al.*, 1983); (B). Short arm–short arm (after Lewis and Cowan, 1990). (C). Long arm–short arm, as proposed. Each model, as conceptualized here, predicts a different amount of intertubule spacing (surface-to-surface), based on the assumption that the long arm extends roughly 50 nm from the microtubule surface (see Fig. 5A), whereas the short arm, comprising only a small portion of the C-terminal domain (see text), extends only a negligible distance from the microtubule surface. In (B), the open circle in the left-hand drawing depicts a bridging protein, and the right-hand drawing depicts only partial microtubule binding. (After Lewis and Cowan, 1990.) The dashed lines indicate incompletely drawn long arms. See Fig. 5 for other labeling conventions.

principal failing of this model is its inability to predict the more heterogeneous nature of the MAP coat surrounding axonal microtubules known to be decorated by both long MAP 1A/1B and short tau protein molecules (Hirokawa *et al.*, 1988b) (see later discussion).

In a more extensive presentation and test of a model of MAP cross-bridging, Lewis *et al.* (1989) emphasize that the lateral extensions serve to bundle microtubules together as well as to space them apart. They propose that cross-bridging by MAP 2 or tau protein involves hydrophobic binding by the relatively short C-terminal "arm" of MAP 2 or tau protein, which is composed of about 70 amino acid residues that extend beyond the microtubule-binding domain to the C-terminus (Fig. 4) (Wiche *et al.*, 1991). They base this hypothesis on the failure of cultured cells to bundle microtubules when transfected with MAP 2 or tau cDNA expression vectors that are missing or that possess a modified version of the nucleotide sequence encoding the short C-terminal arm of either molecule, but not when the vector is missing the sequences encoding different segments of the N-terminal projection arm.

However, their model seems to be strained in its ability to account for microtubule spacing *in vivo*. In a revised version, in fact, the authors had to propose rather unparsimoniously the existence of either a bridging protein between MAP 2 molecules or only partial microtubule binding (see Fig. 6B) in order to approximate even a rather low estimate of 20–40-nm intertubular spacing, while on the other hand leaving the long N-terminal projection arm completely out of the model (Lewis and Cowan, 1990; see comments in Goedert *et al.*, 1991).

A failing of both models is that neither yields an acceptable prediction of intertubule spacing. Recent measurements of isolated MAP 2 molecules prepared by shadowing and examined by high-magnification electron microscopy, which were calibrated against measurements of shadowed single- and double-stranded DNA, estimate the length of the projection domain at about 50 nm (Wille *et al.*, 1992) (Fig. 4), which is well within the range of intertubule spacing (40–100 nm, surface-to-surface) associated with MAP 2 and other HMM MAPs. In the first model, end-to-end interaction between two long projection arms (Sasaki *et al.*, 1983) (Fig. 6A) would produce spacing at two times the length of a long arm, at the high end of the observed range of intertubule spacing. On the other hand, in the second model, the interaction of two short arms (Lewis and Cowan, 1990) (Fig. 6B) would produce spacing much less than the observed range of intertubule spacing. More critically, the interaction of short arms also cannot adequately account for the variation in intertubular spacing when either MAP 2 or tau protein decorate microtubules (see earlier discussion), since the C-terminal short arms of MAP 2 and tau protein are similar if not identical in length (Wiche *et al.*, 1991).

This suggests that the long projection domains of individual MAPs are the predominant components of cross-bridges and that intertubule distance is the equivalent of one long arm (Figs. 6C and 7). How is this likely to be configured? Two possible models are consistent with the observed range of intertubule spacing but include as an additional factor the binding of the long arm of one MAP with some portion of another MAP. In one, the long arm of one MAP extends between two microtubules to bind with the short arm of another MAP (Fig. 6C). This model is based on evidence that segment deletion in the C-terminal short arm of either MAP 2 or tau protein severely diminishes microtubule bundling without affecting microtubule polymerization (Lewis et al., 1989; Lewis and Cowan, 1990; Kanai et al., 1992). The role of the long arm is predicted from the analysis offered earlier that variation in intertubule spacing is only correlated with variation in the length of the long arm. Note, however, that segment deletion in the long arm or even proteolytic elimination of the long arm does not diminish microtubule bundling by MAP 2 (Lewis et al., 1989; Vallee and Borisy, 1977; Friden et al., 1988), although segment deletion in the long arm of tau protein does diminish bundling (Kanai et al., 1992). Moreover, there is no direct evidence as yet for binding between long arms and short arms.

In the second model, the long arm of one MAP extends between two microtubules in antiparallel alignment with the long arm(s) of another MAP(s), involving dimeric and trimeric cross-bridges (Fig. 7A,B). This model is based on recent evidence that isolated MAP 2 molecules, and even MAP 2 chymotryptic microtubule-binding fragments, self-associate in an antiparallel oligomeric alignment, whether they involve entire lengths of different molecules or hairpin-like folds of individual molecules (Wille et al., 1992). Because oligomers (most typically, dimers and trimers) can involve different numbers of MAP 2 molecules overlapping to varying extents, a variety of configurations are possible (Fig. 7B). This would help to account for the variation in spacing and the almost stochastic, lattice-like appearance of MAP 2 cross-bridges (Hirokawa, 1991). Note that there is no direct evidence for a MAP–MAP binding site on the long projection arm of any fibrous MAP that might account for oligomerization. In fact, there is evidence against an essential role in microtubule bundling for any specific segment of the N-terminal long arm of MAP 2 (Lewis et al., 1989). Nevertheless, the tendency of long arms to oligomerize with varying degrees of overlap is consistent with the absence of specific binding sites.

The current model also does not easily account for the supposedly critical role of the C-terminal short arm in bundling predicted in the first model. However, the tendency of C-terminal domains to oligomerize (Wille et al., 1992) may provide a clue. Perhaps the short arm fosters the

A

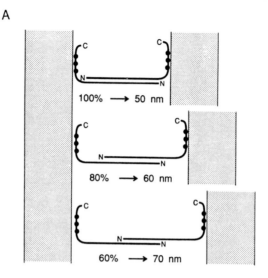

B

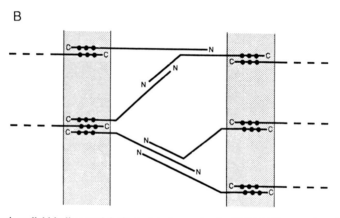

FIG. 7   Antiparallel binding model of intertubule spacing by MAP 2. The drawings illustrate variations in intertubule spacing (A) and cross-bridging pattern (B) attributable to differences in antiparallel oligomeric binding between the long projection arms of MAP 2 molecules. (After Wille *et al.*, 1992.) (A). Decreases in the percentage overlap of two 50-nm long arms lead to effective increases in spacing. (B). Different patterns of oligomeric antiparallel binding (e.g., dimers and trimers involving N-terminal and C-terminal domains and intramolecular bends) result in cross-bridges that branch either close or far from the microtubule wall, are directed at a variety of angles from the wall, and achieve a variety of bridge diameters, all commonly observed *in situ* (Hirokawa, 1991). Intramolecular bends in (B) are confined to approximate positions of the two hinge domains (see Fig. 4). In (B), the orientation of the microtubule binding domain is depicted as across the long axis of microtubules, rather than parallel to the long axis as shown in (A) and in Figs. 5 and 6. No one has as yet established whether the orientation is one way or the other or even another alternative. However, different orientations are broadly compatible with the different models illustrated here and in other figures. See Figs. 5 and 6 for labeling conventions.

antiparallel orientation of C-terminal oligomers in a way that would promote diversely oriented lateral extensions and a more extensively cross-linked and bundled microtubule lattice (Fig. 7B). Of course, if the true length of the short arm turns out to be as disproportionately large as the C-terminal chymotryptic fragment in which it resides (Wille et al., 1992; see also Fig. 4), then all current models of short arm function will require rethinking. However, since a known antigenic sequence near the junction of the microtubule binding domain and the short arm has been mapped to the end of shadowed, proteolytically isolated C-terminal fragments (Wille et al., 1992; see also Fig. 4), it is likely that the disproportionate length of the fragment is attributable to the H_2 region or the microtubule binding domain but probably not to the short arm.

Several authors have questioned the notion of cross-linkage between microtubules by MAPs that is inherent in most of the current models. For example, some have argued that MAPs do not cross-link microtubules but function merely as spacers. This is the essence of the electrostatic model (Sasaki et al., 1983; Brown and Berlin, 1985; Friden et al., 1988) described and discussed earlier. Sato et al. (1988) reached the same conclusion but based it on their observation that microtubules are mechanically elastic, with or without MAPs. The conclusion presumes that cross-linkage would constrain elasticity. Why should this be presumed? MAPs themselves may be as elastic as microtubules. For example, tau has been shown to be elastic, at least when dephosphorylated (Lichtenberg et al., 1988; Hagestedt et al., 1989). Shouldn't an array of elastic polymers be elastic, too? Of course, as Matus (1991) points out, cross-linkage is likely to be weak if it occurs at all, since microtubules in the presence of MAPs do not tend to bundle spontaneously in vitro, only when constrained by cell membranes or centrifugal force. This property is usually cited to support the repulsion model (Brown and Berlin, 1985). However, the concept of cross-linkage need not depend on the premise that individual links between MAPs must be as strong or as avid as those between MAPs and microtubules, demonstrable outside of the cell as well as within. In the biochemical literature of native and recombinant MAPs, too little attention is paid to the possible additive effects of the expression of myriad MAP species and isoforms which one typically finds in vivo (see the following discussion and Sections IV, V,A,3, and VI).

In a somewhat different vein, Chapin et al. (1991a) question whether the purported role of MAPs in bundling microtubules together necessarily requires that they cross-link microtubules, as is commonly assumed. In a critique of the short arm-short arm model of Lewis and Cowan (see Fig. 6B), these authors propose that MAP 2 does not cross-link microtubules but only stabilizes them. They point out that microtubules stabilized by taxol treatment or injection of a guanosine triphosphate (GTP)

analog will bundle in the absence of MAP 2. Thus, they argue, microtubule cross-linking by MAP 2, whether demonstrable or not, is not an
essential factor in bundling. This strikes us as an overly strict interpretation of the Lewis and Cowan model. Although the model specifically
addresses microtubule binding and bundling by MAP 2 and tau, it certainly does not exclude binding and bundling by other MAPs, whether by
the same or other putative mechanisms. For example, Black (1987a) has
shown that taxol-treated microtubules are not necessarily devoid of all
MAPs. Sympathetic neurons in culture still contain both tau and 210-kDa
MAP following taxol treatment. Thus, unpublished results cited by
Chapin et al. (1991a) that taxol-treated 3T3 cells do not contain MAP 2
and tau does not mean that other MAPs, for example, MAP 4 and other
IMM MAPs, are not present and would not bind to and bundle microtubules. We agree with these authors that the effect of direct stabilization
with GTP analogs clearly discounts a *necessary* role for cross-linking in
all instances of bundling. However, this demonstration does not dismiss
a *sufficient* role for cross-linking by MAP 2 and other MAPs in microtubule bundling.

Chapin et al. also propose that a non-MAP endogenous bundling factor
may be at work here, and suggest that dynamin (Shpetner and Vallee,
1989) may be a candidate for such a factor. However, dynamin is also a
MAP, and its bundling properties, like those of fibrous MAPs, appear to
be derived from its direct association with microtubules, not independently of them (Shpetner and Vallee, 1989; Maeda et al., 1992). Other
candidate bundling factors remain to be identified.

Chapin et al. (1991a) also cite both published (Cleveland et al., 1977)
and their own unpublished evidence that tau and MAP 2 behave as
monomers in solution, suggesting that they would not self-associate as
dictated in the model. However, a number of recent studies, some cited
earlier, provide evidence that MAPs, including MAPs 1 and 2 and tau
protein, can either cross-bind or oligomerize and thus have the potential
for cross-linking microtubules through self-association (Furtner and
Wiche, 1987; Lee et al., 1989; Lichtenberg-Kraag and Mandelkow, 1990;
Wille et al., 1992). Ultimately, of course, the significance of self-
association will depend upon demonstrating the nature of MAP–MAP
binding. In the meantime, we suggest that forcing mutually exclusive
distinctions between spacing and cross-linking and bundling is artificial—that there is no reason yet evident why MAPs cannot simultaneously serve these functions within the cytoskeleton.

A new element in the current models (Figs. 6C, 7) that is not easily
accommodated in either of the previous models is that short (tau) and
long (MAP 1A/1B) molecules can bridge in parallel from the same microtubule to different microtubules (Hirokawa et al., 1988b), constrained

only by the distance that their projection arms can extend. The more even distribution of the HMM MAP 1 polypeptides within both dendrites and axons, despite apparent variations in microtubule density between and within dendrites and axons, is a particular challenge to any model of the role of MAPs in cross-bridges and seems to contradict a close association between MAP distribution, microtubule density, and process caliber. However, these MAPs may simply not play as direct a role in regulating microtubule spacing as either MAP 2 or tau protein. MAP 1A colocalizes with MAP 2 in dendrites (Shiomura and Hirokawa, 1987b) and with tau in axons (Hirokawa *et al.*, 1988b). Such colocalization apparently is possible because MAP 1A does not compete for the same microtubule binding sites as either MAP 2 or tau, in the way that the latter compete for the same sites (Hirokawa *et al.*, 1988b). Indeed, the microtubule binding mechanism of the MAP 1s may be quite different from that of MAP 2, tau, or IMM MAPs (Noble *et al.*, 1989; Schoenfeld *et al.*, 1989; Aizawa *et al.*, 1990; Hammarback *et al.*, 1991). Perhaps just as important, the mode or pattern of cytoskeletal interaction or cross-bridging may be quite different for MAP 1A compared with MAP 2 or tau. In axons in particular, it would be impossible for both LMM tau and HMM MAP 1A to cross-link microtubules in precisely the same way, given their different lengths.

Hirokawa *et al.* (1988b) suggest that tau may form short cross-links between closely apposed microtubules, whereas MAP 1A may form longer cross-links between microtubules lying at greater distances from one another. Tau in this case might be the principal regulator of microtubule density and axonal caliber whereas MAP 1A would play a different role in higher order structures of the microtubule latticework, helping to build the highly branched and anastomosing intertubular network seen in fresh, unfixed tissue (Hirokawa *et al.*, 1985). However, there are a number of possible configurations for parallel bridging between axonal microtubules by the MAP 1s and tau that need to be considered and tested (Fig. 8). Better understanding of the relationship between the MAP 1s and tau in axons should also provide clues to the nature of the relationship between the MAP 1s and MAP 2 in dendrites, where the molecules are of comparable size and length but have very different modes of interaction with microtubules (Hammarback *et al.*, 1991).

Although fibrous MAPs may function in a direct way in establishing and maintaining microtubule density, they are unlikely to play a truly regulatory role in controlling the size of dendrites and axons in individual neurons because microtubule density itself appears to be far less of a determinant of process caliber in individual neurons than microtubule number (Hillman, 1979; see also Szaro *et al.*, 1990), with the apparent exception of a basic difference in density between dendrites and axons.

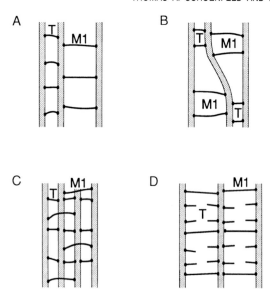

FIG. 8   Variations in the hypothetical cross-bridging of the same set of axonal microtubules by both short tau protein molecules (T) and long MAP 1 molecules (M1). (A) and (B) predict variations in the spacing between microtubules within a particular axon because either tau or MAP 1 dominates between particular pairs (A) or even at particular sites (B). Tau and MAP 1 would not be expected to commingle. On the other hand, (C) and (D) predict more uniform spacing in a particular axon because either tau (C) or MAP 1 (D) dominates throughout the axon. The spacing would be uniformly less with tau dominant (C) than with MAP 1 dominant (D). Moreover, these situations would be expected to involve commingling of tau and MAP 1, with the result that either MAP 1 cross-links only nonadjacent microtubules (C) or tau protein fails to cross-link even adjacent microtubules (D).

In one particular case—spinal motoneuron dendrites—microtubule density appears to increase with the decreasing diameter of each new dendritic branch (Hillman, 1979) but without a corresponding reduction in the density of MAP 2 expression (DeCamilli *et al.*, 1984), illustrating how MAP cross-bridging may be irrelevant to the regulation of local variations in microtubule density and dendritic caliber.

On the other hand, microtubule density, as well as the ratio of microtubules to neurofilaments, varies substantially among different classes of neuron (Hillman, 1979; Szaro *et al.*, 1990). Thus, a more likely role for the putative cross-bridging function of fibrous MAPs may be in the determination and maintenance of class-specific, not individual, variations in cytoskeletal volume and process caliber. One might predict that there are therefore class-specific amalgamations of different MAPs, a possibility that deserves much more attention (Peng *et al.*, 1985; Burgoyne and

Cambray-Deakin, 1988) (see Section IV). Indeed, continued examination of the impact of single MAP species on microtubule spacing (Chen et al., 1992) may only approximate the impact of more complex MAP alliances. Of course, as we have noted, cross-bridging by fibrous MAPs may serve an additional role in neuronal processes, namely, stabilization of microtubule arrays (see Section V).

IV. Typology of Nervous System Cells

Neurons and their supporting cells differentiate into myriad regionally specific classes or types (e.g., pyramidal cells and stellate cells in the cerebral cortex, mitral and granule cells in the olfactory bulb, spiny and aspiny neurons in the basal ganglia, Purkinje cells and basket cells in the cerebellar cortex, fibrous astrocytes in white matter and protoplasmic astrocytes in gray matter). Such differentiation is a hallmark of the richly varied functional specialization that distinguishes each region from the others (Shepherd, 1990). Cells in these classes vary by definition in size, shape, and the ramification of their processes. These are parameters of neuronal form already linked to the expression of MAPs (see Section III), raising the possibility that variations in MAP expression may contribute to the process of class differentiation (see also Goldstein et al., 1983, for the relationship of neurofilament protein expression to cerebellar "neurotypy"). However, in our earlier discussion of the regulation of cellular form by MAPs, we only alluded to the existence of class differences in the expression of various MAPs (see also Matus et al., 1983; Olmsted, 1986). Unfortunately, because very little attention has been given to this topic, there is little more to say, except to highlight some of the class-specific patterns of MAP expression and to raise a host of unanswered questions about the implications of such specificity.

In some respects, the tendency to seek generalizations about individual MAPs in most of the work to date—to show that a particular MAP is "dendritic" or "neuronal" or "juvenile"—has overshadowed evidence that subtle variations in the so-called general patterns do exist for each MAP. These are probably worth paying attention to, since such variations would be multiplied in the amalgamation of different MAPs in different neurons and supporting cells, in ways corresponding to and potentially helping to support the differentiation of cellular form and function.

For example, with few exceptions, MAP 2 has been found only in neuronal cells and not in glia (Papasozomenos and Binder, 1986). Moreover, within neurons, MAP 2 is expressed almost exclusively within somata and dendrites, not axons (see Section III,A,1). However, there are two reports of significant nonsomatodendritic and even non-neuronal staining with anti-MAP 2 antibodies, localized to axons and astrocytes of free-standing axon bundles of the CNS (fimbria) and the PNS (optic nerve, spinal nerves), and to pituicytes of the hypophysis, also a free-standing, unsupported structure (Papasozomenos et al., 1985; Papasozomenos and Binder, 1986). These authors speculate that structures lacking the support provided to white matter tracts embedded within the neuropil of the CNS might be subject to greater mechanical stress and that perhaps MAP 2 acts as a "rigidifying agent" in such cells. However, is it reasonable to view MAP 2 as a "rigidifying agent" in its predominant locale, that is, the somatodendritic compartment of neurons, especially when most such structures are in fact well supported by neuropil? Under what circumstances and in what contexts does MAP 2 help to provide rigidity? Is this the same as microtubule stabilization? Moreover, what about the expression of other MAPs in these same free-standing structures? Is MAP 2 simply added to, or does it replace, axon-typic MAPs such as tau, MAPs 1A and 1B, or MAP 3 in neurons or glial-typic MAP 1s or IMM MAPs in astrocytes? Does their presence, or absence, also contribute to the rigidity function inferred to be essential in free-standing structures?

Other examples abound. MAP 2 reportedly is expressed in the dendrites and perikarya of nearly all principal (long-projecting) and local circuit neurons in the nervous system (see Section III,A,1). Nevertheless, there are significant variations in the intensity of immunoreactivity of different neurons to MAP 2 antibodies. For neocortical pyramidal cells and cerebellar Purkinje cells, the staining is comparable to Golgi impregnation in completeness of staining but, unlike Golgi impregnation, the process involves every neuron present in the section (DeCamilli et al., 1984) (see Section III,A,1). By contrast, staining of hypothalamic multipolar cells is much less intense or differentiating—individual dendritic processes are difficult to resolve and neuronal cell bodies are not stained (Crandall et al., 1989). The same can be said for a number of noncortical neuron classes, for example, brainstem reticular neurons and a variety of neuron types in the thalamus (DeCamilli et al., 1984).

Similar variations can be observed in the distribution of other fibrous MAPs. For example, monospecific antibodies to MAPs 1A and 1B label the dendrites of olfactory bulb mitral and tufted cells to the same degree, whereas the dendrites of olfactory bulb granule cells are labeled, with few exceptions, only by antibodies to MAP 1A (T. A. Schoenfeld, unpublished observations; see Fig. 9). Moreover, two different antigenic sites on the tau protein molecule are apparently not jointly expressed in all classes of axons. Anti-tau antibody Tau-1 (Binder et al., 1985, 1986) stains small-caliber, lightly or unmyelinated axons (e.g., olfactory nerve,

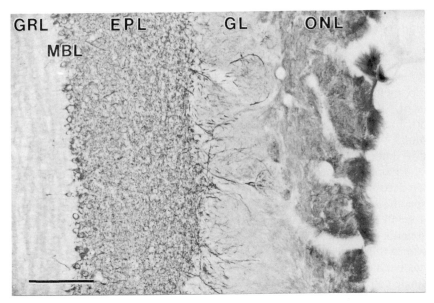

FIG. 9   Distribution of anti-MAP 1B immunoreactivity in the olfactory bulb. Dark reaction product labels in particular the axons of olfactory receptor neurons traveling in the olfactory nerve layer (ONL) of the bulb, although the terminal fields of these axons in the glomerular layer (GL) are not immunolabeled. Cellular labeling is prominent in the somata and dendrites of the mitral body layer (MBL) and external plexiform layer (EPL) but not in the granule cell layer (GRL). Bar = 200 μm.

hippocampal mossy fibers, cerebellar parallel fibers, corticospinal tract) much more intensely and extensively than large-caliber, myelinated axons (e.g., CNS white matter tracts), whereas anti-tau antibody 5E2 (Kowall and Kosik, 1987) produces the opposite staining pattern. Another interesting variation in the expression of tau protein that has recently come to light is the preferential production of the 110-kDa isoform in neurons of the PNS rather than the CNS (Georgieff *et al.,* 1991; Oblinger *et al.,* 1991; Goedert *et al.,* 1992).

The question of MAP specificity also pertains potentially to distinctions between neurons and glial cells. Both MAP 2 and tau are considered to be neuron-specific for the most part, whereas MAPs 3 and 4 and some of the other IMM MAPs are largely glial-specific (Olmsted, 1986). Is it reasonable to think that MAP 2 and tau serve neuron-specific functions, whereas the IMM MAPs serve glial-specific functions? On the other hand, since these MAPs have a number of things in common, including thermostability and microtubule-binding sequences (Aizawa *et al.,* 1990), could they in fact be serving the same kind of function, or filling the same "niche," in different cell types or even in separable regions of the same cell (e.g., MAP 2 in dendrites and tau in axons)?

What about the more ubiquitous MAP 1s, which are found throughout the various compartments of neurons and in glia also (Olmsted, 1986)? Is it significant that they share features such as thermolability and the employment of light chains at the microtubule binding domain (Schoenfeld *et al.*, 1989; Hammarback *et al.*, 1991) that are quite different from the attributes which the previously cited, less ubiquitous MAPs have in common? What function do they serve, or what niche do they fill, that is different from that of the thermostable MAPs?

A complementary approach to seeking patterns of variation and overlap across many cell classes is to focus on the amalgamation of MAPs in specific cell classes by direct, exhaustive analysis of MAP expression in a small set of cell types found either within a particular region or in several different regions of the nervous system. In the rare instances where this sort of thing has been attempted in previous work, the result has been a catalog of MAPs and related cytoskeletal elements found in a single cell type (e.g., cultured sympathetic ganglionic neurons; Black and Kurdyla, 1983; Peng *et al.*, 1985; cerebellar granule cells; Burgoyne and Cambray-Deakin, 1988). Such catalogs are particularly valuable in establishing a framework for further study, in forcing us to think of cells as being composed of, and functionally dependent on, more than the one or two elements that we happen to be specializing in at the moment. Of course, catalogs are only useful in understanding differentiation if they are used comparatively, that is, if they provide evidence for how and why cells are observed to be different, or offer clues that cells should be suspected as being different, if they are not at present observed to be different.

Our preliminary studies of MAP expression in the olfactory system encourage us to think that a comparative, typological approach may prove fruitful in further explicating and uncovering neuronal classes and subclasses. Two recent unpublished observations illustrate cell-specific amalgamations of MAPs that either conform to or predict differentiated classes. Mitral cells and their dendrites show equivalent immunocytochemical expression of MAPs 1A and 1B, whereas granule cells and their dendrites show far greater expression of MAP 1A than MAP 1B (cited earlier in this section). Moreover, tau protein immunoreactivity is prominent only in a small subset of olfactory receptor neuron axons (see Section V,A,3,c; Fig. 2) whereas MAP 1B is prominent in virtually all such axons (Fig. 9). In the former case, the observation conforms to the class distinction between these two neurons and provides a handle for further characterization and understanding of their structural and functional differences (Shepherd and Greer, 1990; Goldstein *et al.*, 1983). In the latter case, the observation provides evidence for the existence of receptor neuron subclasses that are not discernable by classical morphological criteria but which are consistent with recent reports of heterogeneity in olfactory receptor neurons (Hinds *et al.*, 1984; Mackay-Sim and Kittel, 1991). Clearly, it is too early to tell if these observations will prove reliable. However, they illustrate an empirical strategy that is an important new direction for MAP research.

V. Neuronal Development and Stabilization

Since the earliest descriptions of the distribution of MAPs in the nervous system, there has been intense interest in understanding their role in neuronal development (recent reviews in Matus, 1988a,b; Nunez, 1988; Meininger and Binet, 1989; Tucker, 1990). This has been predicated on the evidence that microtubules themselves play an essential role in both neuritic outgrowth (Seeds *et al.*, 1970; Yamada *et al.*, 1970) and the stabilization of adult form (Wuerker and Kirkpatrick, 1972; Hillman, 1979). Initial findings that MAP expression varies with development (Mareck *et al.*, 1980) set the stage for a decade of investigations into the role of MAPs in regulating both microtubule elongation and stabilization during the course of development. More recent evidence that these regulatory roles in growth and stabilization can be influenced by stimulation-dependent phosphorylation and dephosphorylation (Aoki and Siekevitz, 1985; Aletta *et al.*, 1988) now implicates MAPs in mediating the profound effects of experience on neuronal form both during development and in adult learning and memory. Although this constitutes a vast array of putative functions for MAPs, all can be linked to a very basic mechanism, namely the ability of MAPs to regulate the stability of microtubules.

A. Developmental Outgrowth and Stabilization

1. Developmental Changes in MAP Expression

Each MAP isotype shows a distinctive developmental pattern of expression which nevertheless tends to fall into one of two categories. So-called "juvenile" or "early" MAPs are generally expressed at higher levels in the nervous system during fetal and neonatal stages than in adulthood, whereas "adult" or "late" MAPs are generally expressed at higher levels in the adult than in the fetus or neonate (Fig. 10). Early MAPs in brain include the 48-kDa isoform of tau protein (Mareck *et al.*, 1980; Francon *et al.*, 1982; Drubin *et al.*, 1984a), MAP 2c (also known as

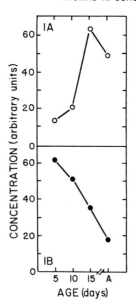

FIG. 10   Opposite changes in MAP 1A and MAP 1B concentration with rat brain development. Data points are based on densitometric analyses of immunoblots of whole brain
tissue, using either anti-MAP 1A (1A, antibody MAP 1A-2) or anti-MAP 1B (1B, antibody
MAP 1B-4). (From Schoenfeld *et al.*, 1989.)

young tau slow, 70-kDa MAP 2 or small MAP 2: Francon *et al.*, 1982;
Couchie and Nunez, 1985; Riederer and Matus, 1985) and MAP 1B
(Riederer *et al.*, 1986; Sato-Yoshitake *et al.*, 1989; Schoenfeld *et al.*,
1989) (Fig. 11). Late MAPs include the heavier isoforms of tau protein
(Mareck *et al.*, 1980; Francon *et al.*, 1982; Drubin *et al.*, 1984a), MAP 2a
(Binder *et al.*, 1984; Burgoyne and Cumming, 1984; Riederer and Matus,
1985) and MAP 1A (Riederer and Matus, 1985; Schoenfeld *et al.*, 1989)
(Fig. 12). Some MAPs are expressed at both early and late stages of
development (e.g., MAP 2b; Binder *et al.*, 1984; Burgoyne and Cumming, 1984; Riederer and Matus, 1985).

There is also evidence that early MAPs are more heavily or completely
phosphorylated at early periods than at late periods of development, that
is, they are more likely to express phosphorylation-dependent epitopes
(e.g., MAP 1B in developing rat cerebellar and cerebral cortex; Sato-
Yoshitake *et al.*, 1989; Schoenfeld *et al.*, 1989; Viereck *et al.*, 1989;
Fischer and Romano-Clarke, 1990; Riederer *et al.*, 1990; Mansfield *et al.*,
1991) or are more resistant to additional phosphorylation by *in vitro* application of radiolabeled adenosine triphosphate (ATP) (e.g., HMM MAP
2 [MAP 2b?] in developing neocortex; Aoki and Siekevitz, 1985). Con-

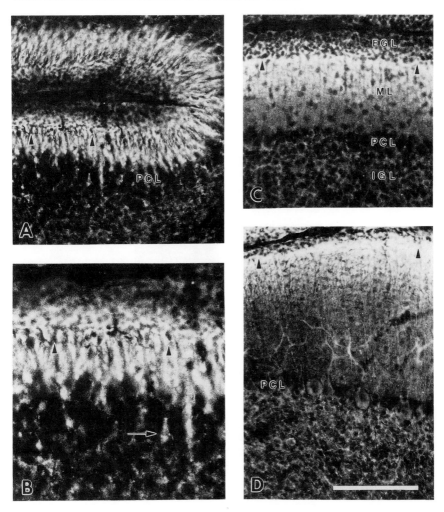

FIG. 11   Anti-MAP 1B-4 immunoreactivity in the rat cerebellum at (A, B), P5; (C), P10; and (D) P20 as seen in sagittal section. Arrowheads indicate the border between the deep, premigratory zone of the external granule layer (EGL) and the superficial part of the molecular layer (ML). (B). A magnified view of a portion of (A); the arrowheads in (A) correspond to the position of the arrowheads in (B). The arrow in (B) points to a labeled granule cell near the end of its migration positioned just deep to the unlabeled profiles of two Purkinje cells. PCL, Purkinje cell layer; IGL, internal granule layer. Bar = 100 μm for (A), (C), and (D); 58 μm for (B). (From Schoenfeld et al., 1989.)

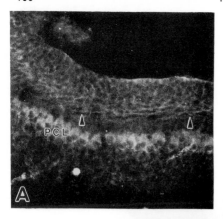

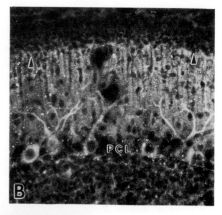

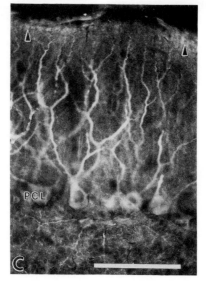

FIG. 12   Anti-MAP 1A-1 immunoreactivity in the rat cerebellum at (A), P5; (B), P10; and (C), P20, as seen in sagittal section. Arrowheads indicate the border between the deep, premigratory zone of the external granule layer and the superficial part of the molecular layer. PCL, Purkinje cell layer. Bar = 100 μm. (From Schoenfeld *et al.*, 1989.)

versely, late MAP isotypes tend to be less phosphorylated in adulthood, although not universally (e.g., MAP 1A; Diaz-Nido *et al.*, 1990).

Since early MAPs tend to be associated with growing processes in the fetal and neonatal nervous system, it might be assumed that they function principally to promote growth or at least derive particular functional significance from an association with growth. However, this assumption may be unwarranted, since there are cases where the expression or lack of expression of early MAPs is irrelevant to growth. For example, regenerating axons express the growth-associated protein, GAP 43, as do developing axons (Skene and Willard, 1981), but fail to express an early MAP found in outgrowing developing axons, namely MAP 1X, a homolog of MAP 1B (Woodhams et al., 1989). On the other hand, the axons of olfactory receptor neurons which have proliferated in the adult, once they have completed outgrowth to the olfactory bulb and show only diminished levels of GAP 43 expression (Verhaagen et al., 1989), nevertheless continue to express MAP 1B at high levels (Schoenfeld et al., 1989; Viereck et al., 1989) (Fig. 9). If the patterns of early MAP expression are not clearly and consistently related to growth per se, what then is their significance?

2. Neurite Outgrowth and Microtubule Assembly

A clue to the functional significance of MAP expression during development comes from the relative ability of MAPs to promote microtubule assembly. The growth of axons and dendrites to adult lengths clearly requires microtubule assembly (Seeds et al., 1970; Yamada et al., 1970; Mitchison and Kirschner, 1988). Since most MAPs are reported to promote microtubule assembly to varying degrees (Murphy and Borisy, 1975; Weingarten et al., 1975; Sloboda et al., 1976; Herzog and Weber, 1978; Mareck et al., 1980; Francon et al., 1982; Riederer et al., 1986), one might have predicted that MAPs influence growth by promoting microtubule assembly to an extent at least equivalent to, if not greater than, that seen in adulthood. However, just the opposite seems to occur. "Early" MAPs, to the extent that they are associated with growth, are actually *less* efficient at promoting microtubule assembly or even binding to microtubules than the "late" MAPs (Mareck et al., 1980; Francon et al., 1982; Bloom et al., 1985; Riederer et al., 1986), with perhaps the notable exception of phosphorylated MAP 1B (Diáz-Nido et al., 1990; Riederer et al., 1990; Mansfield et al., 1991; see later discussion).

This suggests that neuritic growth may not have a simple relationship with microtubule assembly (Letourneau and Ressler, 1984). This is borne out in more direct studies of microtubule assembly and the growth of neurites. For example, the drug taxol is extremely potent in promoting microtubule assembly and protecting against disassembly. However, by promoting the unregulated polymerization and stabilization of nearly all free tubulin (at high taxol concentrations), it apparently sidesteps the vectorial building process that is critical for elongation, effectively stopping growth (Letourneau and Ressler, 1984).

On the other hand, the dynamic instability of MAP-free microtubules actually permits incidences of greater microtubular elongation than observed in MAP-bound microtubules, although net growth is less because of catastrophic disassembly (Mitchison and Kirschner, 1984; Horio and

Hotani, 1986; Keates and Hallett, 1988). It is not known how MAPs moderate a delicate balance between sufficient and excessive promotion of assembly. However, theoretically, net elongation could be achieved and maximized by providing a foundation array of microtubules, stabilized against disassembly through binding and bundling by MAPs, onto which additional tubulin could be added stochastically and from which new microtubules could be extended vectorially and in turn also stabilized (Kirschner and Mitchison, 1986; Mitchison and Kirschner, 1988; Baas et al., 1991). In this way, MAPs may promote microtubule assembly less by "driving" it than by permitting it to occur, through the support of a stabilized microtubular array that is protected against disassembly. That is, they may promote assembly by inhibiting disassembly (Bré and Karsenti, 1990).

This mechanism is illustrated very nicely in studies of tau expression in living cultured cells. At an early stage of neuritic development in cultured cerebellar macroneurons (Ferreira et al., 1989) (see Section III,A,5), when one neurite begins to elongate more rapidly than the others, an immature complement of tau protein isoforms is uniformly expressed in all of the neurites, and thus does not "drive" the outgrowth of the axon specifically. Nevertheless, the presence of tau is critical to the outgrowth of one of the neurites as an axon, for inhibition of tau expression by transfection of tau antisense nucleotides into these neurons prevents any axonal outgrowth (Caceres and Kosik, 1990). Thus, this MAP may play an essential role in stabilizing the already assembled microtubular array against disassembly, but apparently without interfering with, or otherwise participating in, the mechanisms that drive assembly (see discussion in Baas et al., 1991). When tau is introduced into cultured non-neuronal cells which normally lack extended processes, it also promotes microtubule assembly and protects against disassembly. Here, too, tau does not by itself drive the formation and elongation of neurite-like processes (Drubin and Kirschner, 1986). However, if such cells possess mostly free, unpolymerized tubulin to begin with in the presence of a superabundance of tau, then tau may participate over a protracted period in the initial construction of a stabilized microtubular array (Knops et al., 1991).

3. Developmental Plasticity and Stability

Thus, the patterns of early MAP expression may reflect a more general role for MAPs in the regulation of neurite stability. All MAPs apparently stabilize neuronal microtubules to some extent. The question is how far the balance is tipped toward stability and away from instability by the presence of MAPs. Certainly, both the amalgamation of different MAPs within a neurite and their posttranslational modification are potential means for regulating such balance (see the later discussion). What remains to be fully explored and tested experimentally is which MAPs, how they are modified, in what alliances, and with what consequences for both microtubule and neuritic stability and neuronal function overall. We consider some of these issues next.

a. Microtubule Assembly and Bundling Both Contribute to Neurite Stability. Several previous studies using turbidity assays have reported that early MAPs promote less extensive assembly of microtubules than late MAPs (Mareck *et al.*, 1980; Francon *et al.*, 1982; Riederer *et al.*, 1986). Although these data may provide a valid index of the different degrees of stability promoted by different MAPs, there are reasons to doubt that microtubule assembly is the only, or even principal means by which neurite stability is regulated by MAPs. First of all, serial section analysis suggests that individual microtubules are discrete, short segments of discontinuous polymer (Zenker and Hohberg, 1973; Bray and Bunge, 1981; Sasaki *et al.*, 1983; Burton, 1987; Baas and Black, 1990). Thus, a neurite's stability must be tied in part to the stability of the microtubule array, specifically to the cross-links among the microtubules making up the array, since an extended array of discontinuous elements simply could not be established and sustained without effective cross-links.

We have already discussed the role of MAP cross-bridging in the determination of axonal and dendritic caliber through regulation of microtubule spacing (see Section III,B,3). Fibrous MAPs appear to represent the principal components of such cross-bridges, which are formed by domains found within each MAP molecule that project laterally to adjacent microtubules (Dentler *et al.*, 1975; Murphy and Borisy, 1975; Vallee and Borisy, 1977; Herzog and Weber, 1978; see review in Hirokawa, 1991). Since recent work indicates that MAPs may differ as much by the lengths and composition of their projection domains as by the structure and activity of their microtubule binding domains (Aizawa *et al.*, 1990) (see also section III,B,3), variations in MAP function, particularly stabilization, may indeed concern bundling as well as polymerization (Kanai *et al.*, 1992).

Second, the role of reduced assembly in neurite lability may be overstated, and the role of reduced bundling understated, because the turbidity assay in fact cannot distinguish between polymerized individual microtubules and cross-linked microtubule arrays (Fig. 13). This is because turbidity, as a measure of steady-state density or viscosity, is influenced by the density or viscosity of the assembled polymeric arrays, that is, their bundling properties or tendency to gel, not just by assembly from

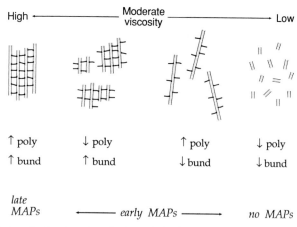

FIG. 13 Variations in the viscosity of microtubules in solution as a function of microtubule polymerization and bundling and the presence of early and late MAPs. Microtubules without MAPs are likely to show low levels of polymerization (poly) and bundling (bund), resulting in low steady-state viscosity or density. At the other extreme, the presence of late MAPs is likely to promote high levels of both, resulting in high viscosity. The presence of early MAPs yields essentially intermediate or moderate levels of viscosity by promoting either more polymerization than bundling, or vice versa, but not both.

free tubulin alone (Lasek, 1988). For example, increased turbidity can actually be a manifestation of increased microtubule bundling even in the face of slightly depressed polymerization (Matus *et al.*, 1987). Thus, when early MAPs are shown by this assay to promote reduced turbidity relative to adult brain MAPs (Francon *et al.*, 1982; Riederer *et al.*, 1986), it is impossible to tell whether the difference is due to reduced polymerization or reduced bundling (Fig. 13).

Histological visualization is especially useful in making such a distinction (Herzog and Weber, 1978; Letourneau and Ressler, 1984; Matus *et al.*, 1987; Lewis *et al.*, 1989; Hirokawa, 1991) and in theory could help to validate functional interpretations of turbidity data. Indeed, in two reports (Jameson *et al.*, 1980; Lindwall and Cole, 1984) in which phosphorylated MAPs were shown to promote less turbidity than unphosphorylated MAPs, electron micrographs (Jameson *et al.*, 1980, Fig. 4; Lindwall and Cole, 1984, Fig. 6) illustrate what appears to be normally assembled but unbundled microtubules formed in the presence of phosphorylated MAPs. Microtubules assembled in the presence of these phosphorylated MAPs were actually twice as long but half as numerous (Jameson *et al.*, 1980; Jameson and Caplow, 1981), resulting in no net change in the mass of assembled microtubules. Thus, the reduced turbidity associated with MAP phosphorylation may be a consequence of reduced bundling as well as altered, if not reduced, assembly.

One can also measure polymerization more directly by assaying sedimentation properties, for example, the proportion of tubulin that sediments as microtubules in the presence of MAPs (Vallee, 1982; Matus *et al.*, 1987) or the tendency of a MAP to fully cosediment with microtubules during purification (Bloom *et al.*, 1985; Diaz-Nido *et al.*, 1990). Whether cross-linked or not, the amount of polymerized tubulin should be directly reflected in the mass of the microtubule pellet. Moreover, the degree of cosedimentation is more likely to depend on the degree of microtubule binding than on the extent of cross-bridging, although this does assume that cross-bridging entails MAP–MAP interactions (Furtner and Wiche, 1987; Lewis *et al.*, 1989) rather than two microtubule binding sites on each MAP molecule. On the other hand, evidence of cosedimentation cannot by itself establish the role of any MAP in the stabilization of microtubule arrays. In any event, these sorts of approaches have not been used in any systematic way to evaluate the differences between early and late MAPs.

b. Early MAPs May Promote Both Lability and Stability. Further complicating this picture is molecular and physiological evidence that early MAPs may in some respects and under some circumstances promote or have the potential to promote both lability and stability, with postranslational phosphorylation a key regulator.

For example, in embryonic and neonatal brain, tau is expressed as a smaller number of isoforms than those found in the full adult complement (1–2 vs. 4–6), which are also lower in molecular weight than those typically found in adult brain (47–49 kDa vs. 52–68 kDa; Francon *et al.*, 1982; Drubin *et al.*, 1984a). In addition, these early-appearing species have fewer microtubule binding sites (predominantly three rather than four imperfect 18-amino acid repeats) than the later-appearing polypeptides (Goedert *et al.*, 1989; Kosik *et al.*, 1989b). The spacing of binding repeats appears to match closely the distance between adjacent tubulin subunits (Lewis *et al.*, 1988), suggesting that one less repeat might directly correspond to one less tubulin subunit linked sequentially along a polymer by a single MAP molecule. Since early tau does not coassemble with tubulin during purification cycles as readily as late tau (Drubin *et al.*, 1984a; Binder *et al.*, 1985), the reduced number of microtubule binding sites may reduce the affinity of early tau for microtubules (Kosik *et al.*, 1989b; Kanai *et al.*, 1992) and in turn lead to a reduction in the nucleation and stabilization of polymerized tubulin.

Early tau also induces quantitatively less bundling of microtubules in cultured cells than do adult tau isoforms, an effect largely attributable to segment deletion in the N-terminal projection domain (Kanai *et al.*, 1992). The smaller number of tau isoforms expressed at early stages of development may also influence the stability of microtubules, although the existence or nature of any potential synergistic or additive activity among the multiple isoforms of this MAP, or any MAP, is unknown (see the following discussion).

On the other hand, there is evidence that tau protein may not be fully phosphorylated *in vivo* at early stages of development, particularly in comparison with other early MAPs (e.g., MAP 1B; see the following discussion), which may foster microtubule stability. Diaz-Nido *et al.* (1990) have shown that intracranial injection of radiolabeled phosphate into 5-day-old rat brains identifies the *in vivo* phosphorylation of MAPs 1B and 2 and β-tubulin but not MAP 1A. Tau is only faintly labeled. In 30-day-old cases, MAP 1A is also radiolabeled but tau isoforms are still only faintly labeled. Kosik *et al.* (1989b) demonstrated that the expression of the α-subunit of calcium–calmodulin-dependent protein kinase, which is more effective than type II cAMP-dependent protein kinase in phosphorylating tau (Yamamoto *et al.*, 1985), coincides with the onset of adult tau expression and the decline in expression of the early isoform. This suggests that early tau lacks a principal means for becoming phosphorylated.

In studies of phosphoproteins in PC12 cells expressed during early neuritic outgrowth following treatment with nerve growth factor (NGF), Greene *et al.* (1983) reported that NGF induces the incorporation of radiolabeled phosphate into an HMM MAP (MAP 1.2, a homolog of MAP 1B) plus a variety of LMM proteins, including tyrosine hydroxylase and several chartin proteins, none of which can be identified as tau protein isoforms (Black *et al.*, 1986) (see also Section V,B,1). Since the turbidity of tubulin coassembled with tau is enhanced by the dephosphorylation of tau (Lindwall and Cole, 1984), these demonstrations suggest that early tau, where dephosphorylated, may be able to promote microtubule binding and/or bundling to a reasonable degree.

On the other hand, Tucker *et al.* (1988a) have shown that phosphatase treatment of sections through the thoracic spinal cord of quail embryos unmasks otherwise faint staining of developing axonal tracts, indicating that tau (or at least the Tau-1 epitope) in these outgrowing axons is largely phosphorylated and would be expected to promote much weaker binding and/or bundling. The basis for this discrepancy is unknown.

MAP 1B, another early MAP, has also been shown to coassemble inefficiently with tubulin, compared with the coassembly efficiency of the late-appearing MAP 1A (Bloom *et al.*, 1985; Noble *et al.*, 1989). It may even coassemble with tubulin less efficiently than early tau, since this isoform of tau at least cosediments with tubulin through the first one to two assembly cycles (Drubin *et al.*, 1984a). Moreover, there is evidence that MAP 1B binds differently to microtubules than tau and MAP 2. The microtubule binding domain of MAP 1B contains shorter repeats than those found in tau or MAP 2 (4 amino acids instead of 18) and they are less regularly spaced and distributed within this domain than those found in tau or MAP 2 (Noble *et al.*, 1989). The irregular spacing and distribution of smaller potential binding sites, even though they are particularly rich in lysine, might be expected to promote weaker bonds between MAP 1B and tubulin than those observed for both tau and MAP 2.

The molecular structure of MAP 1A has recently been determined (Langkopf *et al.*, 1992). Its structure and mode of binding to microtubules seem to be very similar to that of MAP 1B. These MAPs are both composed of lower molecular weight subunits (light chains, 18–34 kDa) that bind both to the N-terminal microtubule binding domain of the heavy chain for each molecule and to tubulin as well (Vallee and Davis, 1983; Kuznetsov and Gelfand, 1987). We have recently demonstrated that MAP 1B and MAP 1A differ in their light chain (LC) content (Schoenfeld *et al.*, 1989). MAP 1B contains only LC1, whereas MAP 1A contains both LC1 and LC2 but with stoichiometrically greater amounts of LC2. Both contain LC3 as well (Fig. 14). LC1 and MAP 1B are actually derived from the same gene product, a polyprotein in which LC1 is an extension of the C-terminal (nontubulin binding) end of MAP 1B before it is cleaved apart (Hammarback *et al.*, 1991). Similarly, LC2 and MAP 1A are derived from a different polyprotein (Langkopf *et al.*, 1992). The origin of LC3 is unknown. The association of LCs with the microtubule binding domain of the heavy chain MAP 1s suggests that they play a role in regulating microtubule binding. It will be of interest to determine whether LC2, in its association with the later-appearing MAP 1A, actually promotes more efficient binding to microtubules than does LC1.

This picture of MAP 1B as a "lability" protein is contrasted with three recent reports suggesting that the efficiency with which MAP 1B coass-

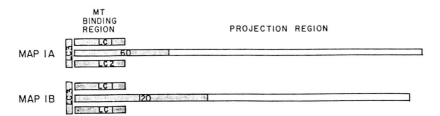

FIG. 14   Schematic representation of MAP 1A and MAP 1B molecules. Each molecule is shown as a complex of one heavy chain plus several light chains (LC1, LC2, LC3) associated with the microtubule binding region of the heavy chain. The drawing identifies microtubule binding fragments of M_r 120,000 for MAP 1B and M_r 60,000 for MAP 1A. (From Schoenfeld *et al.* 1989.)

embles with tubulin is improved by phosphorylation, especially early in development. Diaz-Nido *et al.* (1990) found that *in vivo* phosphorylation of MAP 1B by intracranial injection of radiolabeled phosphate into adult rat brain promotes its differential association with polymerized over free tubulin. Interestingly, a similar result occurred with phosphorylated MAP 1A. This shows that an early and a late MAP (MAPs 1B and 1A, respectively) can have a similar functional relationship with microtubules, although it does not provide evidence that these two MAPs are actually alike in their specific ability to stabilize microtubule arrays. The opposite effect was found with phosphorylated MAP 2, that is, phosphorylated MAP 2 is associated more with unassembled than with assembled tubulin. This is consistent with previous demonstrations (Murthy and Flavin, 1983; Yamamoto *et al.*, 1983; Lindwall and Cole, 1984) that phosphorylation of heat-stable MAPs (including MAP 2 and tau) reduces their promotion of microtubule assembly and/or bundling. It raises the possibility that the thermolabile MAP 1s may differ from thermostable MAPs, as a family, in the functional consequences of phosphorylation.

The authors also report that phosphorylated MAP 1B is predominantly associated with white rather than gray matter in adult rat brain, whereas the opposite is true of phosphorylated MAP 2. Phosphorylated MAP 1A shows no differential association between white and gray matter. The former finding is reminiscent of evidence that MAP 1B readily coassembles with white matter but not gray matter microtubules from calf brain (Bloom *et al.*, 1985), suggesting that the relative inefficiency with which MAP 1B coassembles with gray matter microtubules may reflect its dephosphorylation there.

In a study of MAP 5, a homolog of MAP 1B, in developing cat cerebral cortex, Riederer *et al.* (1990) extended the report by Diaz-Nido *et al.* (1990) in two respects. First, they specifically compared phosphorylated and dephosphorylated (or at least less phosphorylated) isoforms of MAP 5 in their avidity for polymerized or free tubulin and found that MAP 5a, the phosphorylated isoform, is predominantly associated with polymer whereas MAP 5b, the less phosphorylated isoform, is predominantly associated with cytosol. Second, they demonstrate that this differentiation is characteristic only of early development, at least in cortical gray matter, in that MAP 5a expression disappears with development, whereas MAP 5b continues to be expressed, albeit with less intensity. Separate biochemical analyses of developing white matter (corpus callosum) show that MAP 5a is the predominant isoform expressed early in development and that neither is expressed in adulthood. This latter result is contradicted by the findings of Diaz-Nido *et al.* (1990) that phosphorylated MAP 1B is enriched in the corpus callosum and brain stem white matter of adult rat, and those of Sato-Yoshitake *et al.* (1989) that MAP 1B phosphoepitopes continue to be expressed in some axons of adult rat brain and spinal cord.

Mansfield *et al.* (1991) have studied the coassembly properties of phosphorylated MAP 1B in the growth cones and neurites of cultured cerebral cortical neurons. Using an antibody to MAP 1B that recognizes a phosphate-dependent epitope (Ab 150), they found that this phosphoepitope is expressed in distal neurites and growth cones associated with both the soluble, cytosolic pool of tubulin and the insoluble, cytoskeletal fraction. Within growth cones, antibody 150 was more immunoreactive in proximal regions than distal filopodia. Immunoreactivity was also more filamentous in neurites and proximal regions of growth cones than in more distal locations. Such a pattern was more closely correlated with the expression of detyrosinated than tyrosinated α-tubulin. On the other hand, the more widespread expression of a phosphate-independent epitope throughout cultured neurons, in growth cone filopodia, axonal neurites, cell somata, and dendritic neurites, corresponded closely with the distribution of tyrosinated α-tubulin.

Since microtubules are more prominently assembled and bundled in neurites, extending into proximal growth cones, than they are in distal growth cone filopodia (Letourneau, 1985; Gordon-Weeks, 1987), phosphorylated MAP 1B appears to be affiliated largely with stabilized arrays of microtubules located just behind the elongating tips of growing neurites. The expression of detyrosinated α-tubulin here is also consistent with the putative stability of this phospho-MAP 1B-positive domain (Baas and Black, 1990; see the following discussion). Whether phosphorylated MAP 1B is merely associated with stabilized microtubules or actually promotes stabilization remains to be demonstrated, however. Moreover, it is even less clear how MAP 1B may differ from other MAPs, particularly tau and MAP 1A, in the promotion of stabilization.

Another early MAP, MAP 2c, contains the same number and sequence of binding repeats as the late isoform MAP 2b (Papandrikopoulou *et al.*, 1989; Kindler *et al.*, 1990). Thus, unlike 48-kDa tau and MAP 1B, MAP 2c should exhibit the same binding efficiency as the other isoforms in its class, and this has been confirmed (Riederer and Matus, 1985). The N-terminal projection domain of MAP 2c is identical to that of HMM MAP 2 except for the deletion of a sizable internal segment (Papandrikopoulou *et al.*, 1989; Kindler *et al.*, 1990), suggesting that the bundling properties per se ought to be similar as well (Lewis *et al.*, 1989).

However, consider that MAP 2c is not expressed alone in developmentally immature or labile neurons but in tandem with the longer MAP 2b, whereas it is essentially replaced by MAP 2a in mature, stable neurons (Binder *et al.*, 1984; Burgoyne and Cumming, 1984; Riederer and

Matus, 1985), which should be of a length equivalent to that of MAP 2b. Although there may be stoichiometrically somewhat less MAP 2c than 2b expressed in infant brain (Francon *et al.*, 1982; Riederer and Matus, 1985), the mix of 2c with 2b in immature neurons would be heterogeneous nevertheless—a combination of short and long lateral projections—whereas the mix of 2a with 2b in mature neurons would be homogeneous with respect to lateral projections. Perhaps the heterogeneity of MAP 2 isoforms is a critical determinant of the reduced efficiency of cross-bridging activity and hence bundling of microtubules in immature neurons (Kindler *et al.*, 1990).

There has been surprisingly little interest in ascertaining the effect of phosphorylation on MAP 2 function in developing neurons. Unfortunately, the data that do exist appear to be contradictory, although the experimental settings have not really been comparable. Aoki and Siekevitz (1985) reported that HMM MAP 2 from visual cortex of dark-reared kittens cannot be phosphorylated *in vitro* after extraction, whereas it can be phosphorylated *in vitro* if the kittens are exposed to light. Auditory cortex MAP 2 can be phosphorylated *in vitro* in either case. They interpreted this to mean that visual cortex MAP 2 is fully phosphorylated *in vivo* before light exposure and dephosphorylated *in vivo* after light exposure. Auditory cortex MAP 2 of kittens not prevented from having auditory experience is also dephosphorylated *in vivo* by this interpretation. Since visual circuits are more malleable before light exposure (see the discussion in Section V,B,2), these findings are consistent with evidence that MAP 2 phosphorylation is associated with reduced assembly of brain microtubules, reduced binding to brain microtubules, and less cosedimentation with assembled than unassembled brain microtubules (Murthy and Flavin, 1983; Yamamoto *et al.*, 1983; Burns *et al.*, 1984; Diaz-Nido *et al.*, 1990).

On the other hand, Diaz-Nido *et al.* (1990) found that infusion of radiolabeled phosphate into the brains of 5- and 30-day-old rats revealed no age-related reduction in *in vivo* phosphorylation of HMM MAP 2 in whole brain preparations. Although a reduction might have been predicted from the Aoki and Siekevitz study, analysis of whole brain homogenates may have been too crude to reveal regionally specific reductions. Nevertheless, such phosphorylated MAP 2 cosedimented more with unassembled than assembled microtubules.

By contrast, a more recent study (Brugg and Matus, 1991) demonstrates that brain MAP 2 microinjected into cultured fibroblasts must be phosphorylated *in vivo* to a certain extent before binding to microtubules can be observed. This finding seems to contradict the previously cited evidence that phosphorylation of MAP 2 disrupts its association with microtubules. The authors suggest that it may indicate that *in vivo* and *in vitro* phosphorylation affect MAP 2 function differently (Murthy *et al.*, 1985), although the Diaz-Nido *et al.* (1990) data, drawn from an *in vitro* analysis of the effects of *in vivo* phosphorylation, make that hypothesis less tenable.

Variations in the posttranslational modification of tubulin have been shown to correspond to different degrees of stability in microtubules and neurites. For example, axonal microtubules tend to be more acetylated and less tyrosinated than dendritic microtubules and are more resistant to depolymerization (Burgoyne *et al.*, 1982; Cambray-Deakin and Burgoyne, 1987; Ferreira and Caceres, 1989). In axons, detyrosinated microtubules are more resistant to depolymerization than tyrosinated microtubules (Baas and Black, 1990). Moreover, detyrosinated and acetylated α-tubulin does not extend appreciably into the growth cones of growing neurites, which corresponds to the greater lability of the growth cone (Ferreira and Caceres, 1989; Robson and Burgoyne, 1989; Mansfield *et al.*, 1991).

Does the distribution of MAPs in neurons vary with the differential distribution of tubulin isotypes (Meininger and Binet, 1989)? Since modifications like acetylation do not appear to play a causative role in microtubule stabilization (Schulze *et al.*, 1987; Black *et al.*, 1989), do MAPs contribute more directly to the regulation of stability? Until recently, very little attention had been paid to these questions. In cultured cerebellar macroneurons, Ferreira and Caceres (1989) have shown that the fractions of tau and acetylated tubulin that are associated with microtubules are preferentially associated with colchicine-resistant microtubules. MAPs 1A and 2 were not found to show this preferential association with more stable microtubules (see also Robson and Burgoyne, 1989), although another recent study (Gordon-Weeks *et al.*, 1989) has shown that MAP 2 codistributes more completely than tau with taxol-stabilized microtubules in growth cones of 18-hr cultured neurons.

Tau is also expressed in axonal growth cones and dendritic neurites, which express tyrosinated but not acetylated α-tubulin at early stages of development (Ferreira *et al.*, 1989; Ferreira and Caceres, 1989; Gordon-Weeks *et al.*, 1989). However, tau in these processes is more easily solubilized by detergent than is tau in the axonal trunk and cell body, indicating that it is predominantly cytosolic and not bound to the cytoskeleton in axonal growth cones and young dendritic neurites (Ferreira *et al.*, 1989; see also Gordon-Weeks *et al.*, 1989). Since axonal growth cones and early outgrowing dendritic neurites contain a substantial pool of soluble tubulin that is not polymerized into microtubules (Bernhardt and Matus, 1982; Gordon-Weeks, 1987), tau apparently cannot drive polymerization per se in these processes (see also Section V,A,2). Nevertheless, inhibition of tau expression inhibits axonal outgrowth (Caceres and Kosik, 1990), so unlike acetylation, the association of tau with stable microtubules in differentiated axons appears to be more a determinant of stability than a manifestation of it.

Recent work suggests that MAP 1B may also have a distinctive relationship with tubulin isoforms, and one that depends on its phosphorylation. Using primary cultures of rat cerebral cortex, Mansfield *et al.* (1991) have shown that monospecific antibodies to phosphorylated MAP 1B and detyrosinated α-tubulin both stain the proximal portions of axonal growth cones and the associated distal and main segments of the axonal neurite (see previous discussion). An antibody to a phosphorylation-independent epitope of MAP 1B also stains the distal growth cone, particularly filopodia, the soma and proximal axon segments, and dendritic neurites—regions which are also immunoreactive to antibodies recognizing tyrosinated rather than detyrosinated α-tubulin.

The association of phosphorylated MAP 1B with detyrosinated microtubules, and unphosphorylated MAP 1B with tyrosinated microtubules, suggests that phosphorylation may enhance the stabilizing properties of MAP 1B, since detyrosinated microtubules are more resistant to depolymerization than tyrosinated microtubules (Baas and Black, 1990). Recent evidence that phosphorylation enhances the coassembly efficiency of MAP 1B (Diaz-Nido *et al.*, 1990; see previous discussion) is consistent with this inference. However, no one has as yet examined the association of MAP 1B, whether phosphorylated or not, with acetylated microtubules nor whether acetylation and detyrosination, occurring separately or in combination, are associated with microtubules of different stability.

Early MAPs, therefore, can potentially both stabilize and destabilize microtubule arrays, essentially by regulating their disassembly (Fig. 15). As we noted earlier, the balance between stability and lability is probably regulated by different amalgamations of MAPs, varying by both genetic specification and epigenetic modification of binding or bundling functions. An additional regulatory factor may be the availability of binding sites for MAPs on microtubules, manifested perhaps by posttranslational modifications such as detyrosination, acetylation, and phosphorylation. The likely result is a more subtle variation in relative stability that perhaps conforms better to our physiological observations than would the isolated expression of "stability" or "lability" proteins (Fig. 15).

c. Neuritic Lability May Permit Remodeling The weak stabilization conferred by some early MAPs may be necessary to avoid unregulated polymerization of microtubules that might actually stop growth. It may also have another important function, namely, to foster remodeling of neural circuits during development. It is well known that the develop-

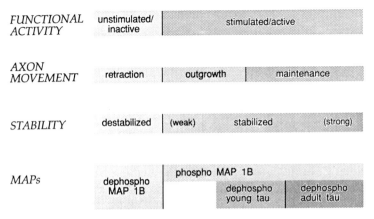

FIG. 15 Relationships between the activity and movement of axons during development, the relative stability of microtubules, and the expression of MAPs, specifically MAP 1B and tau protein, as inferred from the literature reviewed (see text). Lack of stimulation or negligible levels of spontaneous activity lead to the expression of predominantly dephosphorylated MAP 1B, destabilization of the microtubular array, and retraction or even degeneration of the axon. Alternatively, varying degrees of stimulation or activity lead to varying combinations or sequences of MAP expression (phosphorylated MAP 1B, dephosphorylated tau isoforms) that stabilize microtubule arrays to varying degrees in promoting either axonal outgrowth or maintenance.

ment of many connections in the nervous system is completed by a remodeling process that entails selective stabilization and elimination (Wiesel and Hubel, 1963; Changeux and Danchin, 1976; Purves and Lichtman, 1980; Schmidt, 1985; O'Leary, 1989; Constantine-Paton et al., 1990; Bhide and Frost, 1991; Wolpaw et al., 1991). The selection of which axons will be stabilized and which eliminated seems to occur epigenetically: axons that are generally the most active or that exhibit particular patterns of activity are stabilized and retained, whereas the others are lost by retraction and movement to more reactive targets or by degeneration in situ (Schmidt, 1985; see the following discussion). Postsynaptic changes also occur (Schmidt, 1985; Aoki and Siekevitz, 1988; see later discussion) and probably precede and help to determine the presynaptic alterations (Schmidt, 1991). Over time, remodeling becomes more difficult—epigenetic factors are less likely to elicit further adjustments in connectivity (Hubel and Wiesel, 1970).

How might MAPs mediate the process of remodeling? Although there is evidence that the expression of MAPs is stimulation-dependent, it is not yet clear how such expression alters microtubule and neuritic stability (see Section V,B). Certainly, disassembly of microtubules (induced by depolymerizing agents such as colchicine) results in retraction of neurites (Seeds *et al.*, 1970; Yamada *et al.*, 1970). Moreover, the presence of MAPs tends to protect microtubules from depolymerization (Haga and Kurokawa, 1975; Sloboda and Rosenbaum, 1979; Schliwa *et al.*, 1981; Drubin and Kirschner, 1986). However, we do not know whether early and late MAPs differ in the stability they provide to microtubules and neurites in the presence of depolymerizing agents, although there is evidence that microtubules from neonatal brain, presumably under the influence of early MAPs, are less resistant to the depolymerizing effect of cold temperature than those from adult brains (Faivre *et al.*, 1985).

Of course, even early MAPs are needed to promote microtubule assembly and bundling that is sufficient to stabilize outgrowing axons and dendrites behind their growth cones. Phosphorylation of MAP 1B, for example, may promote such stability in distal segments of outgrowing axons. However, this MAP may not provide the degree of stability provided by tau protein, even 48-kDa tau (see earlier discussion). Does this mean that phosphorylated MAP 1B promotes remodeling but 48-kDa tau does not? These differential roles of early MAPs (Fig. 15) remain to be fully explored.

What then is the change in MAPs that promotes disassembly of the sort that would lead to retraction of neurites? This has not been studied. In Fig. 15, we predict that dephosphorylation of MAP 1B might be one means to destabilize the cytoskeleton in an inactive neurite or perhaps a neurite lacking stimulation by neurotrophic factors (see Section V,B,1). Certainly, other variations are possible.

What endogenous agent would initiate disassembly in inactive neurites? Calcium would be an obvious candidate for such an agent, since it causes microtubule disassembly (Schliwa *et al.*, 1981), either directly or through calcium-activated proteases (calpains; Melloni and Pontremoli, 1989). Interestingly, the olfactory nerve of adult rats, whose component axons are believed to regularly degenerate and regrow throughout adult life (Graziadei and Monti Graziadei, 1978; see the following discussion), shows prominent anticalpain immunoreactivity (Siman *et al.*, 1985), which suggests that calpain, and hence calcium, may be an agent in the induction of degeneration there.

On the other hand, the entry of calcium into neurons is more generally associated with synaptic activity than with inactivity (Stanley *et al.*, 1991). Of course, the mechanism or agent leading to disassembly in inactive neurites may not be the same as that which actually signals the inactivity. Interestingly, a cessation of neurite elongation can be brought about by decreases as well as increases in intracellular calcium concentrations outside a narrow range (100–300 nM; Cohan *et al.*, 1987). Although such cellular behavior is consistent with microtubule depolymerization (Seeds *et al.*, 1970; Yamada *et al.*, 1970), it is not known whether and how MAPs might mediate this effect.

At an earlier point in our discussion of development, we cited the enigma of outgrowing axons failing to express early MAPs (regenerating peripheral nerve axons) and of early MAP expression in axons apparently not exhibiting outgrowth (olfactory receptor neuron axons). While these examples illustrate the dissociation of early MAP expression and outgrowth per se, it is difficult to discern what particular quality or function is represented by the presence or absence of MAP expression irrespective of growth. What is different about developing and regenerating axons of the same class of neurons? What, moreover, could be still developmental or immature about the axons of adult neurons that have stopped their outgrowth? In the context of this discussion, perhaps a common denominator is the potential for remodeling. If so, the expression of early MAPs would be a manifestation of that potential.

For example, the lower expression of two early MAPs—MAP 1X (Woodhams et al., 1989) and a 62–kDa isoform of tau (Oblinger et al., 1991)—in regenerating than in developing peripheral nerve axons raises the possibility that regenerating PNS axons are less able or likely to engage in remodeling than are developing PNS axons, despite their outgrowth. Regenerating axons may simply reinnervate vacated synaptic sites without making adjustments as a function of synaptic activity, as developing axons are thought to do (Fawcett and Keynes, 1990).

On the other hand, the continued expression of phosphorylated MAP 1B, an early MAP, in nongrowing olfactory receptor neurons and their axons in adults (Schoenfeld et al., 1989; Viereck et al., 1989) may signal a persistent potential for remodeling even in the absence of growth. It is widely believed that most of these axons, and their parent cell bodies, are destined to degenerate and be replaced in a continuing cycle of persistent receptor neuron proliferation and turnover (Graziadei and Monti Graziadei, 1978). Thus, the association of MAP 1B with the cytoskeleton of these neurons may make it easier for them to degenerate, vacate synaptic sites, and permit innervation by newly generated neurons.

We have recently observed that some olfactory receptor neurons also express tau protein. Anti-Tau-1 immunoreactivity is localized in small, isolated fascicles of olfactory receptor neuron axons in rats and distributed both ventromedially and ventrolaterally in the olfactory nerve layer of the olfactory bulb (Schoenfeld et al., 1993; Fig. 2). We also found anti-tau immunoreactivity in vomeronasal nerve axons, as has been reported (Viereck et al., 1989). In both cases, our evidence suggests that tau-positive axons are also MAP 1B-positive. This additional expression represents an alliance of MAPs that may invest the cytoskeleton of these olfactory receptor neurons with greater stability or resistance to remodeling or turnover than that possessed by neurons that are only MAP 1B-positive.

B. Stimulation-Dependent Stabilization

A solid body of evidence suggests that expression of early and late MAPs is under endogenous, transcriptional control. Studies using cDNA probes for a variety of MAPs have tended to find that the timing of the expression of mRNA transcripts is correlated with the timing of the expression of the MAPs themselves (see review in Tucker, 1990). Nevertheless, it has been known for some time that posttranslational modifications, notably phosphorylation, alter MAP function, especially microtubule assembly and stability (see earlier discussion). While changes in phosphorylation state are reliable features of developmental timetables and may in some way be under endogenous control, recent work is also beginning to show that such changes can be induced by discrete episodes of exogenous stimulation during development and in adulthood. Moreover, even expression of the proteins themselves appears to be responsive to such exogenous factors as hormones, raising the possibility that MAPs are important mediators in the influence of epigenetic events on the development and stabilization of neural networks.

1. Neurotrophic Factors

A number of substances can induce the outgrowth of neurites when they are applied to cultured neurons, including nerve growth factor, estrogen, and gangliosides. Coincident with this outgrowth are changes in the mass and stability of microtubules and in the expression of MAPs.

Studies of the response to NGF have focused on PC12 cells. These were originally established as a clonal line of tumorous adrenal chromaffin (pheochromocytoma) cells that show a reversible response to NGF by differentiating into growing, sympathetic-like neurons (Greene and Tischler, 1976). Distinct short neurites (average 20 μm) appear in some cells within the first 3 days of treatment, but virtually all cells have substantially longer neurites (average 200 μm) by 7–8 days that grow to be very long (about 1 mm) and highly branched within 21 days (Black and Greene, 1982; Burstein et al., 1985; Drubin et al., 1984b, 1985; Jacobs and Stevens, 1986a; Brugg and Matus, 1988). By 8–12 days, the structural appearance of neurites also becomes more regular and stable, and begins to show resistance to microtubule depolymerizing drugs (Jacobs and Stevens, 1986a,b; see also Black and Greene, 1982). The bundling of microtubules within neurites begins even earlier, at 3 days of NGF treatment (Jacobs and Stevens, 1986a; Brugg and Matus, 1988). While the variability of intertubule spacing decreases progressively between 3 and 43 days, the average spacing is remarkably constant (about 69 nm) from the first appearance of bundles at 3 days (Jacobs and Stevens, 1986a).

Nevertheless, all of this is reversible in PC12 cells: neurites disappear within 2–3 days of NGF withdrawal (Greene and Tischler, 1976; Drubin *et al.*, 1985).

Following approximately the same time course of outgrowth (or retraction) consequent to NGF application (or withdrawal) are increases (or decreases) in microtubule mass and the expression (or degradation) of several MAPs, including MAPs 1.2 and 5 (both homologs of MAP 1B; Greene *et al.*, 1983; Aletta *et al.*, 1988; Brugg and Matus, 1988), several isoforms of tau protein (Drubin *et al.*, 1984b, 1985, 1988), MAP 3 (Brugg and Matus, 1988), and 64–80 kDa MAPs now identified as chartins (Greene *et al.*, 1983; Burstein *et al.*, 1985; Black *et al.*, 1986; Greene *et al.*, 1986; Aletta and Greene, 1987; see also Magendantz and Solomon, 1985). Homologs of MAP 1A (MAP 1.1, MAP 1) are not affected by NGF application (Greene *et al.*, 1983; Brugg and Matus, 1988). Increases in MAP 2 following NGF treatment reported by some researchers (Black *et al.*, 1986; Sano *et al.*, 1990) have not been replicated by others (Greene *et al.*, 1983; Brugg and Matus, 1988). Changes in total levels of tubulin over the same time period are not correlated with changes in microtubule mass or with neuritic outgrowth or retraction, although they do increase gradually (Drubin *et al.*, 1985, 1988). These data support the inference that NGF stimulates neurite formation, outgrowth, and stabilization through the promotion of microtubule polymerization, not by directly affecting tubulin but rather by inducing increased expression of MAPs.

NGF may affect both MAP gene transcription and post-translational modification of MAPs in PC12 cells. For example, the induction of neurite outgrowth and expression of MAP 1.2 (MAP 1B) can be blocked by inhibitors of RNA synthesis (Greene *et al.*, 1983). Moreover, the use of cDNA probes for tau mRNA indicates a significant increase in tau mRNA levels in PC12 cells following 5 days of NGF treatment (Drubin *et al.*, 1988). Thus, the measurable increases in the levels of various MAPs following NGF are most likely due to increased protein synthesis and not reduced degradation. NGF also exhibits a particularly striking effect on MAP phosphorylation, inducing the incorporation of phosphate into both MAP 1.2 (MAP 1B) and chartins within several hours of initial treatment (Greene *et al.*, 1983; Aletta *et al.*, 1988). The mechanism for this effect (enhanced kinase activity, reduced phosphatase activity) has not been elucidated.

It is unclear whether the induced changes in any particular MAP or set of MAPs have a specific effect on neurite outgrowth and stabilization in PC12 cells. For example, treatments which specifically affect the phosphorylation of chartins (lithium, forskolin, cholera toxin) have been shown to block NGF-induced neurite growth (Burstein *et al.*, 1985; Greene *et al.*, 1986). While this demonstrates that chartins, particularly the 64-kDa isoform, are necessary mediators for the induction of outgrowth by NGF in PC12 cells, they do not demonstrate either that chartins are sufficient for such growth or that the other MAPs are unnecessary to the induction of such growth. The observation that MAP 5 (MAP 1B) remains diffusely distributed in cytoplasm and largely unbound to microtubules for the first 5 days of NGF treatment, during which time microtubules begin to assemble and bundle together and neurite outgrowth is initiated (Jacobs and Stevens, 1986a; Brugg and Matus, 1988), suggests that MAP 5 (MAP 1B) may not be critical to either microtubule assembly and bundling or to outgrowth, at least in the first 5 days.

A similar observation and inference applies to MAP 3 (Brugg and Matus, 1988), which in fact was never observed to be bound to microtubules in PC12 cells. However, since neither tau nor the chartins have been studied in this way, one cannot conclude from these observations that those MAPs are necessarily more critical to assembly and outgrowth than MAPs 1B and 3 (see the following discussion below). Indeed, in cultured neurons from rat cerebral cortex, both tau and MAP 1B are diffusely expressed in growth cones and neuritic axons and are not in any obvious way bound to microtubules, even after taxol-induced polymerization of free tubulin (Gordon-Weeks *et al.*, 1989; Mansfield *et al.*, 1991). Mansfield *et al.* (1991) propose that the diffuse cytoplasmic expression of phosphorylated MAP 1B may reflect primarily a limitation of tubulin binding sites rather than insufficient binding capacity in this MAP per se, an observation also made to account for the poor cosedimentation properties of MAP 1B isolated from calf brain gray matter (Bloom *et al.*, 1985).

Nevertheless, a demonstration of competency to coassemble with microtubules does not also establish a role for any MAP in actually promoting growth-associated assembly and cross-linkage of microtubules, in part because it is difficult to isolate the functional contribution of any one MAP when other MAPs are expressed in the same outgrowing axonal process. Certainly the evidence suggests that both phosphorylated MAP 1B and dephosphorylated tau may foster the elongation of growing neurites of PC12 cells treated with NGF. However, do they act synergistically or does one in fact dominate in some way? The emergence of resistance to depolymerization by the 7th day of treatment seems to be associated with the accumulation of both tau and MAP 1B (see earlier discussion), so a differentiated role is difficult to discern for either. It is of interest that 125 kDa tau expression precedes that of other MAP isoforms and coincides with the beginning of bundling (Jacobs and Stevens, 1986a). However, the spacing of 69 nm between microtubules corresponds to what would be expected from the projection domains of either an IMM or HMM MAP (see Section III,B,3).

In addition to NGF, both estrogen and gangliosides have been shown to enhance neurite outgrowth in cultured neurons. Treatment of cultured basal hypothalamic neurons with estradiol 17-β (but not estradiol 17-α) leads to an increase in the number of higher order branches of neurites and in the length of neurites that is evident within 3 days and continues to increase through 7 days (Ferreira and Caceres, 1991). Coincident with this growth is a selective increase in tau protein (a 58-kDa isoform) but not in MAPs 1A or 2. Moreover, there is a specific increase in acetylated α-tubulin and in the fraction of polymeric tubulin that is resistant to colchicine depolymerization, which rises to 67% by 7 days. Interestingly, unlike tubulin, MAP 1A or MAP 2, all of the tau expressed in these neurons, whether controls or treated with estradiol 17-β, is associated with the colchicine-resistant fraction, which is evident as early as 3 days after plating. These data suggest that estrogen's neuritogenic effect in cultured hypothalamic neurons is mediated by a specific induction of tau protein expression and an increase in stabilized microtubules.

The effect of gangliosides on neurite outgrowth and MAP expression has been studied in cultured neuroblastoma cells (Ferreira et al., 1990). The principal neuritogenic effect of gangliosides on these cells is to induce the formation of numerous branches and spines, particularly higher order branches which are rare in untreated neuroblastoma cultures. Coincident with the increased incidence of neuritic branching is a specific induction of MAP 2 expression, which also is rare in untreated neuroblastoma cells. An association between MAP 2 expression and neuritic branching is consistent with a likely role for MAP 2 in dendritic branching (see Section III,B,1). As with PC12 cells and cultured hypothalamic neurons, discussed previously, tau expression is closely correlated with the development of microtubule stability in cultured neuroblastoma neurons. Ganglioside treatment enhances somewhat the development of stabilized microtubules and, coincidentally, the expression of tau (Ferreira et al., 1990).

2. Early Sensory Experience

As noted in an earlier discussion (Section V,A,3,c), there is circumstantial evidence that the expression of MAPs early in development may in part foster the remodeling (stabilization and elimination) of neural circuits in a stimulation-dependent manner. Aoki and Siekevitz (1985) have conducted the only direct examination of this hypothesis. They sought to characterize the molecular events that correspond to shifts in visual cortical connectivity and a subsequent decline in cellular plasticity following light exposure in developing kittens. They focused on the in vitro cAMP-dependent phosphorylation of HMM MAP 2 in visual cortex homogenates from kittens either dark reared, exposed to light following dark rearing, or exposed normally to light throughout early development. They interpreted low levels of measurable *in vitro* phosphorylation to reflect a paucity of available phosphorylation sites in the extracted MAP 2 molecules or, in other words, to reflect a high degree of phosphorylation *in vivo*. Likewise, high levels of phosphate binding to MAP 2 *in vitro* were seen to reflect low levels of phosphorylation *in vivo*. In normally reared kittens, Aoki and Siekevitz (1985) found evidence that *in vivo* phosphorylation in visual cortex declines with age. Dark rearing, which is known to prolong visual cortical plasticity, also prolonged the juvenile state of high *in vivo* phosphorylation of MAP 2, whereas even brief exposure to light following prolonged dark rearing was sufficient to bring on the adult state of low *in vivo* phosphorylation. Such changes could not be induced either in the lateral geniculate nucleus (a subcortical part of the visual circuit) or in the auditory cortex. Both areas showed evidence of low *in vivo* phosphorylation of MAP 2.

Although there is some disagreement about how phosphorylation of MAP 2 affects its relationship with microtubules (see Section V,A,3,b), these findings are in accord with a great deal of evidence that phosphorylated thermostable MAPs bind less avidly to microtubules and stimulate a lower degree of microtubule stability than dephosphorylated thermostable MAPs (Jameson *et al.*, 1980; Jameson and Caplow, 1981; Murthy and Flavin, 1983; Yamamoto *et al.*, 1983; Burns *et al.*, 1984; Lindwall and Cole, 1984; Hoshi *et al.*, 1988; Diaz-Nido *et al;* 1990). Phosphorylation of MAP 2 is also reported to weaken its ability to cross-link actin filaments (Sattilaro, 1986; Selden and Pollard, 1986). The greater cytoskeletal lability promoted by such phosphorylation may have a major role in the stimulation-dependent remodeling that characterizes development (see Section V,A,3,c). However, the generality of this mechanism for other varieties of stimulation-dependent developmental processes, whether involving other kinds of visual experience, other sensory modalities, or even more complex phenomena such as social imprinting, remains to be evaluated (see next section).

3. Excitatory Amino Acids

The amino acids glutamate and aspartate are considered to be the principal excitatory neurotransmitters in the CNS. Several recent studies provide evidence that stimulation of neurons with glutamate and its analogs can alter the expression and particularly the phosphorylation of MAPs.

Halpain and Greengard (1990) report that the glutamate analog *N*-methyl-D-aspartate (NMDA) causes a rapid dephosphorylation of MAP 2 when applied to slices of hippocampus and a variety of other forebrain cortical and subcortical regions. Stimulation of non-NMDA glutamate receptors with kainate or quisqualate produces less dephosphorylation of MAP 2. No change in MAP 2 expression per se was observed in any case. Since NMDA receptor activation is believed to play a central role in stimulation-dependent stabilization of neural connections during both development and learning and memory (Wolpaw *et al.*, 1991), these results are consistent with the idea that dephosphorylation of MAP 2 also plays a role in these processes. Indeed, as cited in the previous section, light stimulation during visual development produces a similar effect in visual cortex (Aoki and Siekevitz, 1985).

It is not known whether dephosphorylation under these circumstances leads to a predictable effect on microtubule stability. Thus, it is of interest that another recent study (Bigot and Hunt, 1990) has observed that glutamate agonists, including NMDA, stimulate an increased affiliation of both tubulin and MAP 2 with filamentous structures, presumably microtubules, in the dendrites and somata of cultured spinal cord and cerebral cortical neurons. This is consistent with the inverse relationship between MAP 2 phosphorylation and microtubule assembly noted in most previous reports (Murthy and Flavin, 1983; Yamamoto *et al.*, 1983; Burns *et al.*, 1984; Diaz-Nido *et al.*, 1990). By contrast, glutaminergic stimulation also produced an increase in diffuse anti-Tau-1 immunoreactivity throughout the neuron, including the somatodendritic compartment (see Section III,A,2). While this result also reflects MAP dephosphorylation (Papasozomenos and Binder, 1987), in this case tau did not show a greater association with microtubules, contradicting previous evidence that tau phosphorylation and microtubule assembly are inversely related (Lindwall and Cole, 1984).

Further complicating these overall findings is evidence that glutaminergic stimulation of NMDA receptors in hippocampal cells leads to increased phosphorylation of a 39-kDa protein believed to be a MAP 2 kinase (Bading and Greenberg, 1991). Although the authors cite unpublished observations that such phosphorylation results in increased kinase activity, it remains to be documented whether MAP 2 phosphorylation is specifically increased following NMDA receptor activation in this system, especially in light of previous evidence to the contrary.

VI. Concluding Remarks: Why Such Diversity, and How Is It Regulated?

In reviewing the recent literature on the distribution and function of fibrous MAPs in the nervous system, we have emphasized two broad, recurring themes: that these filamentous molecules, which extend out as side arms from microtubule walls, help to regulate both the form and the stability of nervous system cells. Since they have the ability to bind to and prevent the disassembly of microtubules, as well as to cross-link and bundle together microtubules and other cellular organelles, fibrous MAPs help to regulate the elongation of microtubules and the stabilization of the cytoskeleton. These basic functions underlie the fundamental capability of neurons and glial cells to regulate cellular form. They support the differentiation of neuritic processes into axons and dendrites and help to control the size and branching patterns of these processes. These MAP-dependent functions are also of fundamental importance in regulating cellular stability. They appear to regulate the stimulation-dependent remodeling of neural circuits that occurs during early development and is then sustained in adult form. They also appear to regulate and may even participate in the cascade of molecular events that leads to stabilized alterations in neurons associated with learning and memory.

Running counter to these broad themes, however, is evidence for the molecular diversity of fibrous MAPs. Although there is increasing evidence that the fibrous neuronal MAPs fall into two molecular families which display different thermostability and microtubule-binding strategies (see Section II), individual classes and species within each MAP family come in many forms, differing in overall size, the length of their projection domains, the number and heterogeneity of isotypes, the configuration and properties of their microtubule binding domains, the timing and pattern of developmental expression, and the extent of posttranslational modification. Moreover, we have shown that they are expressed heterogeneously in different cell types and compartments and in varying alliances with each other and other cellular proteins. These variations are far too numerous and often too subtle to be employed merely to regulate the major categorical differences between dendrites and axons, neurons and glial cells, juvenile and adult cells, CNS and PNS cells, and so on, which have been the focus of most research to date. Thus, one may ask: Why is there such diversity, and how is it regulated?

As a preliminary answer to these questions, we have highlighted throughout this chapter an emerging new theme: that MAPs employ diverse molecular strategies to help initiate and sustain variation in cellular form and stability. This diversity itself is regulated diversely through the control of gene expression and intracellular targeting, the effect of stimulation-dependent modifications such as phosphorylation, the synergy of alliances among different MAPs, and other mechanisms. It has been previously suggested that individual MAPs may have singular functions (Vallee *et al.*, 1984). We do not view our current emphasis on MAP alliances as contradicting that earlier view but rather as incorporating it into a more inclusive notion of the diversity of MAP function.

Part of diversity may be the use of a particular MAP for a particular function. For example, a number of unique attributes of MAP 2 (carrying its own kinase, having an affinity for both actin and microtubules, being potentially synthesized in distal dendritic segments, etc.) may give it a unique ability to regulate dendritic branching as a function of afferent stimulation (see Section III,B,1 and Section V,B) that is not dependent on the contribution of any other MAP. However, it is interesting that MAP 2 is not expressed uniformly in all neurons throughout the CNS (DeCamilli *et al.*, 1984)(see Section IV), raising the possibility that not all cells modify the branching of their dendrites once postnatal development is completed, or do so without MAP 2.

As we noted in Section IV, very little time has been devoted to focusing on specific cells and classes of cells in the nervous system, where it might be fruitful to attempt to fully characterize the cytoskeletal amalgamations of MAPs and other proteins and then to test experimentally the role of individual MAPs or groups of MAPs in regulating particular features of cellular function. Some work along this line has begun but there must be much more of this sort of effort if we are to achieve a truly comprehensive understanding of the functions of fibrous MAPs in the nervous system.

Acknowledgments

We are indebted to Dr. Richard Vallee for his mentorship throughout this project. We thank Dr. Jim Crandall for his invaluable help in reviewing the entire manuscript, Donna Santacroce and Taryn Schweitzer for their assistance in reviewing the literature, and Jeri Stolk and Kathy Sutton for their help in preparing the manuscript. We are grateful to the National Institutes of Health, the Worcester Foundation for Experimental Biology, and Clark University for supporting our research; and to the Frances L. Hiatt School of Psychology for supporting this writing project. This work was initiated while both authors were affiliated with the Vallee laboratory and the Worcester Foundation for Experimental Biology.

References

Aizawa, H., Emori, Y., Murofushi, H., Kawasaki, H., Sakai, H., and Suzuki, K. (1990). *J. Biol. Chem.* **265,** 13849–13855.

Aizawa, H., Emori, Y., Mori, A., Murofushi, H., Sakai, H., and Suzuki, K. (1991). *J. Biol. Chem.* **266,** 9841–9846.

Alaimo-Beuret, D., and Matus, A. (1985). *Neuroscience* **14,** 1103–1115.

Aletta, J. M., and Greene, L. A. (1987). *J. Cell Biol.* **105,** 277–290.

Aletta, J. M., Lewis, S. A., Cowan, N.J., and Greene, L.A. (1988). *J. Cell Biol.* **106,** 1573–1581.

Aoki, C., and Siekevitz, P. (1985). *J. Neurosci.* **5,** 2465–2483.

Aoki, C., and Siekevitz, P. (1988). *Sci. Am.* **259,** 56–64.

Baas, P. W., and Black, M. M. (1990). *J. Cell Biol.* **111**, 495–509.

Baas, P. W., Deitch, J. S., Black, M. M., and Banker, G. A. (1988). *Proc. Natl. Acad. Sci. U.S.A.* **85**, 8335–8339.

Baas, P. W., Black, M. M., and Banker, G. A. (1989). *J. Cell Biol.* **109**, 3085–3089.

Baas, P. W., Pienkowski, T. P., and Kosik, K. S. (1991). *J. Cell Biol.* **115**, 1333–1344.

Bading, H., and Greenberg, M. E. (1991). *Science* **253**, 912–914.

Bartlett, W. P., and Banker, G. A. (1984). *J. Neurosci.* **4**, 1954–1965.

Benjamin S., Cambray-Deakin, M. A., and Burgoyne, R. D. (1988). *Neuroscience* **27**, 931–939.

Bernhardt, R., and Matus, A. (1982). *J. Cell Biol.* **92**, 589–593.

Bernhardt, R., and Matus, A. (1984). *J. Comp. Neurol.* **226**, 203–221.

Bernhardt, R., Huber, G., and Matus, A. (1985). *J. Neurosci.* **5**, 977–991.

Bhide, P. G., and Frost, D. O. (1991). *J. Neurosci.* **11**, 485–504.

Bigot, D., and Hunt, S. P. (1990). *Neurosci. Lett.* **111**, 275–280.

Binder, L. I., Frankfurter, A., Kim, H., Caceres, A., Payne, M. R., and Rebhun, L. I. (1984). *Proc. Natl. Acad. Sci. U.S.A.*, **81**, 5613–5617.

Binder, L. I., Frankfurter, A., and Rebhun, L. I. (1985). *J. Cell Biol.* **101**, 1371–1378.

Binder, L. I., Frankfurter, A., and Rebhun, L. I. (1986). *Ann. N.Y. Acad. Sci.* **466**, 145–166.

Black, M. M. (1987a). *J. Neurosci.* **7**, 3695–3702.

Black, M. M. (1987b). *Proc. Natl. Acad. Sci. U.S.A.* **84**, 7783–7787.

Black, M. M., and Baas, P. W. (1989). *Trends Neurosci.* **12**, 211–214.

Black, M. M., and Greene, L. A. (1982). *J. Cell Biol.* **95**, 379–386.

Black, M. M., and Kurdyla, J. T. (1983). *J. Cell Biol.* **97**, 1020–1028.

Black, M. M., Aletta, J. M., and Greene, L. A. (1986). *J. Cell Biol.* **103**, 545–557.

Black, M. M., Baas, P. W., and Humphries, S. (1989). *J. Neurosci.* **9**, 358–368.

Bloom, G. S., Schoenfeld, T. A., and Vallee, R. B. (1984). *J. Cell Biol.* **98**, 320–330.

Bloom, G. S., Luca, F. C., and Vallee, R. B. (1985). *Proc. Natl. Acad. Sci. U.S.A.* **82**, 5404–5408.

Bray, D., and Bunge, M. B. (1981). *J. Neurocytol.* **10**, 589–605.

Bré, M. H., and Karsenti, E. (1990). *Cell Motil. Cytoskeleton* **15**, 88–98.

Brion, J. P., Guilleminot, J., Couchie, D., Flament-Durand, J., and Nunez, J. (1988). *Neuroscience*, **25**, 139–146.

Brown, P. A., and Berlin, R. D. (1985). *J. Cell Biol.* **101**, 1492–1500.

Bruckenstein, D. A., and Higgins, D. (1988a). *Dev. Biol.* **128**, 324–336.

Bruckenstein, D. A., and Higgins, D. (1988b). *Dev. Biol.* **128**, 337–348.

Bruckenstein, D., Johnson, M. I., and Higgins, D. (1989). *Dev. Brain Res.* **46**, 21–32.

Bruckenstein, D. A., Lein, P. J., Higgins, D., and Fremeau, R. T., Jr. (1990). *Neuron* **5**, 809–819.

Brugg, B., and Matus, A. (1988). *J. Cell Biol.* **107**, 643–650.

Brugg, B., and Matus, A. (1991). *J. Cell Biol.* **114**, 735–743.

Burgoyne, R. D. (1991). In "The Neuronal Cytoskeleton" (R. D. Burgoyne, ed.), pp. 75–91. Wiley-Liss, New York.

Burgoyne, R. D., and Cambray-Deakin, M. A. (1988). *Brain Res. Rev.* **13**, 77–101.

Burgoyne, R. D., and Cumming, R. (1984). *Neuroscience* **11**, 157–167.

Burgoyne, R. D., Gray, E. G., Sullivan, K., and Barron, J. (1982). *Neurosci. Lett.* **31**, 81–85.

Burns, R. G., Islam, K., and Chapman, R. (1984). *Eur. J. Biochem.* **141**, 609–615.

Burstein, D.E., Seeley, P.J., and Greene, L.A. (1985). *J. Cell Biol.* **101**, 862–870.

Burton, P. R. (1987). *Brain Res.* **409**, 71–78.

Burton, P. R. (1988). *Brain Res.* **473**, 107–115.

Burton, P. R., and Paige, J. L. (1981). *Proc. Natl. Acad. Sci. U.S.A.* **78**, 3269–3273.

Caceres, A., and Kosik, K. S. (1990). *Nature (London)* **343**, 461–463.

Caceres, A., Payne, M. R., Binder, L. I., and Steward, O. (1983). *Proc. Natl. Acad. Sci. U.S.A.* **80**, 1738–1742.

Caceres, A., Banker, G. A., Steward, O., Binder, L., and Payne, M. (1984a). *Dev. Brain Res.* **13**, 314–318.

Caceres, A., Binder, L. I., Payne, M. R., Bender, P., Rebhun, L., and Steward, O. (1984b). *J. Neurosci.* **4**, 394–410.

Caceres, A., Banker, G. A., and Binder, L. (1986). *J. Neurosci.* **6**, 714–722.

Caceres, A., Potrebic, S., and Kosik, K. S. (1991). *J. Neurosci.* **11**, 1515–1523.

Cambray-Deakin, M. A., and Burgoyne, R. D. (1987). *J. Cell Biol.* **104**, 1569–1574.

Cambray-Deakin, M. A., Norman, K.-M., and Burgoyne, R. D. (1987). *Dev. Brain Res.* **34**, 1–7.

Chamak, B., Fellous, A., Glowinski, J., and Prochiantz, A. (1987). *J. Neurosci.* **7**, 3163–3170.

Changeux, J.-P., and Danchin, A. (1976). *Nature (London)* **264**, 705–712.

Chapin, S. J., Bulinski, J. C., and Gundersen, G. G. (1991a). *Nature (London)* **349**, 24.

Chapin, S.J., Bulinski, J.C., and Matus, A. (1991b). *J. Cell Biol.* **115**, 339a.

Chen, J., Kanai, Y., Cowan, N.J., and Hirokawa, N. (1992). *Nature (London)* **360**, 674–677.

Cleveland, D. W., Hwo, S.-Y., and Kirschner, M. W. (1977). *J. Mol. Biol.* **116**, 227–247.

Cohan, C. S., Connor, J. A., and Kater, S. B. (1987). *J. Neurosci.* **7**, 3588–3599.

Constantine-Paton, M., Cline, H. T., and Debski, E. (1990). *Annu. Rev. Neurosci.* **13**, 129–154.

Coss, R. G., and Perkel, D. H. (1985). *Behav. Neural Biol.* **44**, 151–185.

Couchie, D., and Nunez, J. (1985). *FEBS Lett.* **188**, 331–335.

Crandall, J. E., Tobet, S. A., Fischer, I., and Fox, T. O. (1989). *Brain Res. Bull.* **22**, 571–574.

Davies, P. (1988). *J. Clin. Psychiatry* **49**, suppl., p. 23.

DeCamilli, P., Harris, S. M., Huttner, W. B., and Greengard, P. (1983). *J. Cell Biol.* **96**, 1355–1373.

DeCamilli, P., Miller, P. E., Navone, F., Theurkauf, W. E., and Vallee, R. B. (1984). *Neuroscience* **11**, 819–846.

deMey, J., Aerts, F., De Raeymacker, M., Daneels, G., Moeremans, M., DeWever, B., Vandre, D. D., Vallee, R. B., Borisy, G. G., and de Brabander, M. (1987). *Prog. Zool.* **34**, 187–206.

Dentler, W. L., Grannett, S., and Rosenbaum, J. L. (1975). *J. Cell Biol.* **65**, 237–241.

Diaz-Nido, J., Serrano, L., Hernandez, M. A., and Avila, J. (1990). *J. Neurochem.* **54**, 211–222.

Dingus, J., Obar, R., Hyams, J., Goedert, M., and Vallee, R.B. (1991). *J. Biol. Chem.* **266**, 18854–18860.

Dinsmore, J. H., and Solomon, F. (1991). *Cell* **64**, 817–826.

Dotti, C. G., and Banker, G. A. (1987). *Nature (London)* **330**, 254–256.

Dotti, C. G., Banker, G. A., and Binder, L. I. (1987). *Neuroscience* **23**, 121–130.

Dotti, C. G., Sullivan, C. A., and Banker, G. A. (1988). *J. Neurosci.* **8**, 1454–1468.

Drubin, D. G., and Kirschner, M. W. (1986). *J. Cell Biol.* **103**, 2739–2746.

Drubin, D. G., Caput, D., and Kirschner, M. W. (1984a). *J. Cell Biol.* **98**, 1090–1097.

Drubin, D., Kirschner, M., and Feinstein, S. (1984b). *In* "Molecular Biology of the Cytoskeleton" (G. G. Borisy, D. W. Cleveland, and D. B. Murphy, eds.), pp. 343–355. Cold Spring Harbor Lab. Press, Cold Spring Harbor, New York.

Drubin, D. G., Feinstein, S. C., Shooter, E. M., and Kirschner, M. W. (1985). *J. Cell Biol.* **101**, 1799–1807.

Drubin, D., Kobayashi, S., Kellogg, D., and Kirschner, M. (1988). *J. Cell Biol.* **106,** 1583–1591.

Escobar, M. I., Pimienta, H., Caviness, V. S., Jr., Jacobson, M., Crandall, J. E., and Kosik, K. S. (1986). *Neuroscience* **17,** 975–989.

Faivre, C., Legrand, C., and Rabie, A. (1985). *Int. J. Dev. Neurosci.* **3,** 559–565.

Fawcett, J. W., and Keynes, R. J. (1990). *Annu. Rev. Neurosci.* **13,** 43–60.

Ferreira, A., and Caceres, A. (1989). *Dev. Brain Res.* **49,** 205–213.

Ferreira, A., and Caceres, A. (1991). *J. Neurosci.* **11,** 392–400.

Ferreira, A., Busciglio, J., and Caceres, A. (1989). *Dev. Brain Res.* **49,** 215–228.

Ferreira, A., Busciglio, J., Landa, C., and Caceres, A. (1990). *J. Neurosci.* **10,** 293–302.

Fischer, I., and Romano-Clarke, G. (1990). *J. Neurochem.* **55,** 328–333.

Fischer, I., Kosik, K. S., and Sapirstein, V. S. (1987). *Brain Res.* **436,** 39–48.

Francon, J., Lennon, A. M., Fellows, A., Moreck, A., Pierre, M., and Nunez, J. (1982). *Eur. J. Biochem.* **129,** 465–471.

Friden, B., Nordh, J., Wallin, M., Deinum, J., and Norden, B. (1988). *Biochim. Biophys. Acta* **955,** 135–142.

Friede, R. L., and Samorajski, T. (1970). *Anat. Rec.* **167,** 379–388.

Friedrich, P., and Aszódi, A. (1991). *FEBS Lett.* **295,** 5–9.

Furtner, R., and Wiche, G. (1987). *Eur. J. Cell Biol.* **45,** 1–8.

Garner, C. C., Tucker, R. P., and Matus, A. (1988). *Nature (London)* **336,** 674–677.

Georgieff, I. S., Liem, R. K. H., Mellado, W., Nunez, J., and Shelanski, M. L. (1991). *J. Cell Sci.* **100,** 55–60.

Goedert, M., Spillantini, M. G., Potier, M. C., Ulrich, J., and Crowther, R. A. (1989). *EMBO J.* **8,** 393–399.

Goedert, M., Crowther, R. A., and Garner, C. C. (1991). *Trends Neurosci.* **14,** 193–199.

Goedert, M., Spillantini, M. G., and Crowther, R. A. (1992). *Proc. Natl. Acad. Sci. U.S.A.* **89,** 1983–1987.

Goldstein, M. E., Sternberger, L. A., and Sternberger, N. H. (1983). *Proc. Natl. Acad. Sci. U.S.A.* **80,** 3101–3105.

Gordon-Weeks, P. R. (1987). *Neuroscience* **21,** 977–989.

Gordon-Weeks, P. R., Mansfield, S. G., and Curran, I. (1989). *Dev. Brain Res.* **49,** 305–310.

Goslin, K., and Banker, G. (1989). *J. Cell Biol.* **108,** 1507–1516.

Graziadei, P. P. C., and Monti Graziadei, G. A. (1978). *In* "Handbook of Sensory Physiology. Vol. IX: Development of Sensory Systems" (M. Jacobson, ed.), pp. 55–83. Springer-Verlag, Berlin.

Greene, L. A., and Tischler, A. S. (1976). *Proc. Natl. Acad. Sci. U.S.A.* **73,** 2424–2428.

Greene, L. A., Liem, R. K. H., and Shelanski, M. L. (1983). *J. Cell Biol.* **96,** 76–83.

Greene, L. A., Drexler, S. A., Connolly, J. L., Rukenstein, A., and Green, S. H. (1986). *J. Cell Biol.* **103,** 1967–1978.

Grundke-Iqbal, I., and Iqbal, K. (1989). *Prog. Clin. Biol. Res.* **317,** 745–753.

Haga, T., and Kurokawa, M. (1975). *Biochim. Biophys. Acta* **392,** 335–345.

Hagestedt, T., Lichtenberg, B., Wille, H., Mandelkow, E. M., and Mandelkow, E. (1989). *J. Cell Biol.* **109,** 1643–1651.

Halpain, S., and Greengard, P. (1990). *Neuron* **5,** 237–246.

Hammarback, J. A., Obar, R. A., Hughes, S. M., and Vallee, R. B. (1991). *Neuron* **7,** 129–139.

Hernandez, M. A., Avila, J., Moya, F., and Alberto, C. (1989). *Neuroscience* **29,** 471–477.

Herzog, W., and Weber, K. (1978). *Eur. J. Biochem.* **92,** 1–8.

Hillman, D. E. (1979). *In* "The Neurosciences: Fourth Study Program" (F. O. Schmitt and F. G. Worden, eds.), pp. 477–498. MIT Press, Cambridge, MA.

Hinds, J. W., Hinds, P. L., and McNelly, N. A. (1984). *Anat. Rec.* **210**, 375–383.
Hirokawa, N. (1991). *In* "The Neuronal Cytoskeleton" (R. D. Burgoyne, ed.), pp. 5–74. Wiley-Liss, New York.
Hirokawa, N., Bloom, G. S., and Vallee, R. B. (1985). *J. Cell Biol.* **101**, 227–239.
Hirokawa, N., Hisanaga, S.-I., and Shiomura, Y. (1988a). *J. Neurosci.* **8**, 2769–2779.
Hirokawa, N., Shiomura, Y., and Okabe, S. (1988b). *J. Cell Biol.* **107**, 1449–1459.
Hoffman, P. N., Griffin, J. W., and Price, D. L. (1984). *J. Cell Biol.* **99**, 705–714.
Horio, T., and Hotani, H. (1986). *Nature (London)* **321**, 605–607.
Hoshi, M., Akiyama, T., Shinohara, Y., Miyata, Y., Ogawara, H., Nishida, E., and Sakai, H. (1988). *Eur. J. Biochem.* **174**, 225–230.
Hubel, D. H., and Wiesel, T. N. (1970). *J. Physiol. (London)* **206**, 419–436.
Huber, G., and Matus, A. (1984a). *J. Cell Biol.* **98**, 777–781.
Huber, G., and Matus, A. (1984b). *J. Neurosci.* **4**, 151–160.
Huber, G., and Matus, A. (1990). *J. Cell Sci.* **95**, 237–246.
Huber, G., Alaimo-Beuret, D., and Matus, A. (1985). *J. Cell Biol.* **100**, 496–507.
Jacobs, J. R., and Stevens, J. K. (1986a). *J. Cell Biol.* **103**, 895–906.
Jacobs, J. R., and Stevens, J. K. (1986b). *J. Cell Biol.* **103**, 907–915.
Jameson, L., and Caplow, M. (1981). *Proc. Natl. Acad. Sci. U.S.A.* **78**, 3413–3417.
Jameson, L., Frey, T., Zeeberg, B., Dalldorf, F., and Caplow, M. (1980). *Biochemistry* **19**, 2472–2479.
Job, D., Pabion, M., and Margolis, R. (1985). *J. Cell Biol.* **101**, 1680–1689.
Kanai, Y., Chen, J., and Hirokawa, N. (1992). *EMBO J.* **11**, 3953–3961.
Keates, R. A., and Hallett, F. R. (1988). *Science* **241**, 1642–1645.
Kim, H., Jensen, C. G., and Rebhun, L. I. (1986). *Ann. N.Y. Acad. Sci.* **466**, 218–239.
Kindler, S., Schulz, B., Goedert, M., and Garner, C. C. (1990). *J. Biol. Chem.* **265**, 19679–19684.
Kirschner, M., and Mitchison, T. (1986). *Cell* **45**, 329–342.
Kleiman, R., Banker, G., and Steward, O. (1990). *Neuron* **5**, 821–830.
Klunk, W. E., and Abraham, D. J. (1988). *Psychiatr. Dev.* **6**, 121–152.
Knops, J., Kosik, K. S., Lee, G., Pardee, J. D., Cohen-Gould, L., and McConlogue, L. (1991). *J. Cell Biol.* **114**, 725–733.
Kosik, K. S. (1989). *J. Gerontol.* **44**, B55–B58.
Kosik, K. S. (1991). *Trends Neurosci.* **14**, 218–219.
Kosik, K. S. (1992). *Science* **256**, 780–783.
Kosik, K. S., and Finch, E. A. (1987). *J. Neurosci.* **7**, 3142–3153.
Kosik, K. S., Crandall, J. E., Mufson, E. J., and Neve, R. L. (1989a). *Ann. Neurol.* **26**, 352–361.
Kosik, K. S., Orecchio, L. D., Bakalis, S., and Neve, R. L. (1989b). *Neuron* **2**, 1389–1397.
Kotani, S., Murofushi, H., Maekawa, S., Aizawa, H., and Sakai, H. (1988). *J. Biol. Chem.* **263**, 5385–5389.
Kowall, N. W., and Kosik, K. S. (1987). *Ann. Neurol.* **22**, 639–643.
Kuznetsov, S. A., and Gelfard, V. I. (1987). *FEBS Lett.* **212**, 145–148.
Langkopf, A., Hammarback, J. A., Müller, R., Vallee, R. B., and Garner, C. C. (1992). *J. Biol. Chem.* **267**, 16561–16566.
Lasek, R. J. (1988). *In* "Intrinsic Determinants of Neuronal Form and Function" (R. J. Lasek and M. M. Black, eds.), pp. 3–58. Alan R. Liss, New York.
Lee, G., Neve, R. L., and Kosik, K. S. (1989). *Neuron* **2**, 1615–1624.
Letourneau, P. C. (1985). *In* "Molecular Bases of Neural Development" (G. M. Edelman, W. E. Gall, and W. M. Cowan, eds.), pp. 269–293. Wiley, New York.
Letourneau, P. C., and Ressler, A. N. (1984). *J. Cell Biol.* **98**, 1355–1362.
Lewis, S. A., and Cowan, N. (1990). *Nature (London)* **345**, 674.

Lewis, S. A., Wang, D. H., and Cowan, N. J. (1988). *Science* **242**, 936–939.

Lewis, S. A., Avonov, I. E., Lee, G.-H., and Cowan, N. J. (1989). *Nature (London)* **342**, 498–505.

Lichtenberg, B., Mandelkow, E.-M., Hagestedt, T., and Mandelkow, E. (1988). *Nature (London)* **334**, 359–362.

Lichtenberg-Kraag, B., and Mandelkow, E.-M. (1990). *J. Struct. Biol.* **105**, 46–53.

Lien, L. L., Boyce, F. M., Kleyn, P., Brzustowicz, L. M., Menninger, J., Ward, D. C., Gilliam, T. C., and Kunkel, L. M. (1991). *Proc. Natl. Acad. Sci. U.S.A.* **88**, 7873–7876.

Lindwall, G., and Cole, R. D. (1984). *J. Biol. Chem.* **259**, 5301–5305.

Luca, F. C., Bloom, G. S., and Vallee, R. B. (1986). *Proc. Natl. Acad. Sci. U.S.A.* **83**, 1006–1010.

Mackay-Sim, A., and Kittel, P. W. (1991). *Eur. J. Neurosci.* **3**, 209–215.

Maeda, K., Nakata, T., Noda, Y., Sato–Yoshitake, R., and Hirokawa, N. (1992). *Mol. Biol. Cell* **3**, 1181–1194.

Magendantz, M., and Solomon, F. (1985). *Proc. Natl. Acad. Sci. U.S.A.* **82**, 6581–6585.

Mansfield, S. G., Diaz-Nido, J., Gordon–Weeks, P. R., and Avila, J. (1991). *J. Neurocytol.* **20**, 1007–1022.

Mareck, A., Fellous, A., Francon, J., and Nunez, J. (1980). *Nature (London)* **284**, 353–355.

Matus, A. (1988a). *Annu. Rev. Neurosci.* **11**, 29–44.

Matus, A. (1988b). *In* "The Making of the Nervous System" (J. G. Parnavelos, C. D. Stern, and R. V. Stirling, eds.), pp. 421–433. Oxford Univ. Press, Oxford.

Matus, A. (1991). *J. Cell Sci., Suppl.* **15**, 61–67.

Matus, A., and Riederer, B. (1986). *Ann. N.Y. Acad. Sci.* **466**, 167–179.

Matus, A., Bernhardt, R., and Hugh-Jones, T. (1981). *Proc. Natl. Acad. Sci. U.S.A.* **78**, 371–389.

Matus, A. I., Ackermann, M., Pehling, G., Byers, H. R., and Fujiwara, K. (1982). *Proc. Natl. Acad. Sci. U.S.A.* **79**, 7590–7594.

Matus, A., Huber, G., and Berhnardt, R. (1983). *Cold Spring Harbor Symp. Quant. Biol.* **48**, Part 2, 775–782.

Matus, A., Bernhardt, R., Bodmer, R., and Alaimo, D. (1986). *Neuroscience* **17**, 371–389.

Matus, A., Riederer, B., and Huber, G. (1987). *J. Neurochem.* **49**, 714–720.

Matus, A., Delhaye-Bouchaud, N., and Mariani, J. (1990). *J. Comp. Neurol.* **297**, 435–440.

McKerracher, L., Vallee, R.B., and Aguayo, A.J. (1989). *Visual Neurosci.* **2**, 349–356.

Meininger, V., and Binet, S. (1989). *Int. Rev. Cytol.* **114**, 21–79.

Melloni, E., and Pontremoli, S. (1989). *Trends Neurosci.* **12**, 438–444.

Migheli, A., Butler, M., Brown, K., and Shelanski, M.L. (1988). *J. Neurosci.* **8**, 1846–1851.

Miller, P., Walker, U., Theurkauf, W. E., Vallee, R. B., and DeCamilli, P. (1982). *Proc. Natl. Acad. Sci. U.S.A.* **79**, 5562–5566.

Mitchison, T., and Kirschner, M. (1984). *Nature (London)* **312**, 237–242.

Mitchison, T., and Kirschner, M. (1988). *Neuron* **1**, 761–772.

Morales, M., and Fifkova, E. (1989). *Cell Tissue Res.* **256**, 447–456.

Murofushi, H., Kotani, S., Aizawa, H., Hisanaga, S., Hirokawa, N., and Sakai, H. (1986). *J. Cell Biol.* **103**, 1911–1919.

Murofushi, H., Suzuki, M., Sakai, H., and Kobayashi, S. (1989). *Cell Tissue Res.* **255**, 315–322.

Murphy, D. B., and Borisy, G. G. (1975). *Proc. Natl. Acad. Sci. U.S.A.* **72**, 2696–2700.

Murthy, A. S. N., and Flavin, M. (1983). *Eur. J. Biochem.* **137**, 37–46.

Murthy, A. S. N., Bramblett, G. T., and Flavin, M. (1985). *J. Biol. Chem.* **260**, 4364–4370.

Niinobe, M., Maeda, N., Ino, H., and Mikoshiba, K. (1988). *J. Neurochem.* **51**, 1132–1139.

Noble, M., Lewis, S. A., and Cowan, N. J. (1989). *J. Cell Biol.* **109**, 3367–3376.

Novikoff, A. B., Novikoff, P. M., Quintana, N., and Davis, C. (1972). *J. Histochem. Cytochem.* **20**, 745–749.
Nunez, J. (1988). *Trends Neurosci.* **11**, 477–479.
Oblinger, M. M., Argasinski, A., Wong, J., and Kosik, K. S. (1991). *J. Neurosci.* **11**, 2453–2459.
Okabe, S., and Hirokawa, N. (1989). *Proc. Natl. Acad. Sci. U.S.A.* **86**, 4127–4131.
Okabe, S., Shiomura, Y., and Hirokawa, N. (1989). *Brain Res.* **483**, 335–346.
O'Leary, D. D. M. (1989). *Trends Neurosci.* **12**, 400–406.
Olmsted, J. B. (1986). *Annu. Rev. Cell Biol.* **2**, 421–457.
Papandrikopoulou, A., Doll, T., Tucker, R. P., Garner, C. C., and Matus, A. (1989). *Nature* (*London*) **340**, 650–652.
Papasozomenos, S. C., and Binder, L. I. (1986). *J. Neurosci.* **6**, 1748–1756.
Papasozomenos, S. C., and Binder, L. I. (1987). *Cell Motil. Cytoskeleton* **8**, 210–226.
Papasozomenos, S. C., Binder, L. I., Bender, P., and Payne, M. R. (1985). *J. Cell Biol.* **100**, 74–85.
Parysek, L. M., Asnes, C. F., and Olmsted, J. R. (1984). *J. Cell Biol.* **99**, 1309–1315.
Parysek, L. M., del Cerro, M., and Olmsted, J. B. (1985). *Neuroscience* **15**, 869–875.
Peng, I., Binder, L. I., and Black, M. M. (1985). *Brain Res.* **361**, 200–211.
Peng, I., Binder, L. I., and Black, M. M. (1986). *J. Cell Biol.* **102**, 252–262.
Peters, A., Palay, S., and Webster, H. deF. (1976). "The Fine Structure of the Nervous System," p. 81. Saunders, Philadelphia.
Purves, D., and Lichtman, J. W. (1980). *Science* **210**, 153–157.
Ramon y Cajal, S. (1895). "Les nouvelles idées sur la structure du système nerveux chez l'homme et chez les vertébrés." Reinwald, Paris.
Riederer, B., and Matus, A. (1985). *Proc. Natl. Acad. Sci. U.S.A.* **82**, 6006–6009.
Riederer, B., Cohen, R., and Matus, A. (1986). *J. Neurocytol.* **15**, 763–775.
Riederer, B. M., Guadano-Ferraz, A., and Innocenti, G. M. (1990). *Dev. Brain Res.* **56**, 235–243.
Robson, S. J., and Burgoyne, R. D. (1989). *Cell Motil. Cytoskeleton* **12**, 273–282.
Sano, M., Katoh-Semba, R., Kitajima, S., and Sato, C. (1990). *Brain Res.* **510**, 269–276.
Sasaki, S., Stevens, J. K., and Bodick, N. (1983). *Brain Res.* **259**, 193–206.
Sasaki-Sherrington, S. E., Jacobs, J. R., and Stevens, J. K. (1984). *J. Cell Biol.* **98**, 1279–1290.
Sato, M., Schwartz, W. H., Selden, S. C., and Pollard, T. D. (1988). *J. Cell Biol.* **106**, 1205–1211.
Sato-Yoshitake, R., Shiomura, Y., Miyasaka, H., and Hirokawa, N. (1989). *Neuron* **3**, 229–238.
Sattilaro, R. F. (1986). *Biochemistry* **25**, 2003–2009.
Sattilaro, R. F., Dentler, W. L., and LeCluyse, E. L. (1981). *J. Cell Biol.* **90**, 467–473.
Scheibel, M. E., and Scheibel, A. B. (1978). *In* "Neuroanatomical Research Techniques" (R. T. Robertson, ed.), pp. 89–114. Academic Press, New York.
Schliwa, M., Euteneuer, U., Bulinski, J. C., and Izant, J. G. (1981). *Proc. Natl. Acad. Sci. U.S.A.* **78**, 1037–1041.
Schmidt, J. T. (1985). *In* "Molecular Bases of Neural Development" (G. M. Edelman, W. E. Gall, and W. M. Cowan, eds.), pp. 453–480. Wiley, New York.
Schmidt, J. T. (1991). *Ann. N.Y. Acad. Sci.* **627**, 10–25.
Schoenfeld, T. A., McKerracher, L., Obar, R., and Vallee, R. B. (1989). *J. Neurosci.* **9**, 1712–1730.
Schoenfeld, T. A., Meltser, H. M., and May, A. L. (1993). *Chem. Senses* **18**, 624–625.
Schulze, E., Asai, D. J., Bulinski, J. C., and Kirschner, M. (1987). *J. Cell Biol.* **105**, 2167–2177.

Seeds, N. W., Gilman, A. G., Amano, T., and Nirenburg, M. W. (1970). *Proc. Natl. Acad. Sci. U.S.A.* **66**, 160–167.

Selden, S. C., and Pollard, T. D. (1986). *Ann. N.Y. Acad. Sci.* **466**, 803–812.

Selkoe, D. J. (1986). *Neurobiol. Ageing* **7**, 425–430.

Selkoe, D. J., Podlisny, M. B., Gronbeck, A., Mammen, A., and Kosik, K. S. (1990). *Adv. Neurol.* **51**, 171–179.

Shepherd, G. M., ed. (1990). "The Synaptic Organization of the Brain," 3rd Ed. Oxford Univ. Press, New York.

Shepherd, G. M., and Greer, C. A. (1990). *In* "The Synaptic Organization of the Brain" (G. M. Shepherd, ed.), 3rd Ed., pp. 133–169. Oxford Univ. Press, New York.

Shiomura, Y., and Hirokawa, N. (1987a). *J. Neurosci.* **7**, 1461–1469.

Shiomura, Y., and Hirokawa, N. (1987b). *J. Cell Biol.* **104**, 1575–1578.

Shiomura, Y., and Hirokawa, N. (1989). *Brain Res.* **502**, 356–364.

Shpetner, H. S., and Vallee, R. B. (1989). *Cell* **59**, 421–432.

Siman, R., Gall, C., Perlmutter, L. S., Christian, C., Baudry, M., and Lynch, G. (1985). *Brain Res.* **347**, 399–403.

Sims, K. B., Crandall, J. E., Kosik, K. S., and Williams, R. S. (1988). *Brain Res.* **449**, 192–200.

Skene, J. H. P., and Willard, M. (1981). *J. Cell Biol.* **89**, 96–103.

Sloboda, R. D., and Rosenbaum, J. L. (1979). *Biochemistry* **18**, 48–55.

Sloboda, R. D., Rudolph, S. A., Rosenbaum, J. L., and Greengard, P. (1975). *Proc. Natl. Acad. Sci. U.S.A.* **72**, 177–181.

Sloboda, R. D., Dentler, W. L., and Rosenbaum, J. L. (1976). *Biochemistry* **15**, 4497–4505.

Stanley, E. F., Nowycky, M. C., and Triggle, D. J., eds. (1991). *Ann. N.Y. Acad. Sci.* **635**, 1–506.

Sternberger, L. A., and Sternberger, N. H. (1983). *Proc. Natl. Acad. Sci. U.S.A.* **80**, 6126–6130.

Szaro, B. G., Whitmall, M. H., and Gainer, H. (1990). *J. Comp. Neurol.* **302**, 220–255.

Trojanowski, J. Q., Schuck, T., Schmidt, M. L., and Lee, V. M. (1989a). *J. Histochem. Cytochem.* **37**, 209–215.

Trojanowski, J. Q., Schuck, T., Schmidt, M. L., and Lee, V. M. (1989b). *J. Neurosci. Methods* **29**, 171–180.

Tucker, R. P. (1990). *Brain Res. Rev.* **15**, 101–120.

Tucker, R. P., and Matus, A. I. (1987). *Development* **101**, 535–546.

Tucker, R. P., and Matus, A. I. (1988). *Dev. Biol.* **130**, 423–434.

Tucker, R. P., Binder, L. I., and Matus, A. I. (1988a). *J. Comp. Neurol.* **271**, 44–55.

Tucker, R. P., Binder, L. I., Viereck, C., Hemmings, B. A., and Matus, A. I. (1988b). *J. Neurosci.* **8**, 4503–4512.

Tucker, R. P., Garner, C. C., and Matus, A. I. (1989). *Neuron* **2**, 1245–1256.

Tytell, M., Brady, S. T., and Lasek, R. J. (1984). *Proc. Natl. Acad. Sci. U.S.A.* **81**, 1570–1574.

Vallee, R. (1980). *Proc. Natl. Acad. Sci. U.S.A.* **77**, 3206–3210.

Vallee, R. B. (1982). *J. Cell Biol.* **92**, 435–444.

Vallee, R. B. (1985). *Biochem. Biophys. Res. Commun.* **133**, 128–133.

Vallee, R. B., and Bloom, G. S. (1984). *Mod. Cell Biol.* **3**, 21–76.

Vallee, R. B., and Bloom, G. S. (1991). *Annu. Rev. Neurosci.* **14**, 59–92.

Vallee, R. B., and Borisy, G. G. (1977). *J. Biol. Chem.* **252**, 377–382.

Vallee, R. B., and Davis, S. E. (1983). *Proc. Natl. Acad. Sci. U.S.A.* **80**, 1342–1346.

Vallee, R. B., and Shpetner, H. S. (1990). *Annu. Rev. Biochem.* **59**, 909–932.

Vallee, R. B., DiBartolomeis, M. J., and Theurkauf, W. E. (1981). *J. Cell Biol.* **90**, 568–576.

Vallee, R. B., Bloom, G. S., and Luca, F. C. (1984). *In* "Molecular Biology of the Cytoskeleton" (G. G. Borisy, D. W. Cleveland, and D. B. Murphy, eds.), pp. 111–130. Cold Spring Harbor Lab. Press, Cold Spring Harbor, New York.

Verhaagen, J., Oestreicher, A. B., Gispen, W. H., and Margolis, F. L. (1989). *J. Neurosci.* **9**, 683–691.

Viereck, C., Tucker, R. P., and Matus, A. (1989). *J. Neurosci.* **9**, 3547–3557.

Voter, W. A., and Erickson, H. P. (1982). *J. Ultrastruct. Res.* **80**, 374–382.

Weingarten, M., Lockwood, A., Hwo, S., and Kirschner, M. (1975). *Proc. Natl. Acad. Sci. U.S.A.* **72**, 1858–1862.

Wiche, G. (1989). *Biochem. J.* **259**, 1–12.

Wiche, G., Briones, E., Hirt, H., Krepler, R., Artlieb, V., and Denk, H. (1983). *EMBO J.* **2**, 1915–1920.

Wiche, G., Oberkamins, C., and Himmler, A. (1991). *Int. Rev. Cytol.* **124**, 217–273.

Wiesel, T. N., and Hubel, D. (1963). *J. Neurophysiol.* **26**, 1003–1017.

Wille, H., Mandelkow, E.-M., Dingus, J., Vallee, R.B., Binder, L.I., and Mandelkow, E. (1992). *J. Struct. Biol.* **108**, 49–61.

Wolpaw, J. R., Schmidt, J. T., and Vaughn, T. M., eds. (1991). *Ann. N.Y. Acad. Sci.* **627**, 1–399.

Wood, J. G., Mirra, S., Pollack, N. J., and Binder, L. I. (1986). *Proc. Natl. Acad. Sci. U.S.A.* **83**, 4040–4043.

Woodhams, P. L., Calvert, R., and Dunnett, S. B. (1989). *Neuroscience* **28**, 49–59.

Wuerker, R. B., and Kirkpatrick, J. B. (1972). *Int. Rev. Cytol.* **33**, 45–75.

Yamada, K. M., Spooner, B. S., and Wessels, N. K. (1970). *Proc. Natl. Acad. Sci. U.S.A.* **66**, 1206–1212.

Yamamoto, H., Fukunaga, K., Tanaka, E., and Miyamoto, E. (1983). *J. Neurochem.* **41**, 1119–1125.

Yamamoto, H., Fukunaga, K., Goto, S., Tanaka, E., and Miyamoto, E. (1985). *J. Neurochem.* **44**, 759–768.

Yee, V. C., Pestronk, A., Alderson, K., and Yuan, C. M. (1988). *J. Neurocytol.* **17**, 649–656.

Zenker, W., and Hohberg, E. (1973). *J. Neurocytol.* **2**, 143–148.

Cholinesterases in Avian Neurogenesis

Paul G. Layer and Elmar Willbold
Technical University of Darmstadt, Institute for Zoology, 64287 Darmstadt,
Germany

I. New Frontiers for Cholinesterase Research

The best-known member of the cholinesterase family is acetylcholinesterase, which degrades the neurotransmitter acetylcholine at the postsynaptic membrane of cholinergic synapses. Other cholinesterases include butyrylcholinesterase (BChE, nonspecific cholinesterase, pseudocholinesterase), and propionylcholinesterase. BChE is enriched in the liver, heart, brain, and especially in serum; and it is able to hydrolyze higher cholinesters as well as acetylcholine. However, compared with AChE, it does so only at a reduced rate and shows no substrate inhibition. BChE has no known physiological function in adult organisms. The pharmacological inhibition of BChE and naturally occurring mutations that lead to an enzymatically inactive protein have no obvious effect on normal development and health (Goedde *et al.*, 1967).

The expression of different types of cholinesterase genes at odd times and places, such as in red cell membranes, migrating neural crest cells, retinal pigmented epithelium, or intestinal epithelial cells (Drews, 1975; Miki and Mizoguti, 1982a; Layer and Kaulich, 1991; Sine *et al.*, 1991; Salceda and Martinez, 1992), has been known for a long time, but it is not well understood (Ozaki, 1974; Silver, 1974; Graybiel and Ragsdale, 1982; Robertson, 1987). Research in the following areas has revived interest in these enzymes, since it suggests that additional physiological and developmental functions of cholinesterases are highly possible.

1. Cholinesterases could exert additional noncholinergic enzymatic functions; for example, an arylacylamidase has been detected. Moreover, a peptidase and substance P hydrolyzing activity have been reported, yet they could not be demonstrated unequivocally with the cholinesterase

molecule (George and Balasubramanian, 1981; Lockridge, 1982; Balasubramanian, 1984; Chatonnet and Masson, 1985; Small *et al.*, 1986; Boopathy and Balasubramanian, 1987; Rao and Balasubramanian, 1990).

2. AChE and BChE are coded for by separate genes located on different chromosomes (Soreq *et al.*, 1987; Zakut *et al.*, 1989; Arpagaus *et al.*, 1990; Gaughan *et al.*, 1991; Allerdice *et al.*, 1991; Getman *et al.*, 1992). Therefore the hypothesis that BChE is a protein precursor of AChE can no longer be sustained (Koelle *et al.*, 1976, 1977a,b, 1979a,b). Rather, another function for BChE is needed. By now, the amino acid sequences of BChE and AChE (Doctor *et al.*, 1990; Jbilo and Chatonnet, 1990) as well as the nucleotide sequences of the corresponding genes (Sikorav *et al.*, 1987; Maulet *et al.*, 1990; Li *et al.*, 1991) are known for a number of species. As expected, they show a high degree of homology (Prody *et al.*, 1987; McTiernan *et al.*, 1987; Dreyfus *et al.*, 1988; Soreq and Prody, 1989).

3. Modern recombinant DNA technology has revealed complex cholinesterase homologies with a variety of other esterases as well as with nonhydrolase proteins like thyroglobulin (Schumacher *et al.*, 1986), the *Drosophila* tactins (Barthalay *et al.*, 1990; De la Escalera *et al.*, 1990; Krejci *et al.*, 1991) or lipases (Christie *et al.*, 1991). These sequence homologies indicate a close evolutionary relationship within a new family of proteins, including the cholinesterases. It is supposed that the more specialized AChE emerged from a duplicated copy of the "older" BChE gene. Yet, the fact that BChE has not been lost during the subsequent evolution indicates that it must play some indispensable function.

4. Despite the existence of independent genes for AChE and BChE, observations from different research areas indicate that the close evolutionary relationship between AChE and BChE is also reflected in complex temporal and regulatory interdependencies during ontogenesis (Koelle *et al.*, 1976, 1977a,b, 1979a,b; Edwards and Brimijoin, 1982; Berman *et al.*, 1987; Adler and Filbert, 1990; Layer *et al.*, 1992). Either one or both cholinesterases show significantly reduced or increased activity in a series of neurological disorders and tumors (Topilko and Caillou, 1988; Rakonczay, 1988). Both types of cholinesterases are also implicated in developmental anomalies such as Down's syndrome (Price *et al.*, 1982) and neural tube defects (Smith *et al.*, 1979), and in dementias (Atack *et al.*, 1985; Wright *et al.*, 1993).

Because some of these abnormalities arise during embryonic development, it is reasonable to search for functions of the cholinergic system during normal embryogenesis. A close relationship between the expression of specific cholinesterases and important developmental events such as the end of proliferation and the beginning of differentiation could be revealed (for reviews see Drews, 1975 and Layer, 1990, 1991a).

5. AChE is secreted in the substantia nigra and possibly acts as a neuromodulator in the adult brain (Greenfield, 1984, 1991; Greenfield *et al.*, 1984; Jones and Greenfield, 1991).

In this chapter we present our findings on the expression of BChE and AChE during avian neurogenesis, with particular emphasis on our most recent studies, which reveal new functions for both BChE and AChE.

II. Acetylcholinesterase and Butyrylcholinesterase in Developing Nervous Tissues

A. Two Periods of Expression

The production of cholinesterases is developmentally regulated in a complex manner. The enzymes appear in different molecular forms in specific cell types and specific tissue parts (for example, inside and outside the neural tube during various stages of embryonic development). It is therefore advisable to subdivide the entire history of embryonic cholinesterase development into two main periods—embryonic and synaptic.

The first period consists of the early expression of cholinesterases— during neurogenesis when cells of the neuroepithelium leave the mitotic cycle and start to differentiate. We call these cholinesterases "embryonic cholinesterases" [while the term "embryonic cholinesterase" was introduced earlier, at that time BChE and AChE were not considered separately; compare Drews (1975)].

This early "morphogenetic" period is distinguished from the "synaptic" period of cholinesterase expression—the time when neurofibrillar laminae and then synapses are established. Only then do the tissues reach their physiological function. Cholinesterases that remain at cholinergic synapses then will perform their conventional synaptic functions.

In our studies we have concentrated on the embryonic cholinesterases because they are the candidates most likely to fulfill morphogenetic functions. One has to keep in mind that in reality both periods will overlap in time and space.

B. Expression of Cholinesterases in Proliferation and Differentiation

Our studies on avian embryos revealed a close relationship between the expression of BChE and the onset of differentiation, and between the regulation of AChE and BChE (Layer, 1983; Weikert *et al.*, 1990; Willbold and Layer, 1992a,b; Layer *et al.*, 1992).

Typically, both enzymes show a mutually exclusive expression. In the early neural tube of an HH 14–20-day chicken embryo (staged after Hamburger and Hamilton, 1951), a diffuse staining along the ventricular side of the neural tube indicates BChE that is transiently produced by neuroblasts during their last cell cycle (Layer, 1983). Soon after they have finished their last mitosis, the neuroblasts rapidly accumulate AChE while migrating to the mantle surface of the brain (Miki and Mizoguti, 1982b; Layer, 1983; Mizoguti and Miki, 1985; Layer *et al.*, 1987; Willbold and Layer, 1992a). For this reason the expression of AChE marks one of the earliest postmitotic events in the life of early neurons (see Fig. 1).

Cell bodies of the neural tube undergo a characteristic inside-outside movement during their proliferation period. Cell division always occurs on the ventricular side of the neuroepithelium (Sauer, 1935). Since BChE is localized along the ventricular side, whereas AChE appears along the opposite side, it was reasonable to search for a relationship between the expression of these enzymes and the transition of the cells from proliferation to differentiation. Comparable observations have been described in cell lines *in vitro:* BChE is predominantly expressed in mouse ependymoblastoma cells, with maximum activity during monolayer formation. Afterward, BChE activity is inversely proportional to the rate of cell division (Chudinovskaia, 1978).

During the transition, four different classes of cells with increasing AChE activities have been distinguished. Full expression of AChE is considered an early marker for differentiation (Zacks, 1952; Miki and Mizoguti, 1982b; Mizoguti and Miki, 1985).

In our studies on brains of young chicken embryos as well as on retinal *in vitro* reaggregates, we compared the appearance of both enzymes with cell proliferation by using a double-labeling technique (Vollmer and Layer, 1986b, 1987). The proliferation zones were diffusely stained by BChE but not by AChE. Patches intensely stained for BChE accompanied clusters of cells in the final stages of mitosis and on their way to the differentiation zone, where they began expressing AChE.

It is noteworthy that cells expressing AChE are fully separated from the mitotically active cells. Thus a histological border seems to separate these two compartments (Willbold and Layer, 1992a). By applying varying [^{3}H]thymidine pulses, we identified an 11-hr period from the last thymidine uptake to full AChE expression. Thus, the decrease of cell proliferation is closely followed by a decrease in BChE activity, whereas AChE increases at a rate inverse to that of BChE (Fig. 1, bottom).

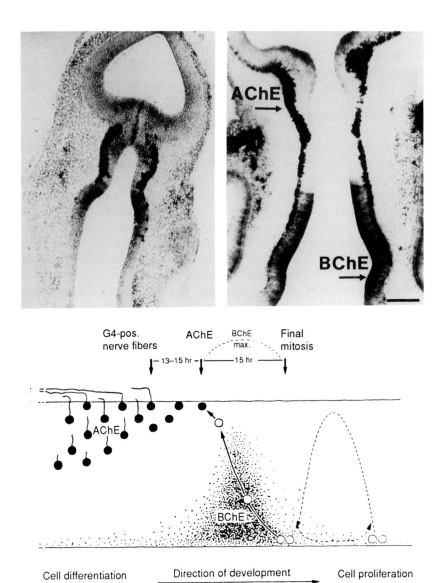

FIG. 1   Distribution of both classes of cholinesterases during development of the chicken brain. *Top.* Frozen horizontal sections of HH 16 and HH 20-stage (after Hamburger and Hamilton, 1951) chick embryos showing the diencephalic and midbrain regions. AChE heavily labels cell bodies on the outer surface of the neuroepithelium; BChE is indicated by the diffuse dark staining on the ventricular side. Note that during development, BChE patches spread in front of AChE-activated cells. The sections were stained by using acetylthiocholine as substrate (Karnovsky and Roots, 1964), thus revealing both classes of cholinesterases. The specificity of each type of cholinesterase can be checked separately by using appropriate inhibitors. Bar = 20 μm. *Bottom.* Scheme depicting the expression of AChE and BChE in relation to the final stage of cell proliferation, the onset of differentiation, and neurite outgrowth. For further description see the text. "Direction of Development" indicates the spatial progression of the AChE-differentiation front.

III. Molecular Polymorphism of Cholinesterases during Brain Development

A. Early Neurogenesis

To date, the primary sequences of *Torpedo* AChE, fetal bovine AChE, *Drosophila* AChE, and of human BChE have been identified (Schumacher *et al.*, 1986; McTiernan *et al.*, 1987; Sikorav *et al.*, 1987; Doctor *et al.*, 1990; Ekstrom *et al.*, 1993; Vellom *et al.*, 1993). While only a few different organisms permit a direct comparison of AChE and BChE, a close homology of 38–54% between the two enzymes becomes evident. In particular, the sequence Gly–Glu–Ser–Ala–Gly in the active center has been highly conserved. Despite the similarity to the center of serine proteases (Gly–Asp–Ser–Gly–Gly), cholinesterases present no further similarities with this group of enzymes. Instead, cholinesterases form a separate molecular family that includes thyroglobulin, rabbit liver esterase, and esterase-6 from *Drosophila* (Chatonnet and Lockridge, 1989).

Depending on the species, the tissues, and the specific time of their expression, cholinesterases can exist in a multitude of distinct molecular forms. They represent complex associations of catalytic and structural subunits. Globular and asymmetric forms and membrane- or extracellular matrix-associated forms can be distinguished. The globular forms exist as hydrophilic and soluble or as amphiphilic, membrane-bound enzymes. In particular cases, the enzymes can be released into the extracellular spaces.

In vertebrates, the globular forms predominate. Generally, in blood we find exclusively globular forms. Here AChE occurs as dimer, while serum BChE is present as tetramer (Silman and Futerman, 1987). Noticeably in chicks, erythrocyte AChE is low, whereas in serum all three globular forms of BChE are present. Here, a significant age-dependancy of BChE contents has been noticed (Bennett and Bennett, 1991). Globular cholinesterases are also found in other tissues, such as the liver, the urogenital sytem, the digestive tract, the placenta, and also in some endocrine and exocrine glands. In chick liver, predominantly monomeric and tetrameric BChE and only minor amounts of AChE are present.

Asymmetric forms of cholinesterases are particularly prevalent at skeletal neuromuscular junctions, but they are also found in heart muscle and peripheral ganglia. In brain, globular forms (G1, monomeric; G2, dimeric; G4, tetrameric) dominate their asymmetric counterparts. In the embryonic chick brain, the globular forms make up 95% of the total AChE activity. Moreover, collagen-tailed asymmetric forms, which typically are associated with glycosaminoglycans of the extracellular matrix, are found in extracts of 10-day-old chick retina and tectum (Villafruela *et al.*, 1981), but seem to make no major contribution to the total AChE present in chicken embryonic brain.

Molecular forms of BChE have been described that are precise counterparts of the AChE forms in rat tissues (Vigny *et al.*, 1978), chick embryonic muscle (Lyles *et al.*, 1979; Allemand *et al.*, 1981), chick serum (Lyles *et al.*, 1980), *Torpedo marmorata* electric organ (Toutant *et al.*, 1985), and postmortem human brain (Atack *et al.*, 1986). In human fetus, only G4 BChE has been found, whereas several forms of AChE are expressed (Zakut *et al.*, 1985). The functional significance of the different forms expressed at different developmental stages and in different species and tissues is not known.

B. Small Molecular Forms in Embryonic Nervous Tissues

We have used sucrose gradient analysis to study the expression of specific molecular forms of the two cholinesterases in the chicken nervous system during early neurogenesis of retinal and brain tissues (Layer *et al.*, 1987). Besides other minor constituents, both enzymes are present in two major molecular forms. AChE exists as G2 and G4 globular forms, whereas BChE is represented by G1 and G4 globular forms. During development, a continuous increase of G4 over G2 AChE is noted.

Both in retina and tectum, the processes of proliferation, cell migration, and laminae formation take place before E10–E12 (for review see Jacobson, 1978), whereas synaptogenesis starts only after E12 (Glees and Sheppard, 1964; Meller, 1964; LaVail and Cowan, 1971a,b; Cantino and Daneo, 1972; Rager, 1976). From our data, two discrete developmental phases of cholinesterase expresssion have emerged, which are linked by a more or less pronounced intermediate phase. Cholinesterases expressed during morphogenetic periods show a higher percentage of the low-molecular-weight species, whereas those being expressed during synaptogenetic periods are more of the G4 type.

The main constituents of the molecular pool of cholinesterase in the young chicken brain thus have turned out to be simple, with only a monomeric BChE, a dimeric AChE, and a tetrameric form for both of them. This parallels the situation described for chick muscle (Lyles *et al.*, 1979) and for human fetal brain (Muller *et al.*, 1985). AChE shows a continuous shift from the dimeric into the tetrameric form (compare Marchand *et al.*, 1977), which could be related to the reported preferential association of the G4 form with axons (Couraud and DiGiamberardino, 1980).

C. Butyrylcholinesterase in Mature Retina

In the retina, G4-AChE reaches about 30% of the total AChE. This relatively low level of AChE corresponds to a high value of G4-BChE at late stages. In contrast, G4-BChE in older brain is low. In other words, the content of G4-BChE in brain remains almost constant at 45%, but in retina there is a drastic shift from 35% G4-BChE before E5 to 70% G4-BChE at E7. It is possible that the high amounts of G4-AChE found in brain could correlate with a higher percentage of efferent cholinergic connections. In the mature inner plexiform layer of the retina, there are synaptic sublaminae expressing BChE almost exclusively (unpublished observations); these are expected to contain the G4 form of BChE prevailing in retina. It will be interesting to find out if they are associated with afferent connections projecting back from the brain to the retina.

As discussed before, the BChE expressed early is associated with cellular mitotic activity. Conversely, BChE that is expressed in the late period of maturation may participate in some unknown synaptic function. Significant amounts of the total cholinesterase activity in adult chick muscle end plate are attributable to BChE (Jedrzejczyk et al., 1984). During the very beginning of development of the myotubes, a significant fraction of the cholinesterases consists of BChE (Layer et al., 1988b). As a corollary, hybrid molecules between AChE and BChE have been reported in embryonic muscle (Tsim et al., 1988a,b), the significance of which is unknown.

D. Embryonic Brain Cholinesterases and Membrane Association

Detergents are needed to solubilize cholinesterases that are membrane bound. Two ways of anchoring to membranes have been reported. Experiments with phosphatidylinositol-specific phospholipase C (PIPL C) from *Staphylococcus aureus* revealed that AChE from erythrocytes is solubilized by this enzyme (Low and Finean, 1977). In the meantime, a number of other cell surface glycoproteins, including cell adhesion molecules and other ectoenzymes, have been identified that similarly anchor to the cell membrane, for example, the F11 fasciculation molecule (Wolff et al., 1989), alkaline phosphatase, renal dipeptidase, or 5'-nucleotidase (Zimmermann, 1992).

Interestingly, only the G2 forms of AChE from different tissues are sensitive to PIPL C. The hydrophobic forms of AChE from mammalian central nervous system (CNS) tissue are insensitive to PIPL C (Futerman et al., 1985a,b). Instead, a catalytically active and soluble enzyme can be released through limited proteolytic digestion. In the brain, a 20 kDa subunit seems to mediate anchoring to the membrane (Gennari *et al.*, 1987; Inestrosa *et al.*, 1987; Fuentes *et al.*, 1988). Similarly, in the embryonic and the adult chicken brain, we found that both AChE and BChE are insensitive to PIPL C digestion (Treskatis, 1990; Treskatis *et al.*, 1992). Rather, 85% of the embryonic and 60% of the adult membrane-bound BChE is released by limited proteolysis following a combined protocol of proteinase K and pronase digestion.

Most strikingly, AChE activity from embryonic or adult chicken brain tissue was not releasable under these circumstances. A detailed analysis showed that not only the membrane anchor but also AChE activity itself is much more sensitive to such a protease treatment than BChE. Is membrane-associated BChE more resistant to protease attack, and if so, why? This finding should be followed up, since it could bear on the functional significance of both enzymes.

E. Membrane-Bound Brain Cholinesterases

The distribution of both types of cholinesterases in the various subcellular compartments of embryonic and adult chicken brain has been studied using subcellular fractionation (Treskatis, 1990). Major fractions—more than 70%—of both AChE and BChE are associated with membranes. In the embryonic brain, the soluble fraction of BChE—36%—was remarkably high. This hints at a possible release of BChE in the embryonic brain, a suspicion supported by our own cell culture experiments. Otherwise, the soluble fraction for both cholinesterases was less than 20%. Cholinesterases within nuclei (less than 0.5% of AChE and BChE) and the microsomal fractions (less than 2% for both) are negligible.

F. Brain and Serum Butyrylcholinesterase

Butyrylcholinesterase from chicken serum has been purified to homogeneity with more than 250 units/mg specific activity (Treskatis *et al.*, 1992) and used to produce monoclonal antibodies (Ebert, 1988). These BChE-specific antibodies also recognize BChE from brain tissue. Therefore we could use them to isolate the enzymes from embryonic and adult brain, although they occur only in minute amounts.

As revealed by sodium dodecyl sulfate-polyacrylamide (SDS–PAGE) gel electrophoresis, the serum enzyme is represented by a double band of 79–82 kDa, while the brain enzyme has a size of only 74 kDa. Limited digestion by V8-protease of the serum and brain preparations leads to similar peptide patterns, indicating the close similarity at the protein level of both types of molecules. Furthermore, enzymatic deglycosylation has shown that their core proteins consist of 59 kDa subunits. Their different molecular weights thus are due to different glycosylation patterns (see Fig. 2).

Since BChE from different tissue sources shows different degrees of glycosylation (Liao *et al.*, 1991, 1992; Brodbeck and Liao, 1992), it seems possible that the glycoparts could help to define regional specificity and possibly even contribute to various functions of the molecule. We therefore have started to investigate the precise glycosylation patterns of BChE from embryonic and adult serum and brain. For these various sources, we not only find differences in the distribution of the adhesion-relevant HNK-1 epitope (see below), but more generally significant differences in their overall glycosylation patterns.

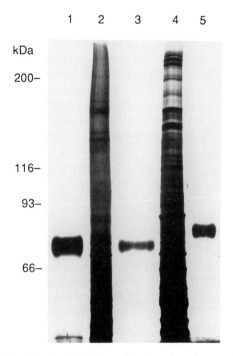

FIG. 2   The subunit size of brain BChE is smaller than that of serum BChE, as shown by SDS-PAGE. mAb 6-immunoisolated BChE fractions are shown from: adult brain membranes (1), a water-soluble fraction of adult brains (3), and serum (5). The original brain membranes and the water-soluble supernatant are shown in lanes (2) and (4), respectively. Note the one-step efficiency of the immunoaffinity chromatography on mAb 6 to isolate both serum and brain BChEs. (From Treskatis *et al.*, 1992.)

G. Biosynthesis and Regulation of Cholinesterase Expression

Considering the many forms of cholinesterases that exist in various tissues, the question of their biosynthesis and regulation arises.

The regulation of the biosynthesis of cholinesterases is only partially understood. In particular areas of the nervous system, denervation leads to a decrease of AChE activity. A postulated release factor may stimulate the biosynthesis of AChE. This factor could be acetylcholine itself, but other factors, including cyclic adenosine monophosphate (cAMP) and cyclic guanosine monophosphate (cGMP), have been considered as well (Silver, 1974; Massoulié and Bon, 1982). More recently, attention has been drawn to the dipeptide Gly-Gln, which can be produced posttranslationally from β-endorphin. This dipeptide can increase the biosynthesis of AChE (Koelle, 1988; Koelle et al., 1988a,b; Haynes, 1989).

It is likely that the heterogeneity of the various molecular forms is determined before translation into their respective protein sequences. Using in vitro translation in Xenopus oocytes, heterogeneous RNA molecules of cholinesterases were found coding for the various forms of proteins (Soreq et al., 1984, 1989). In the case of Torpedo AChE, soluble and hydrophobic dimers are achieved through alternative splicing, creating varying C-terminals with otherwise identical sequences (Gibney et al., 1988; Sikorav et al., 1988).

In muscle cells, a dimeric globular form of AChE is found within cells that is also secreted. The tetrameric globular form is associated mainly with the membrane. The rate of synthesis was about 6-fold higher for the dimeric than for the tetrameric forms. A mutual conversion of both forms could not be observed. Similar to the globular soluble or membrane-associated forms, the asymmetric forms of AChE are formed independently in the cell and, after a period of maturation in the Golgi apparatus, are secreted into the extracellular space (Rotundo and Fambrough, 1980; Rotundo, 1984).

IV. Functional Analysis of Cholinesterase Actions

A. In Vitro Models to Investigate Functional Mechanisms of Cholinesterase Action during Development

Our data strongly support the hypothesis that cholinesterases may act as important cues during development of the avian nervous system. In particular, AChE seems to have more functions than simply degrading acetylcholine at the postsynaptic membrane. Drews (1975) has suggested that during embryogenesis AChE is associated with migratory cells. In order to determine the migratory behavior of embryonic cells in the AChE active state, he studied the differentiation of limb bud mesenchyme into cartilage nodules *in vitro*. Once the cells acquire cholinesterase activity, their locomotory behavior changes. They start to pile up and adhere together, leading to the formation of cell aggregates (Drews and Drews, 1973). Other examples of AChE expression in migratory cells are the blastomeres and epithelial cells of *Ciona intestinalis* (Minganti and Falugi, 1980) or cells of the adrenal gland of postmetamorphic *Rana esculenta* (Grassi Milano *et al.*, 1985).

However, these observations do not tell us about the underlying molecular mechanisms, nor do they identify the structures that the enzymes act upon. How can they influence the normal neurogenesis of an avian brain? Interestingly, neurotransmitters themselves or neurotransmitter-related molecules can be directly involved as morphogens in embryogenesis (Buznikov *et al.*, 1964; Buznikov, 1971, 1990; MacMahon, 1974). Growing axons respond to the application of serotonin *in vitro* (Haydon *et al.*, 1984) and acetylcholine is released by growing growth cones after electric stimulation (Sun and Poo, 1987). Thus all molecules that are able to interact with neurotransmitters may be involved in the regulation of such possible functions during embryogenesis. This would be one mechanism by which cholinesterases could act on the development of a brain.

We have explored these questions further through a number of different culture techniques. Depending upon the experiment and the test system, we either added purified enzyme preparations or alternatively, specific anticholinesterases or antibodies to cells of embryonic retinal or tectal origin. We determined their effects on (1) mutual regulation of cholinesterase expression, (2) cell proliferation, (3) neurite outgrowth patterns, and (4) their aggregation behavior and ability to generate histotypic structures. Mainly, the following three systems were used:

1. Embryonic retinal and tectal cells were enzymatically and mechanically dissociated and cultured in a conventional monolayer system. This technique is very suitable for investigating the release of substances into the supernatant and the alteration of cell and fiber morphology. In this system, cells are isolated from each other; they tend to dedifferentiate and begin to lose characteristic properties.

2. We used aggregation culture systems (Vollmer *et al.*, 1984; Vollmer and Layer, 1986a,b, 1987; Layer and Willbold, 1989; Layer *et al.*, 1990; Wolburg *et al.*, 1991) to organize the cells more histotypically. In these systems, the cells are prevented from settling by continuous rotation of the culture dishes. Thus the isolated cells aggregate and form three-

dimensional spheroids floating freely in the culture medium. This technique offers numerous advantages compared with monolayer cultures. The cells retain many of their normal properties, such as the ability to proliferate, to migrate, and to differentiate. The histological and biochemical development of spheroids *in vitro* often closely resembles the development of tissue *in vivo*.

At one point we used the classical retinal reaggregation system which uses retinal cells from the central part of the eye (Moscona, 1952; Sheffield and Moscona, 1969; Fujisawa, 1973). When only retinal cells are used, nuclear layers develop in a reversed orientation, forming so-called rosetted spheroids. Alternatively, we introduced a second *in vitro* retinal regeneration system by mixing dissociated pigmented epithelial and retinal cells from the eye margin (Vollmer *et al.*, 1984; Layer and Willbold, 1989) (see Fig. 3A). In both systems, after primary aggregation, cells multiply about 10-fold *in vitro* and then start to undergo histogenesis, including the formation of all main nuclear and plexiform layers (Layer and Willbold, 1989). Noticeably in the latter system—which we call stratospheroids—the layers develop in a correct orientation when pigmented cells from the eye margin are added (Vollmer *et al.*, 1984; Wolburg *et al.*, 1991).

The expression and the cellular localization of cholinesterases in these spheroids follows patterns almost identical to those observed *in vivo* (Layer *et al.*, 1992). AChE is found in the inner half of the inner nuclear layer and in ganglion cells. Again, the formation of synaptic sublaminae within the inner plexiform layer is indicated by weak BChE and then strong AChE staining (see Fig. 3C,D).

3. In order to follow neurite growth patterns, we used the so-called stripe assay method (Walter *et al.*, 1987). In this assay, embryonic retinas are cut into strips and the outgrowing axons of the ganglion cells are allowed to grow on strips pretreated with suitable substrates.

B. Downregulation of Butyrylcholinesterase during in Vitro Regeneration of Chicken Retina

In higher vertebrates such as birds or mammals, the capacity for regeneration of the central nervous system is often restricted to certain parts or to specific embryonic stages (Egar and Singer, 1972; Goss, 1974; Anderson and Waxman, 1981, 1983; Cowan and Finger, 1982). The neural retina—after injury or complete removal—exhibits a more-or-less pronounced capacity to regenerate (Orts-Llorca and Genis-Galvez, 1960; Stroeva, 1960; Coulombre and Coulombre, 1965, 1970; Park and Hollenberg, 1989; Willbold and Layer, 1992b). In the chick, retinal regeneration can occur in at least two different ways. It can occur in the center part of the eye either

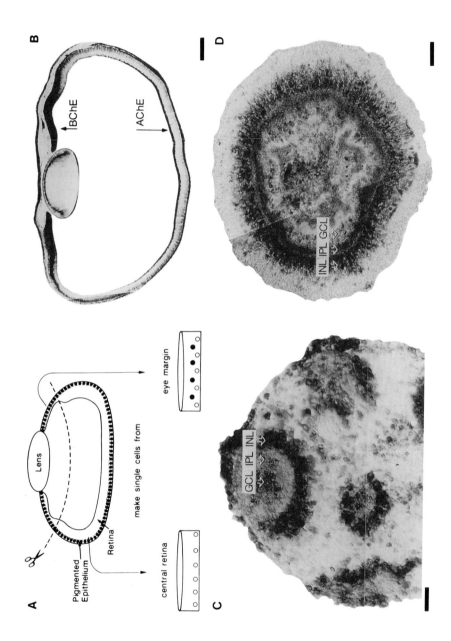

A

Lens

Pigmented
Epithelium

Retina

make single cells from

eye margin

central retina

B

BChE

AChE

C

GCL IPL INL

D

INL IPL GCL

by transdifferentiation from adjacent cells of the retinal pigmented epithelium or by proliferation of stem cells located in the eye margin. The ability to regenerate a retina is lost around embryonic days 4 or 5 (McKeehan, 1961; Coulombre and Coulombre, 1965).

Cells of the eye margin, which give rise to retinal regenerates, are distinctly different from the cells in the central parts of the eye; for example, their ability to take up [³H]dihydroxyphenylalanine (DOPA) or [³H]thymidine differs (for a review see Stroeva and Mitashov, 1983), as well as the distribution of Na^+/K^+ pumps (Burke et al., 1991). Interestingly, this eye region, which fosters cells with regenerative capacity, expresses high amounts of BChE over most of the embryonic period (Layer, 1983; Willbold and Layer, 1992b) (see Fig. 3B, Fig. 4). This is not surprising since the transition of mitotic cells into a differentiated state may be accompanied by BChE expression shortly before and possibly during the transition itself. When cells from this eye region are transferred into rotation culture, BChE is downregulated to low levels over a period of 7 days (see Fig. 5). Concomitantly, the expression of AChE increases and reflects the continuing differentiation of the retinal regenerate.

These data demonstrate that the change from BChE to AChE expression is halted in the in vivo eye margin. We have suggested that the BChE-producing cells represent a population of stem cells that have been arrested before undergoing their final division. Similar to other BChE-producing cell populations with mitogenic capacity (Dubovy and Haninec, 1990; Layer, 1991a), the high BChE activity in cells of the ciliary margin may thus delineate the ciliary stem cell population that retains some capacity to regenerate under appropriate in vivo (after retinectomy) or in vitro conditions (formation of stratospheroids; see Fig. 6). Only then will

---

FIG. 3   (A). Scheme showing the location of the cell populations responsible for the production of histotypic regenerates (so-called retinospheroids). When kept in rotation culture, E6 chicken cells of the central retina form so-called rosetted spheroids (left; see also C); cells from the eye margin, including neuroepithelial and pigmented cells, form so-called stratospheroids (right; see also D). (B). A frozen section of an E6 chicken eye stained for both AChE and BChE. Eyes from this stage provide the starting tissues for the production of retinospheroids. Note that the transmitotic marker BChE heavily labels the eye margin, whereas the postmitotic marker AChE is first expressed in ganglion cells along the inner surface of the retina. (C, D). Frozen sections of typical examples of rosetted spheroids (C) and of stratospheroids (D) after 14 days in culture. Their histotypic organization is revealed by AChE staining: Similar to an E10 retina, cells of the inner half of the inner nuclear layer (INL) and of the ganglion cell layer (GCL) plus a subband of the inner plexiform layer (IPL) are stained in stratospheroids (D). In rosetted spheroids (C), the corresponding AChE-positive cells of the inner half of the inner nuclear layer surround an inner plexiform layer-like circular space which includes single stained ganglion cells as well as unorganized material. Bars = 250 μm (B), 25 μm (C), 50 μm (D). (From Willbold and Layer, 1992b.)

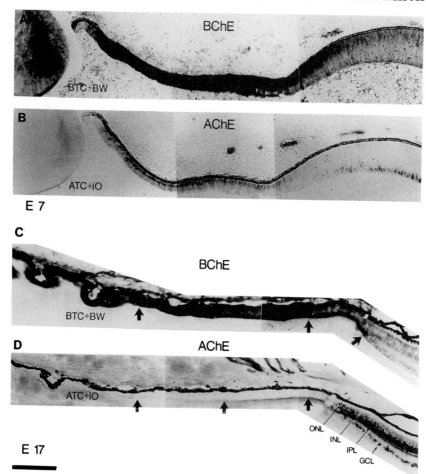

FIG. 4   High BChE and low AChE activity along the ciliary margin of chicken eyes over an extended embryonic period, indicative of its nondifferentiated state. Comparative histochemical staining of the ciliary margins of an E7 (A, B) and an E17 eye (C, D) specifically for BChE (A, C) and AChE (B, D). Parallel frozen sections of each eye were incubated according to the Karnovsky-Roots technique modified by Kugler (Karnovsky and Roots, 1964; Kugler, 1987). The solutions contained butyrylthiocholine (A, C) plus BW284C51 (a specific inhibitor for AChE) or acetylthiocholine (B, D) plus iso-OMPA (a specific inhibitor for BChE), respectively. The arrows in C delineate high BChE activity along the basal membrane of the marginal neuroepithelium. In the ciliary margin, a minimal amount of AChE is detectable at E7. Note the total absence of AChE in the ciliary margin at E17 (arrows in D) and the typical staining pattern of AChE within the functional retina. AChE is localized to cell bodies in the inner half of the inner nuclear layer and in specific sublaminae of the inner plexiform layer. Reaction times were 10 hr at 37° C and overnight at room temperature for BChE (A, C); 4 hr at 37° C for AChE (B, D). Lens is to the left. Bar = 200 μm. (From Willbold and Layer, 1992b.)

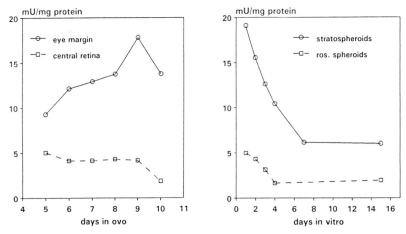

FIG. 5   BChE is high in cells of the ciliary margin (calculated on protein basis, left), but is downregulated after transfer into rotation culture (right), paralleling the accelerating phase of its cell proliferation (cf. Fig. 6). Only in the eye margin does BChE activity remain high over a long period of *in vivo* retinogenesis (left), which is indicative of cells locked between proliferation and differentiation. When brought into *in vitro* conditions, BChE drops within the first 7 days. The time courses of quantitative development [measured according to the method of Ellman *et al.* (1961)] of specific BChE activities for homogenates from central retina and eye margin (left) are compared with their *in vitro* counterparts (rosetted spheroids and stratospheroids, right). (According to Willbold and Layer, 1992b.)

they go through their final cell divisions, to subsequently downregulate their BChE and differentiate into a new retinal structure.

C.   Butyrylcholinesterase and Rate of Cell Proliferation

We determined the action of BChE on cell proliferation by adding either purified BChE from horse serum, or alternatively, specific anticholinesterases to monolayer cultures of retinal and tectal single cells. A highly purified fraction of BChE causes a 2 to 3-fold increase of [³H]thymidine uptake/mg protein in a concentration-dependent manner. Pretreatment of BChE by diisopropylfluorophosphate (DFP) reduces the esteratic activity to less than 1%; concomitantly, the stimulatory potential of this inactivated enzyme is drastically reduced. This indicates that most of the stimulatory effect is due to the esterolytic enzyme activity of BChE (Layer, 1991b).

  Specific inhibitors of both classes of cholinesterases applied with the medium affect cell proliferation accordingly. General anticholinesterases, including phenylmethylsulfonylfluoride, methylsulfonylfluoride, eserine,

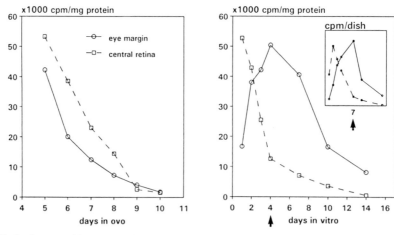

FIG. 6   Strong self-renewal of cells from the eye margin *in vitro*, as demonstrated by a prolonged period of cell proliferation (right). Time courses of [³H]thymidine uptake (calculated on protein basis) into cells from the central retina and the eye margin, respectively, are compared in normal embryonic retinas (left) with their dissociated E6 cells, which form retinospheroids under *in vitro* conditions (right). At the appropriate stages 30 μCi [³H]thymidine plus 20 μl 10⁻² unlabeled 2'-deoxythymidine were either injected into the eye or added to the culture medium. Embryos and culture dishes were then further incubated for 16 hr. Insert in right panel shows the thymidine uptake data plotted as cpm/dish. Note that uptake rates increase only for 2 days in rosetted spheroids, but 7 days in stratospheroids. (According to Willbold and Layer, 1992b.)

and neostigmine all inhibit [³H]thymidine uptake by about 20–40%. In the 5–10 μM range, specific inhibitors of BChE, for example, ethopropazine and tetraisopropylpyrophosphoramide (iso-OMPA), diminish the uptake rate, again indicating a stimulatory function for the system's inherent BChE. In contrast, BW284C51—a specific inhibitor of AChE—shows no effect on cell proliferation. From our short term experiments it is not clear whether BChE acts as a growth factor, whether it releases a growth factor by acting on some unknown substrate, or whether it speeds up the last cell cycle.

D. Butyrylcholinesterase Activity and Acetylcholinesterase Expression

An important and crucial question is the relationship between BChE and AChE. On the one hand there exists no constant quantitative relationship between the two enzymes and so far no physiological function for BChE is apparent. The absence of BChE activity in adults has no obvious effects.

On the other hand, some kind of interrelationship between the two enzymes does exist, possibly originating from their evolutionary relationship. A temporal and regulatory interdependence of AChE and BChE first noticed by Koelle *et al.* is becoming more and more evident (Koelle *et al.*, 1976, 1977a,b, 1979a,b). We have been able to show that BChE has a regulatory role in AChE expression. Adding the irreversible BChE inhibitor, iso-OMPA, to reaggregating retinal cells in rotary culture not only leads to a full inhibition of BChE, but unexpectedly, iso-OMPA also suppresses the expression of AChE by 35–60%. The diminished AChE activity is due to a reduced expression of AChE in fiber rich areas (Layer *et al.*, 1992) (see Fig. 7).

Control   +iso–OMPA

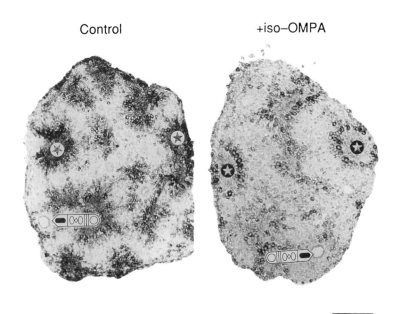

FIG. 7   The BChE inhibitor iso-OMPA downregulates AChE expression in plexiform areas (marked by stars) of retinospheroids, indicating a regulatory role for BChE in AChE expression. Two frozen sections from control (left) and iso-OMPA-treated retinospheroids (right; both 10 days in culture) have been stained for AChE activity. Note the strong and typical staining pattern in the control: a ring of cells corresponding to the AChE positive cells in the inner half of the inner nuclear layer plus the corresponding inner plexiform layer-like internal areas are AChE positive. Note that staining in these neurite-containing areas is drastically suppressed in the presence of iso-OMPA (right, stars). The pictogram indicates the sequence of retinal cell types with their inverted laminar arangement in retinospheroids: black oval, photoreceptor; single line, outer plexiform layer (OPL); triple symbol, horizontal, bipolar and amacrine cells of inner nuclear layer (INL); triple line, inner plexiform layer (IPL); circle, ganglion cell. Bar = 100 μm. (From Layer *et al.*, 1992.)

Even more pronounced, the release of AChE into the media of stationary tectal cell cultures is inhibited by iso-OMPA and reduced to less than
20% (see Fig. 8). This effect is not due to a direct cross-inhibition of AChE
by iso-OMPA. In contrast, treatment of retinal reaggregates with the
reversible AChE inhibitor, BW284C51, slightly stimulates the expression
of AChE (during the viable period of the aggregates).

We have concluded that the cellular expression of AChE is regulated by
both the amount of active BChE and active AChE within neuronal tissues.
This finding is most remarkable considering that BChE is expressed during
the time of final mitosis. Does BChE itself trigger the expression of AChE,
which is so strictly associated with the differentiated state of cells? If
BChE triggers the final transition of mitotic cells into the postmitotic
compartment, then this enzyme would indeed play a key role in neurogenesis.

E. Cholinesterases as Cell Adhesion Molecules

AChE is expressed shortly before neurites extend. Do cholinesterases
somehow affect neurite growth? In this section we describe (1) a close

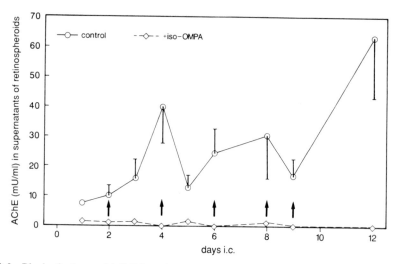

FIG. 8   Block of release of AChE into the culture medium by the BChE inhibitor iso-OMPA
(cf. Fig. 7), indicating a regulatory relationship. Supernatants from four separate plates were
determined in duplicate according to the Ellman method. Note that the activities before a
change of media (indicated by arrows) are high due to accumulation. Standard deviation is
indicated. A direct inhibition of AChE by iso-OMPA at the given concentration range is not
detectable (data not shown). (According to Layer et al., 1992.)

association of the expression of cholinesterases with that of the HNK-1 epitope, a glycopart typically found on cell adhesion molecules, and (2) then show that anticholinesterase drugs change neurite morphology under *in vitro* conditions.

1. Developmental Regulation of the HNK-1 Epitope during Formation of Cranial Nerves

The expression of cholinesterases during the establishment of the peripheral nervous system is more complex than that for the neural tube. Soon after closure of the neural tube, the cells of the neural crest begin to migrate to specific positions of the body to give rise to a variety of different structures, for example, the pigmented cells of the epidermis; the medullary cells of the adrenal gland; or the neurons and supporting glial cells of the sensory, sympathetic, and parasympathetic nervous systems.

The cranial nerves are complex systems in terms of their spatial routes and their contributing cell lineages. Both the motor roots and the sensory roots of cranial nerves, namely, the trigeminal nerve (V), facialis nerve (VII), glossopharyngeal nerve (IX), and vagus nerve (X), are coarranged in an orderly manner with specific subdivisions of the early hindbrain called rhombomeres (Gräper, 1913; Lumsden and Keynes, 1989; Keynes *et al.,* 1990; Layer and Alber, 1990; Lumsden, 1990). Their sensory cells, which are located within cranial ganglia and which receive sensory input from placodal cells, are derived from neural crest cells, whereas cells within the distal ganglia are supplied by the placodes (for a review see Noden 1988). An exception is seen in the trigeminal ganglion, which also contains placodal cells in its proximal part (Kupffer, 1906; Johnston, 1966; Narayanan and Narayanan, 1980).

Interestingly, about 90% of all migrating neural crest cells express AChE (Drews, 1975) (see Fig. 9). However, the usefulness of this activity as a marker has been questioned since not all neural crest cells express the enzyme (Cochard and Coltey, 1983), possibly because of their different states of differentiation. On the assumption that it is an early and more reliable marker, the expression of the HNK-1 epitope has been introduced for marking neural crest cells (Rickman *et al.,* 1985; Bronner-Fraser, 1986).

We directly compared cholinesterase and HNK-1 expression through double-labeling studies. During the development of cranial nerves V–X in the hindbrain of chick and quail, we could show that the neural crest cells transiently express AChE activity during their emigration. The closer they come to their target fields, the more the enzyme decreases. These target areas as well as the migratory tracks in the cranial mesenchyme show elevated BChE activity, foreshadowing the course of later projecting

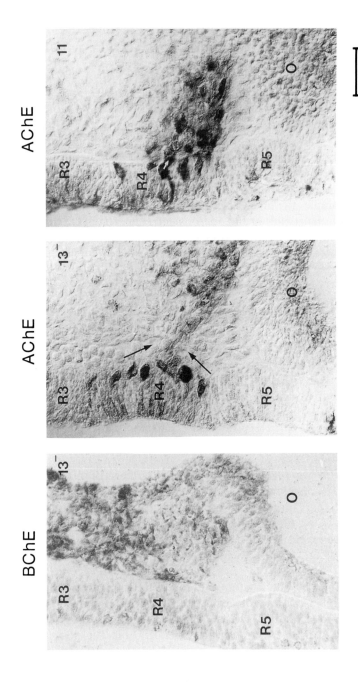

FIG. 9 When neural crest cells emigrate from the neural tube, they express high amounts of AChE (middle, right). At the same time BChE is present in the surrounding mesenchyme and on ectodermal placodes (left). Frozen sections of HH 11 and HH13⁻ chick hindbrains are shown. The small white arrow (right) marks a cell just emigrating from the outer edge of the tube from rhombomere R4. Note the absence of BChE in the neural tube and its expression in mesenchymal and placodal cells (left) adjacent to the rhombomeres R3 and R4. AChE-positive cells in rhombomere R4 (middle, arrows indicate their point of emigration from the tube) have just reached the mantle surface. Some of them will leave the neuroepithelium on defined tracks. O, otocyst. Bar = 50 μm. (From Layer and Kaulich, 1991.)

cranial nerves. BChE increases strongly in cells that may represent immature glial Schwann cells (see Fig. 10). Glial cells are supposed to play important roles as guiding and support structures. They are often characterized by high amounts of BChE (Barth and Ghandour, 1983).

As AChE decreases, and before neurites begin to extend, an HNK-1-positive scaffold of extracellular matrix becomes established. At present we cannot decide whether the HNK-1 epitope is expressed on AChE-and/or on BChE-positive cells. Since this epitope might be an independent mediator of adhesive function (Riopelle *et al.*, 1986; Cole and Schachner, 1987; Künemund *et al.*, 1988), its expression could support the aggregation and stabilization of neuronal contacts. Concomitantly, tunnel-shaped HNK-1 matrices are established within which cranial neurites begin to extend (see Fig. 11).

These results suggest that HNK-1 may be spatiotemporally regulated on cholinesterase-positive cells of neural crest origin. It seems most likely that cholinesterases with and without the HNK-1 epitope serve different functions, since AChE is related to their migration, and HNK-1 to their aggregation and to the formation of an extracellular scaffold. As another example of the sensitive regulation of HNK-1 on cholinesterases, it is worthwhile mentioning that both AChE and BChE, but not HNK-1, are expressed in the ectodermal placodes (Layer and Kaulich, 1991).

Cholinesterases are not the only molecules that precede neurite growth, but their expression occurs very early. HNK-1-positive neural crest cells (Bronner-Fraser, 1986), cytotactin (a cell–substrate adhesion molecule) (Grumet *et al.*, 1985; Hoffman *et al.*, 1988), tenascin (a mesenchymal–extracellular matrix glycoprotein) (Mackie *et al.*, 1988; Chiquet-Ehrismann *et al.*, 1986; Aufderheide *et al.*, 1987; Chiquet, 1989); or molecules such as laminin, fibronectin, N-CAM, or N-cadherin (for a comprehensive review, see Keynes *et al.*, 1990) are found in the rostral sclerotome of the trunk, whereas the peanut-agglutinin (PNA)-lectin binding protein is expressed in the caudal part of the sclerotome (Stern *et al.*, 1986). This asymmetric distribution of molecules that normally mediate cell–cell or cell–matrix interactions seems to be indispensable for directing growing motor axons and for establishing proper synaptic contacts in the target regions.

BChE can be detected earlier than any of the other molecules mentioned. In fact, we have seen cases where the future somite within the mesodermal segmental plate which has not yet split off shows elevated BChE asymmetry (unpublished observations). One may speculate that the expression of BChE must represent some valuable information. Because BChE is considered to be an evolutionarily "old" and relatively unspecific molecule with presumably more than one catalytic function, one could imagine that it originally served multiple functions. Among other

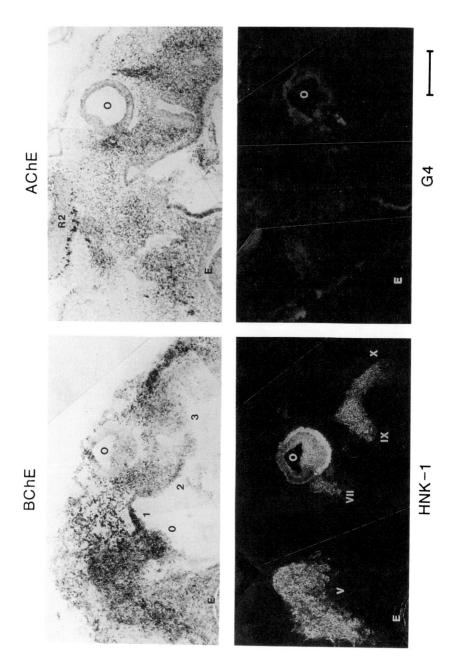

AChE

BChE

G4

HNK−1

processes, it could have been involved in mediating cell–cell or cell–matrix contacts. During subsequent evolution this task was taken over by more specialized molecules.

2. Cholinesterases as Cues for Pathfinding and Targeting Outgrowing Axons

Outgrowing cranial nerves represent a highly dynamic system in which cholinesterases are expressed according to a stringent spatiotemporal pattern: The migrating cells and nuclei from which long efferent fibers originate express high amounts of AChE, whereas their target areas and the future fiber tracts are foreshadowed by high amounts of BChE (see Fig. 12). Sometimes the target areas are not characterized by high BChE but by high (sometimes transient) AChE activity.

This scheme seems to be a general one during the establishment of specific nerve connections because examples from other systems reveal highly comparable patterns. Weikert *et al.* (1990) could show for the entire embryonic chick brain that the formation of long efferent fiber tracts [detected by the fiber-specific G4 antibody (Rathjen *et al.*, 1987)] is preceded by the expression of AChE in the originating cells (Weikert *et al.*, 1990; see also Wilson *et al.*, 1990). Comparable patterns have been detected in other species. In the lizard *Uromastix hardwickii,* all nuclei show intense AChE activity, those of the hypothalamus exclusively, whereas the enzymatic activity in the fiber tracts is totally negative for AChE and intensively positive for BChE (Sethi and Tewari, 1976). In rats, in the ventrobasal complex, the onset of AChE activity precedes ingrowth of extrinsic afferents (Kristt, 1989), and developing primary sensory thalamocortical neurons transiently express AChE and transport it via their thalamocortical projections to the target areas in the sensory cortex (Robertson *et al.*, 1988, Robertson, 1991). In cats, afferents from the

---

FIG. 10   Neurite growth is a late process preceded by BChE, AChE, and HNK-1 expression. AChE-positive cells (top right) migrate within BChE-positive areas (top left). This migration precedes the formation of an HNK-1-positive extracellular matrix (bottom left). Here parasagittal parallel frozen sections of the cranial periphery of a stage HH 18 chicken embryo are shown. The formation of the cranial nerve is revealed by the G4 antibody [specific for long projecting neurites; Rathjen *et al.* (1987)] and has just started in nerves V, VII, and VIII (bottom right). (V, trigeminal nerve; VII, facioacoustic nerve; IX, glossopharyngeal nerve; X, vagus nerve). Note staining of BChE in mesenchymal and placodal areas, within which AChE-producing cells are migrating on more restricted tracks. Spatially, extracellular HNK-1 expression coincides largely with BChE activity. The future branching of the major nerves is foreshadowed by all three markers at this stage (arrows). Branchial arches (placodes stained strongly by BChE, and weakly by AChE and G4) are numbered 0–3: 0, maxillary process; 1, mandibular arch; 2, hyoid arch; 3, branchial arch III; E, eye; O, otocyst. Bar = 200 μm. (From Layer and Kaulich, 1991.)

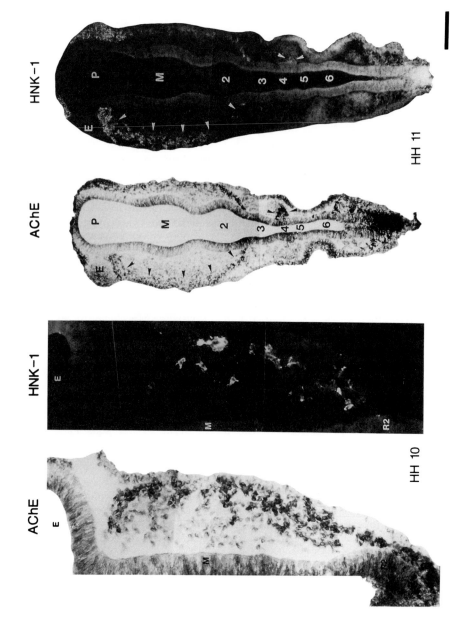

HNK-1

AChE

P

M

2

3

4

5

6

E

HH 11

HNK-1

E

M

R2

AChE

E

M

R2

HH 10

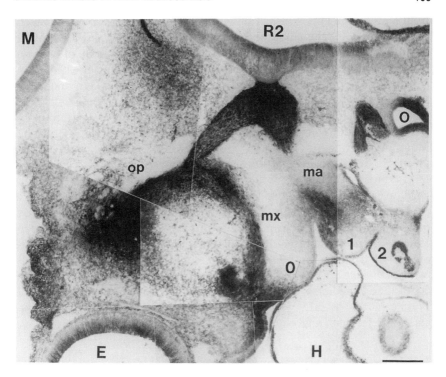

FIG. 12 Cells expressing BChE outline the pathway of growing peripheral nerves. Here BChE staining of the trigeminal nerve branches is revealed on a frozen sagittal section of a stage HH 21 quail embryo. Note that the maxillary placode (near O) also expresses large amounts of BChE. E, eye; H, heart; O, otocyst; M, mesencephalon; R2, rhombomere 2; op, ophthalmic branch; mx, maxillar branch; ma, mandibular branch of trigeminal nerve; 0, maxillar arch; 1, mandibular arch; 2, hyoid arch. Bar = 100 μm. (From Layer and Kaulich, 1991.)

FIG. 11 Coexpression of AChE and HNK-1. Note that neural crest cells *first* express AChE activity, and *then* the HNK-1 epitope as they approach their destination fields. Here the cephalic region of an HH 10⁻ and an HH 11 embryo is shown. *Left*. A population of AChE-positive cells migrates from rhombomere R2 to the vicinity of the eye stalk (E). The expression of HNK-1 just begins. *Right*. Similar sections at lower magnification. Note that with the advancing developmental state of the cells, expression of AChE decreases and that of HNK-1 increases. The migratory pathways are marked by arrows. P, prosencephalon; M, mesencephalon; 2–6, numbers of rhombomeres; O, otocyst. Bar = 50 μm (left pair of figures), 210 μm (right pair of figures). (According to Layer and Kaulich, 1991.)

frontal cortex and the substantia nigra project onto AChE-rich patches of the superior colliculus (Illing and Graybiel, 1985), and in *Drosophila* AChE is present in retinula cells during axonal growth and formation of terminations within the developing optic lobes (Wolfgang and Forte, 1989).

During development of the segmented trunk motor system of the chick, the motoneuronal cell bodies of the neural tube and the myotomes express AChE in a rostrocaudal wave, whereas BChE-positive cells are found in the early dermomyotome and within the rostral half of the sclerotome. Motor axons begin to grow out from the AChE-positive motoneurons through a BChE-positive sclerotomal space toward the anlage of the myotome (Layer *et al.*, 1988a). Could BChE here act as target-secreted factor? Data from other systems support this hypothesis. In *Tupaia*, BChE is expressed diffusely throughout the lateral geniculate nucleus and its expression is independent from retinal input (Horn and Carey, 1988). In chicks, *in vitro* studies showed that the density and complexity of fiber outgrowth from the metencephalic trigeminal (V) motor nucleus is significantly enhanced when the cells are kept in the appropriate jaw-muscle conditioned medium, which is their normal target tissue (Heaton and Wayne, 1986). BChE, which is often a secreted protein, is a possible candidate for such a growth-promoting and fiber attracting factor.

3. Alteration of Neurite Growth by Anticholinesterases

The formation of the nervous system is a complex process governed by cell proliferation, differentiation, and migration, as well as the establishment and stabilization of complex networks of neuronal connections. These contacts are mediated through different mechanisms by a variety of molecules present in the extracellular matrix, or on the membranes of neuronal or glial cells.

One important group of such molecules consists of the so-called cell adhesion molecules (CAM). There are three major groups of these molecules which are localized on axonal surfaces: the cadherins, the integrins, and the members of the immunoglobulin superfamily (for a comprehensive review see Rathjen, 1991). In particular, fasciculation molecules such as N-CAM, L1, G4, and TAG-1 are indispensable for the proper outgrowth of neurites and for keeping them together within organized fiber bundles. However, cell adhesion-mediating molecules seem to have more complex biological roles than simply mediating cell adhesion. Some of them are coexpressed, indicating a high redundancy and overlapping scopes of action.

What is the evidence that cholinesterases may be involved in cell–cell or cell–matrix communication? First, cholinesterases share sequence homologies with the adhesion molecules neurotactin and glutactin from *Dro-*

sophila (Barthalay et al., 1990; De la Escalera et al., 1990; Krejci et al., 1991). Another widespread feature of cell adhesion molecules is their modification by covalently linked carbohydrates. Carbohydrates are powerful and probably the original mediators of cell recognition. Membrane-bound carbohydrates are the linkage targets for many hormones, toxins, viruses, bacteria, or cells (for review see Sharon and Lis, 1993).

Among the cell adhesion molecules of the immunoglobulin superfamily, the so-called L2/HNK-1 sugar epitope (Kruse et al., 1984; Chou et al., 1986; Tucker et al., 1988; Linnemann and Bock, 1989) seems to be a specific marker. There is evidence that, at least in particular cases, this epitope itself is an independent mediator of adhesive function (Riopelle et al., 1986; Cole and Schachner, 1987; Künemund et al., 1988). Therefore a hypothesis has been put forward suggesting that the mere presence of this epitope on a glycoprotein may indicate its adhesive nature. AChE from Torpedo (Bon et al., 1987) and highly purified BChE from chicken embryonic brain (Treskatis et al., 1992; unpublished data) bear the HNK-1 epitope, whereas BChE from chicken serum shows only weak expression or none (Weikert, 1993).

In a similar fashion, different monoclonal antibodies that recognize specific N-linked carbohydrates originally raised against AChE from Torpedo (Liao et al., 1991, 1992) bind to muscle and brain AChE from different fishes and mammals but not to their erythrocyte AChE. This indicates that the recognized carbohydrate epitopes are highly conserved during vertebrate evolution. Moreover, because the different forms of AChE arise from one gene by differential splicing, AChE at least undergoes different posttranslational modifications. Additional data support such a hypothesis of the modification and modulation of the action of cholinesterase molecules by different residues. After sialic acids were removed from neuronal and glial cells in culture, increased AChE and BChE activity was observed (Stefanovic et al., 1975). Although it is not clear whether the activity of cholinesterases can be trapped or regulated by a direct linkage of sialic acid to specific parts of the AChE molecule, this systems resembles the modification of adhesion molecules by sialic acid during embryogenesis. N-CAM, a widespread cell adhesion molecule in the nervous system, also carries this moiety. Enzymatic removal of sialic acid caused an increase in the aggregation of membrane vesicles and in the thickness of neurite bundles, indicating that sialic acid regulates the action of N-CAM (Rutishauser et al., 1985).

All these data support the hypothesis that cholinesterases can have properties similar to cell adhesion molecules and that they may be involved in processes of cell-cell and cell–environment contacts as well as in regulating and guiding neurite growth. However, it is not clear whether this hypothetical property is mediated by the esteratic function, that is, by the active site of the molecule, by other noncholinergic sites, or by their carbohydrate calyx.

Recently, we provided the first direct evidence showing that neurite growth *in vitro* from various neuronal tissues of the chicken embryo can be modified by some anticholinesterase agents. By growing axonal processes from retinal explant cultures on striped laminin carpets (Walter *et al.,* 1987), we observed definite morphological changes, such as defasciculation of neurite bundles by BW284C51 and bambuterol. Since both drugs are thought to block exclusively AChE and BChE, respectively, these results demonstrate that both cholinesterases can affect adhesive mechanisms (Fig. 13). Moreover, we quantified these effects by measuring the expression of the neurite-specific G4 antigen in tectal cell cultures in the presence of increasing concentrations of various anticholinesterase drugs. The effects were then compared directly with their cholinesterase inhibitory action. BW284C51 and ethopropazine, which inhibit AChE and BChE, respectively, strongly decrease neurite growth in a dose-dependent manner. It surprised us when we found that echothiopate, an agent that inhibits both cholinesterases, does not change neuritic growth (Layer *et al.,* 1993). Interestingly, similar effects were also obtained by treating retinospheroids with anticholinesterases. Both iso-OMPA and BW284C51 have definite morphological and biochemical effects on histogenesis (Layer *et al.,* 1992; unpublished data), while treatment of spheroids with echothiopate has no obvious effects on viabilty and histogenesis (Willbold, 1991; unpublished data).

How can these seemingly contradictory results be reconciled? These data strongly suggest that (1) both cholinesterases can regulate axonal growth, and (2) the regulation is not the result of enzyme activity per se, since at least one drug was found that inhibits all cholinesterase activities but not neurite growth. Therefore, a secondary site on cholinesterase molecules must be responsible for adhesive functions. This result then draws our attention away from the active site of cholinesterases to molecular territories that are still unknown.

V. Conclusions

Taken together, the results presented have shown that cholinesterases are expressed in specific spatial and temporal patterns and that they may have important functions during early neurogenesis:

> AChE and BChE show a high degree of homology, yet both enzymes are genetically independent proteins coded for by different genes located on different chromosomes.

+ iso-OMPA   Control   + BW 284C51

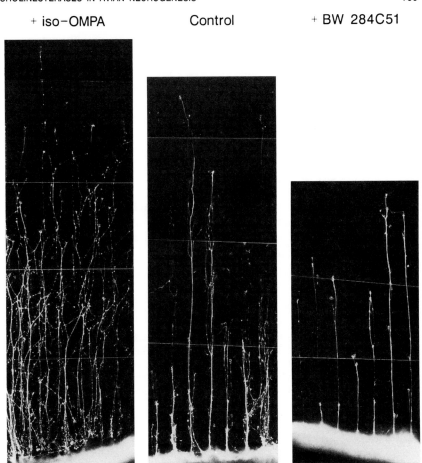

FIG. 13   Anticholinesterases change the morphology of retinal explant neurites when grown on the laminin stripe assay (Walter *et al.*, 1987). Strips of E6 retinae cut in a naso-temporal orientation (explants are visible at the lower edge of the micrographs) are explanted on coverslips that present 40 to 60 μm-wide parallel laminin strips in a perpendicular direction. Specific inhibitors of AChE (BW284C51, 50 μM, right) and of BChE (iso-OMPA, 100 μM, left) are added; cultures are shown after 1 day. Neurites are visualized by binding of the G4 antibody plus rhodamine–isothiocyanate–coupled second antibody. Note that BW284C51 leads to shorter and thinner neurite bundles. Iso-OMPA slightly increases neurite growth with longer and less organized bundles. Bar = 150 μm. (From Layer *et al.*, 1993.)

Due to different glycosylation patterns, BChE isolated from brain is smaller than that from serum. In younger nervous tissues, the membrane-associated and small molecular forms prevail.

BChE is mainly expressed during the transition from proliferation to differentiation and is possibly involved in activating the expression of AChE. Accordingly, BChE can interfere with the rate of cell proliferation.

Migrating cells with efferent fibers often express high amounts of AChE, whereas their target areas and the future fiber tracts are foreshadowed by high amounts of BChE or AChE.

Cholinesterases can bear the cell adhesion molecule-specific epitope HNK-1. Cholinesterases are possibly involved in the pathfinding and targeting of outgrowing axons.

In vitro, both enzymes can regulate neurite growth, probably by means of a nonenzymatic mechanism.

The implications of these findings are far-reaching. Not only developmental aspects of the cholinergic system, but most likely aspects of their role in neurological diseases may appear in a new light. Indeed, new frontiers are open for cholinesterase research.

Acknowledgments

We wish to thank our colleagues R. Alber, Ch. Ebert, G. Fischer von Mollard, H. Frank, R. Girgert, S. Kaulich, S. Kotz, L. Liu, P. Mansky, F. G. Rathjen, S. Rommel, O. Sporns, S. Treskatis, G. Vollmer, and T. Weikert, who have contributed to the progress of this project. We also thank A. Gierer, A. S. Gordon, F. Hucho, H. Meinhardt, and L. Puelles for many helpful discussions and R. Groemke-Lutz for photolaboratory work.

References

Adler, M., and Filbert, M. G. (1990). Role of butyrylcholinesterase in canine tracheal smooth muscle function. *FEBS Lett.* **267,** 107–110.

Allemand, P., Bon, S., Massoulié, J., and Vigny, M. (1981). The quaternary structure of chicken acetylcholinesterase and butyrylcholinesterase; effect of collagenase and trypsin. *J. Neurochem.* **36,** 860–867.

Allerdice, P. W., Gardner, H. A. R., Galutira, D., Lockridge, O., La Du, B. N., and McAlpine, P. J. (1991). The cloned butyrylcholinesterase (BChE) gene maps to a single chromosome site, 3q26. *Genomics* **11,** 452–454.

Anderson, M. J., and Waxman, S. G. (1981). Morphology of regenerated spinal cord in *Sternachus albifrons. Cell Tissue Res.* **219,** 1–8.

Anderson, M. J., and Waxman, S. G. (1983). Regeneration of spinal neurons in inframammalian vertebrates: Morphological and developmental aspects. *J. Hirnforsch.* **24,** 371–398.

Arpagaus, M., Kott, M., Vatsis, K. P., Bartels, C. F., La Du, B. N., and Lockridge, O. (1990). Structure of the gene for human butyrylcholinesterase. Evidence for a single copy. *Biochemistry* **29,** 124–131.

Atack, J. R., Perry, E. K., Perry, R. H., Wilson, I. D., Bober, M. J., Blessed, G., and Tomlinson, B. E. (1985). Blood acetyl- and butyrylcholinesterases in senile dementia of Alzheimer type. *J. Neurol. Sci.* **70,** 1–12.

Atack, J. R., Perry, E. K., Bonham, J. R., Candy, J. M., and Perry, R. H. (1986). Molecular forms of acetylcholinesterase and butyrylcholinesterase in the aged human central nervous system. *J. Neurochem.* **47**, 263–277.

Aufderheide, E., Chiquet-Ehrismann, R., and Ekblom, P. (1987). Epithelial-mesenchymal interactions in the developing kidney lead to the expression of tenascin in the mesenchyme. *J. Cell Biol.* **105**, 599–608.

Balasubramanian, A. S. (1984). Have cholinesterases more than one function. *TINS* **7**, 467–468.

Barth, F., and Ghandour, M. S. (1983). Cellular localisation of butyrylcholinesterase in adult rat cerebellum determined by immunofluorescence. *Neurosci. Lett.* **39**, 149–153.

Barthalay, Y., Hipeau-Jacquotte, R., De la Escalera, S., Jimenez, F., and Piovant, M. (1990). *Drosophila* neurotactin mediates heterophilic cell adhesion. *EMBO J.* **9**, 3603–3609.

Bennett, R. S., and Bennett, J. K. (1991). Age-dependent changes in activity of mallard plasma cholinesterases. *J. Wild. Dis.* **27**, 116–118.

Berman, H. A., Decker, M. M., and Sangmee, J. (1987). Reciprocal regulation of acetyl-cholinesterase and butyrylcholinesterase in mammalian skeletal muscle. *Dev. Biol.* **120**, 154–161.

Bon, S., Méflah, K., Musset, F., Grassi, J., and Massoulié, J. (1987). An immunoglobulin M monoclonal antibody, recognizing a subset of acetycholinesterase molecules from electric organ of *Electrophorus* and *Torpedo,* belongs to the HNK-1 anti-carbohydrate family. *J. Neurochem.* **49**, 1720–1731.

Boopathy, R., and Balasubramanian, A. S. (1987). A peptidase activity exhibited by human serum pseudocholinesterase. *Eur. J. Biochem.* **162**, 191–197.

Brodbeck, U., and Liao, J. (1992). Subunit assembly and glycosylation of mammalian brain acetylcholinesterase. *In* "Multidisciplinary Approaches to Cholinesterase Functions" (A. Shafferman and B. Velan, eds.), pp. 33–38. Plenum Publishing, New York and London.

Bronner-Fraser, M. (1986). Analysis of the early stages of trunk neural crest migration in avian embryos using monoclonal antibody HNK-1. *Dev. Biol.* **115**, 44–55.

Burke, J. M., Mckay, B. S., and Jaffe, G. J. (1991). Retinal-pigment epithelial-cells of the posterior pole have fewer Na/K adenosine-triphosphatase pumps than peripheral cells. *Invest. Ophthalmol. Vis. Sci.* **32**, 2042–2046.

Buznikov, G. A. (1971). The role of nervous system mediators in individual development. *Ontogenez* **2**, 5–13.

Buznikov, G. A. (1990). Neurotransmitters in embryogenesis. *In* "Soviet Scientific Reviews, Supplement Series, Physiology and General Biology" (T. M. Turpaev, ed.). Trans. by A. Bastow. Harwood Academic Publishers, Chur, Switzerland.

Buznikov, G. A., Chudakova, I. W., and Zwezdina, N. D. (1964). The role of neurohumours in early embryogenesis. I. Serotonin content of developing embryos of sea urchin and loach. *J. Embryol. Exp. Morph.* **12**, 563–573.

Cantino, D., and Daneo, L. S. (1972). Synaptic junctions in the developing chick optic tectum. *Experientia* **29**, 85–87.

Chatonnet, A., and Lockridge, O. (1989). Comparison of butyrylcholinesterase and acetyl-cholinesterase. *Biochem. J.* **260**, 625–634.

Chatonnet, A., and Masson, P. (1985). Study of the peptidasic site of cholinesterase, prelimi-nary results. *FEBS Lett.* **182**, 493–498.

Chiquet, M. (1989). Tenascin/J1/Cytotactin: The potential function of hexabrachion proteins in neural development. *Dev. Neurosci.* **11**, 266–275.

Chiquet-Ehrismann, R., Mackie, E. L., Pearson, C. A., and Sakakura, T. (1986). Tenascin: An extracellular matrix protein involved in tissue interaction during fetal development and oncogenesis. *Cell* **47**, 131–139.

Chou, D. K. H., Ilyas, A. A., Evans, J. E., Costello, C., Quarles, R. H., and Jungalwala, F. B. (1986). Structure of sulfated glucuronyl glycolipids in the nervous system reacting with

HNK-1 antibody and some IgM paraproteins in neuropathy. *J. Biol. Chem.* **261,** 11717–11725.

Christie, D. L., Cleverly, D. R., and O'Connor, C. J. (1991). Human milk bile-salt stimulated lipase. Sequence similarity with rat lysophospholipase and homology with the active site region of cholinesterases. *FEBS Lett.* **278,** 190–194.

Chudinovskaia, N. V. (1978). Aktivnost' kholesterazy v perevivaemykh tkanevykh kul'turakh. *Arkh. Anat. Gistol. Embriol.* **74,** 93–98.

Cochard, P., and Coltey, P. (1983). Cholinergic traits in the neural crest: Acetylcholinesterase in crest cells of the chick embryo. *Dev. Biol.* **98,** 221–238.

Cole, G. J., and Schachner, M. (1987). Localization of the L2 monoclonal antibody binding site on chicken neural cell adhesion molecule (NCAM) and evidence for its role in N-CAM-mediated cell adhesion. *Neurosci. Lett.* **78,** 227–232.

Coulombre, J. L., and Coulombre, A. J. (1965). Regeneration of neural retina from pigmented epithelium in the chick embryo. *Dev. Biol.* **12,** 79–92.

Coulombre, J. L., and Coulombre, A. J. (1970). Influence of mouse neural retina on regeneration of chick neural retina from chick embryonic pigmented epithelium. *Nature (London)* **228,** 559–560.

Couraud, J. Y., and DiGiamberardino, L. (1980). Axonal transport of the molecular forms of acetylcholinesterase in chick sciatic nerve. *J. Neurochem.* **35,** 1035–1066.

Cowan, W. M., and Finger, T. E. (1982). Regeneration and regulation in the developing central nervous system, with special reference to the reconstitution of the optic tectum of the chick following removal of the mesencephalic alar plate. In "Neuronal Development" (N. C. Spitzer, ed.), pp. 377–415, Plenum Publishing, New York.

De la Escalera, S., Bockamp, E. O., Moya, F., Piovant, M., and Jiménez, F. (1990). Characterization and gene cloning of neurotactin, a *Drosophila* transmembrane protein related to cholinesterases. *EMBO J.* **9,** 3593–3601.

Doctor, B. P., Chapman, T. C., Christner, C. E., Deal, C. D., De-La-Hoz, D. M., Gentry, M. K., Ogert, R. A., Rush, R. S., Smyth, K. K., and Wolfe, A. D. (1990). Complete amino acid sequence of fetal bovine serum acetylcholinesterase and its comparison in various regions with other cholinesterases. *FEBS Lett.* **266,** 123–127.

Drews, U. (1975). Cholinesterase in embryonic development. *Progr. Histochem. Cytochem.* **7,** 1–53.

Drews, U., and Drews, U. (1973). Cholinesterase in der Extremitätenentwicklung des Hühnchens. II. Fermentaktivität und Bewegungsverhalten der präsumptiven Knorpelzellen *in vitro. Roux's Arch. Dev. Biol.* **173,** 208–227.

Dreyfus, P., Zevin-Sokin, D., Seidman, S., Prody, C., Zisling, R., Zakut, H., and Soreq, H. (1988). Cross-homologies and structural differences between human cholinesterases revealed by antibodies against cDNA-produced human butyrylcholinesterase peptides. *J. Neurochem.* **51,** 1858–1867.

Dubovy, P., and Haninec, P. (1990). Non-specific cholinesterase activity of the developing peripheral nerves and its possible function in cells in intimate contact with growing axons of chick embryo. *Int. J. Dev. Neurosci.* **8,** 589–602.

Ebert, C. (1988). Biochemische, immunologische und histologische Untersuchungen zur Butyrylcholinesterase des Hühnchens. Dissertation der Fakultät für Biologie der Eberhard-Karls-Universität Tübingen, Tübingen, Germany.

Edwards, J. A., and Brimijoin, S. (1982). Divergent regulation of acetylcholinesterase and butyrylcholinesterase in tissues of the rat. *J. Neurochem.* **38,** 1393–1403.

Egar, M., and Singer, M. (1972). The role of ependyma in spinal cord regeneration in the Urodele, *Triturus. Exp. Neurol.* **37,** 422–430.

Ekstrom, T. J., Klump, W. M., Getman, D., Karin, M., and Taylor, P. (1993). Promoter elements and transcriptional regulation of the acetylcholinesterase gene. *DNA Cell Biol.* **12,** 63–72.

Ellman, G. L., Courtney, D. K., Anders, V., and Featherstone, R. M. (1961). A new and rapid colorimetric determination of acetylcholinesterase activity. *Biochem. Pharmacol.* **7,** 88–95.

Fuentes, M. E., Rosenberry, T. L., and Inestrosa, N. (1988). A 13 kDa fragment is responsible for the hydrophobic aggregation of brain G4 acetylcholinesterase. *Biochem. J.* **256,** 1047–1050.

Fujisawa, H. (1973). The process of reconstruction of histological architecture from dissociated retinal cells. *Roux's Arch. Dev. Biol.* **171,** 312–330.

Futerman, A. H., Fiorini, R. M., Roth, E., Low, M. G., and Silman, I. (1985a). Physicochemical behaviour and structural characteristics of membrane-bound acetylcholinesterase from *Torpedo*: Effect of phophatidylinositol-specific phospholipase C. *Biochem. J.* **226,** 369–377.

Futerman, A. H., Low, M. G., Michaelson, D. M., and Silman, I. (1985b). Solubilization of membrane-bound acetylcholinesterase by a phosphatidylinositol-specific phospholipase C. *J. Neurochem.* **45,** 1487–1494.

Gaughan, G., Park, H., Priddle, J., Craig, I., and Craig, S. D. (1991). Refinement of the localization of human butyrylcholinesterase to chromosome 3q26.1-q26.2 using a PCR-derived probe. *Genomics* **11,** 455–458.

Gennari, K., Brunner, J., and Brodbeck, U. (1987). Tetrameric detergent-soluble acetylcholinesterase from human caudate nucleus: Subunit composition and number of active sites. *J. Neurochem.* **49,** 12–18.

George, S. T., and Balasubramanian, A. S. (1981). The aryl acylamidases and their relationship to cholinesterases in human serum, erythrocytes and liver. *Eur. J. Biochem.* **121,** 177–186.

Getman, D. K., Eubanks, J. H., Camp, S., Evans, G. A., and Taylor, P. (1992). The human gene encoding acetylcholinesterase is located on the long arm of chromosome-7. *Am. J. Hum. Genet.* **51,** 170–177.

Gibney, G., MacPhee-Quigley, K., Thompson, B., Vedvick, T., Low, M. G., and Taylor, P. (1988). Divergence in primary structure between the molecular forms of acetylcholinesterase. *J. Biol. Chem.* **263,** 1140–1145.

Glees, P., and Sheppard, B. L. (1964). Electron microscopical studies of the synapse in the developing chick spinal chord. *Z. Zellforsch. Mikrosk. Anat.* **62,** 356–362.

Goedde, H. W., Doenicke, A., and Altland, K. (1967). "Pseudocholinesterasen. Pharmakogenetik, Biochemie, Klinik." Springer-Verlag, Berlin.

Goss, R. J. (1974). "Regeneration. Probleme-Experimente-Ergebnisse." Georg Thieme Verlag, Stuttgart.

Gräper, L. (1913). Die Rhombomeren und ihre Nervenbeziehungen. *Arch. Mikrosk. Anat.* **83,** 380–429.

Grassi Milano, E., Arizzi, M., Bracci, M. A., and Manelli, H. (1985). Active cell movements and embryonic cholinesterase in the postmetamorphic completion of the adrenal morphogenesis in frogs. *Exp. Cell Biol.* **53,** 328–334.

Graybiel, A. M., and Ragsdale, C. W. (1982). Pseudocholinesterase staining in the primary visual pathway of the macaque monkey. *Nature (London)* **299,** 439–442.

Greenfield, S. A. (1984). Acetylcholinesterase may have novel functions in the brain. *TINS* **7,** 364–368.

Greenfield, S. A. (1991). A noncholinergic action of acetylcholinesterase (AChE) in the brain: From neuronal secretion to the generation of movement. *Cell. Mol. Neurobiol.* **11,** 55–77.

Greenfield, S. A., Chubb, I. W., Grunewald, R. A., Henderson, Z., May, J. M., Portnoy, S., Weston, J., and Wright, M. C. (1984). A non-cholinergic function for acetylcholinesterase in the substantia nigra: Behavioural evidence. *Exp. Brain Res.* **54,** 513–520.

Grumet, M., Hoffman, S., Crossin, K. L., and Edelman, G. M. (1985). Cytotactin, an extracellular matrix protein of neural and non-neural tissues that mediates glia-neuron interaction. *Proc. Natl. Acad. Sci. U. S. A.* **82**, 8075–8079.

Hamburger, V., and Hamilton, H. L. (1951). A series of normal stages in the development of the chick embryo. *J. Morphol.* **88**, 49–92.

Haynes, L. (1989). Regulation of AChE by a dipeptide. *Trends Pharmacol. Sci.* **10**, 136.

Haydon, P. G., McCobb, D. P., and Kater, S. B. (1984). Serotonin selectively inhibits growth cone motility and synaptogenesis of specific identified neurons. *Science* **226**, 561–564.

Heaton, M. B., and Wayne, D. B. (1986). Specific responsiveness of chick trigeminal motor nucleus explants to target-conditioned media. *J. Comp. Neurol.* **243**, 381–387.

Hoffman, S., Crossin, K. L., and Edelman, G. M. (1988). Molecular forms, binding functions, and developmental expression patterns of cytotactin and cytotactin-binding proteoglycan, an interactive pair of extracellular matrix molecules. *J. Cell Biol.* **106**, 519–532.

Horn, K. M., and Carey, R. G. (1988). Origin of butyrylcholinesterase in the lateral geniculate nucleus of tree shrew. *Brain Res.* **448**, 386–390.

Illing, R. B., and Graybiel, A. M. (1985). Convergence of afferents from frontal cortex and substantia nigra onto acetylcholinesterase-rich patches of the cats superior colliculus. *Neuroscience* **14**, 455–482.

Inestrosa, N. C., Roberts, W. L., Marshall, T. L., and Rosenberry, T. L. (1987). Acetylcholinesterase from bovine caudate nucleus is attached to membranes by a novel subunit distinct from those of acetylcholinesterase in other tissues. *J. Biol. Chem.* **262**, 4441–4444.

Jacobson, M. (1978). "Developmental Neurobiology." 2nd ed., Plenum Publishing, New York.

Jbilo, O., and Chatonnet, A. (1990). Complete sequence of rabbit butyrylcholinesterase. *Nucleic Acids Res.* **18**, 3990.

Jedrzejczyk, J., Silman, I., Lai, J., and Barnard, E. A. (1984). Molecular forms of acetylcholinesterase in synaptic and extrasynaptic regions of avian tonic muscle. *Neurosci. Lett.* **46**, 283–289.

Johnston, M. C. (1966). A radioautographic study of the migration and fate of cranial neural crest cells in the chick embryo. *Anat. Rec.* **156**, 143–156.

Jones, S. A., and Greenfield, S. A. (1991). Behavioural correlates of the release and subsequent action of acetylcholinesterase secreted in the substantia nigra. *Eur. J. Neurosci.* **3**, 292–295.

Karnovsky, M. J., and Roots, L. J. (1964). A "direct-coloring" thiocholine method for cholinesterases. *J. Histochem. Cytochem.* **12**, 219–221.

Keynes, R., Cook, G., Davies, J., Lumsden, A., Norris, W., and Stern, C. (1990). Segmentation and the development of the vertebrate nervous system. *J. Physiol. (Paris)* **84**, 27–32.

Koelle, G. B. (1988). Enhancement of acetylcholinesterase synthesis by glycyl-L-glutamine: An example of a small peptide that regulates differential transcription? *Trends Pharmacol. Sci.* **9**, 318–321.

Koelle, W. A., Koelle, G. B., and Smyrl, E. G. (1976). Effects of persistent selective suppression of ganglionic butyrylcholinesterase on steady-state and regenerating levels of acetylcholinesterase: Implications regarding function of butyrylcholinesterase and regulation of protein synthesis. *Proc. Natl. Acad. Sci. U. S. A.* **73**, 2936–2938.

Koelle, G. B., Koelle, W. A., and Smyrl, E. G. (1977a). Effects of inactivation of butyrylcholinesterase on steady state and regenerating levels of ganglionic acetylcholinesterase. *J. Neurochem.* **28**, 313–319.

Koelle, W. A., Smyrl, E. G., Ruch, G. A., Siddons, V. E., and Koelle, G. B. (1977b). Effects of protection of butyrylcholinesterase on regeneration of ganglionic acetylcholinesterase. *J. Neurochem.* **28**, 307–311.

Koelle, G. B., Koelle, W. A., and Smyrl, E. G. (1979a). Steady state and regenerating levels of acetylcholinesterase in the superior cervical ganglion of the rat following selective inactivation of propionylcholinesterase. *J. Neurochem.* **33**, 1159–1164.

Koelle, G. B., Rickard, K. K., and Ruch, G. A. (1979b). Interrelationships between ganglionic acetylcholinesterase and nonspecific cholinesterase of the cat and rat. *Proc. Natl. Acad. Sci. U. S. A.* **76**, 6012–6016.

Koelle, G. B., Massoulié, J., Eugene, D., and Melone, M. A. (1988a). Effects of glycyl-L-glutamine *in vitro* on the molecular forms of acetylcholinesterase in the preganglionically superior cervical ganglion of the cat. *Proc. Natl. Acad. Sci. U. S. A.* **85**, 1686–1690.

Koelle, G. B., Skau, K. A., Thampi, N. S., Hymel, D. M., and Han, M. S. (1988b). Maintenance by glycyl-L-glutamine *in vivo* of molecular forms of acetylcholinesterase in the preganglionically denervated superior cervical ganglion of the cat. *Proc. Natl. Acad. Sci. U. S. A.* **85**, 6215–6217.

Krejci, E., Duval, N., Chatonnet, A., Vincens, P., and Massoulié, J. (1991). Cholinesterase-like domains in enzymes and structural proteins: Functional and evolutionary relationships and identification of a catalytically essential aspartic acid. *Proc. Natl. Acad. Sci. U. S. A.* **88**, 6647–6651.

Kristt, D. A. (1989). Acetylcholinesterase in immature thalamic neurons: Relation to afferentiation, development, regulation and cellular distribution. *Neuroscience* **29**, 27–43.

Kruse, J., Mailhammer, R., Wernecke, H., Faissner, A., Sommer, I., Goridis, C., and Schachner, M. (1984). Neural cell adhesion molecules and myelin-associated glycoprotein share a common moiety recognized by monoclonal antibodies to L2 and HNK-1. *Nature (London)* **311**, 153–155.

Kugler, P. (1987). Improvement of the method of Karnovsky and Roots for the histochemical demonstration of acetylcholinesterase. *Histochemistry* **86**, 531–532.

Kupffer, C. v. (1906). "Die Morphogenie des Centralnervensystems." *In* Hertwigs Handbuch der vergleichenden und experimentellen Entwickelungslehre der Wirbeltiere" (O. Hertwig, ed.), Band 2, Teil 3. Gustav Fischer, Jena, Germany.

Künemund, V., Jungalwala, F. B., Fischer, G., Chou, D. K. H., Keilhauer, G., and Schachner, M. (1988). The L2/HNK-1 carbohydrate of neural cell adhesion molecules is involved in cell interactions. *J. Cell Biol.* **106**, 213–223.

LaVail, J. H., and Cowan, W. M. (1971a). The development of the chick optic tectum. I. Normal morphology and cytoarchitectonic development. *Brain Res.* **28**, 391–419.

LaVail, J. H., and Cowan, W. M. (1971b). The development of the chick optic tectum. II. Autoradiographic studies. *Brain Res.* **28**, 421–441.

Layer, P. G. (1983). Comparative localization of acetylcholinesterase and pseudocholinesterase during morphogenesis of the chick brain. *Proc. Natl. Acad. Sci. U. S. A.* **80**, 6413–6417.

Layer, P. G. (1990). Cholinesterases preceding major tracts in vertebrate neurogenesis. *BioEssays* **12**, 415–420.

Layer, P. G. (1991a). Cholinesterases during development of the avian nervous system. *Cell. Mol. Neurobiol.* **11**, 7–33.

Layer, P. G. (1991b) Expression and possible functions of cholinesterases during chicken neurogenesis. *In* "Cholinesterases. Structure, Function, Mechanism, Genetics, and Cell Biology" (J. Massoulié, F. Bacou, E. Barnard, A. Chatonnet, B. Doctor, and D. M. Quinn, eds.), pp. 350–357. ACS-Books, Conference Proceeding Series, American Chemical Society, Washington DC.

Layer, P.G., and Alber, R. (1990). Patterning of early chick brain vesicles as revealed by peanut agglutinin and cholinesterases. *Development* **109**, 613–624.

Layer, P. G., and Kaulich, S. (1991). Cranial nerve growth in birds is preceded by cholinesterase expression during neural crest cell migration and the formation of an HNK-1 scaffold. *Cell Tissue Res.* **265**, 393–407.

Layer, P. G., and Willbold, E. (1989). Embryonic chicken retinal cells can regenerate all cell layers *in vitro*, but ciliary pigmented cells induce their correct polarity. *Cell Tissue Res.* **258**, 233–242.

Layer, P. G., Alber, R., and Sporns, O. (1987). Quantitative development and molecular forms of acetyl- and butyrylcholinesterase during morphogenesis and synaptogenesis of chick brain and retina. *J. Neurochem.* **49,** 175–182.

Layer, P. G., Alber, R., and Rathjen, F. G. (1988a). Sequential activation of butyrylcholinesterase in rostral half somites and acetylcholinesterase in motoneurones and myotomes preceding growth of motor axons. *Development* **102,** 387–396.

Layer, P. G., Rommel, S., Bülthoff, H., and Hengstenberg, R. (1988b). Independent spatial waves of biochemical differentiation along the surface of chicken brain as revealed by the sequential expression of acetylcholinesterase. *Cell Tissue Res.* **251,** 587–595.

Layer, P. G., Alber, R., Mansky, P., Vollmer, G., and Willbold, E. (1990). Regeneration of a chimeric retina from single cells *in vitro*: Cell-lineage-dependent formation of radial cell columns by segregated chick and quail cells. *Cell Tissue Res.* **259,** 187–198.

Layer, P. G., Weikert, T., and Willbold, E. (1992). Chicken retinospheroids as developmental and toxicological *in vitro* models: Acetylcholinesterase is regulated by its own and by butyrylcholinesterase activity. *Cell Tissue Res.* **268,** 409–418.

Layer, P. G., Weikert, T., and Alber, R. (1993). Cholinesterases regulate neurite growth of chick nerve cells *in vitro* by means of a non-enzymatic mechanism. *Cell Tissue Res.* **273,** 219–226.

Li, Y., Camp, S., Rachinsky, T. L., Getman, D., and Taylor, P. (1991). Gene structure of mammalian acetylcholinesterase. Alternative exons dictate tissue-specific expression. *J. Biol. Chem.* **266,** 23083–23090.

Liao, J., Heider, H., Sun, M. C., Stieger, S., and Brodbeck, U. (1991). The monoclonal antibody 2G8 is carbohydrate-specific and distinguishes between different forms of vertebrate cholinesterases. *Eur. J. Biochem.* **198,** 59–65.

Liao, J., Heider, H., Sun, M. C., and Brodbeck, U. (1992). Different glycosylation in acetylcholinesterase from mammalian erythrocytes. *J. Neurochem.* **58,** 1230–1238.

Linnemann, D., and Bock, E. (1989). Cell adhesion molecules in neural development. *Dev. Neurosci.* **11,** 149–173.

Lockridge, O. (1982). Substance P hydrolysis by human serum cholinesterase. *J. Neurochem.* **39,** 106–110.

Low, M. G., and Finean, J. B. (1977). Non-lytic release of acetylcholinesterase from erythrocytes by phosphatidylinositol-specific phospholipase C. *FEBS Lett.* **82,** 143–146.

Lumsden, A. (1990). The cellular basis of segmentation in the developing hindbrain. *TINS* **13,** 329–335.

Lumsden, A., and Keynes, R. (1989). Segmental patterns of neuronal development in the chick hindbrain. *Nature (London)* **337,** 424–428.

Lyles, J. M., Silman, I., and Barnard, E. A. (1979). Developmental changes in levels and forms of cholinesterases in muscles of normal and dystrophic chickens. *J. Neurochem.* **33,** 727–738.

Lyles, J. M., Barnard, E. A., and Silman, I. (1980). Changes in the levels and forms of cholinesterases in the blood plasma of normal and dystrophic chickens. *J. Neurochem.* **34,** 978–987.

MacMahon, D. (1974). Chemical messengers in development: A hypothesis. *Science* **185,** 1012–1021.

Mackie, E. J., Tucker, R. P., Halfter, W., Chiquet-Ehrismann, R., and Epperlein, H. H. (1988). The distribution of tenascin coincides with pathways of neural crest cell migration. *Development* **102,** 237–250.

Marchand, A., Chapouthier, G., and Massoulié, J. (1977). Developmental aspects of acetylcholinesterase activity in chick brain. *FEBS Lett.* **78,** 233–236.

Massoulié, J., and Bon, S. (1982). The molecular forms of cholinesterase and acetylcholinesterase in vertebrates. *Ann. Rev. Neurosci.* **5,** 57–106.

Maulet, Y., Camp, S., Gibney, G., Rachinsky, T., Ekstrom, T. J., and Taylor, P. (1990).

Single gene encodes glycophospholipid-anchored and asymmetric acetylcholinesterase forms: Alternative coding exons contain inverted repeat sequences. *Neuron* **4**, 289–301.

McKeehan, M. S. (1961). The capacity for lens regeneration in the chick embryo. *Anat. Rec.* **141**, 227–230.

McTiernan, C., Adkins, S., Chatonnet, A., Vaughan, T., Bartels, C. F., Kott, M., Rosenberry, T. L., La Du, B. N., and Lockridge, O. (1987). Brain cDNA clone for human cholinesterase. *Proc. Natl. Acad. Sci. U. S. A.* **84**, 6682–6686.

Meller, K. (1964). Elektronenmikroskopische Befunde zur Differenzierung der Rezeptorzellen und Bipolarzellen der Retina und ihrer synaptischen Verbindungen. *Z. Zellforsch. Mikrosk. Anat.* **64**, 733–750.

Miki, A., and Mizoguti, H. (1982a). Acetylcholinesterase activity in the myotome of the early chick embryo. *Cell Tissue Res.* **227**, 23–40.

Miki, A., and Mizoguti, H. (1982b). Proliferating ability, morphological development and acetylcholinesterase activity of the neural tube cells in early chick embryos. An electron microscopic study. *Histochemistry* **76**, 303–314.

Minganti, A., and Falugi, C. (1980). An epithelial localization of acetylcholinesterase in the ascidian *Ciona intestinalis* embryos and larvae. *Acta Embryol. Morphol. Exp.* **1**, 143–155.

Mizoguti, H., and Miki, A. (1985). Interrelationship among the proliferating ability, morphological development and acetylcholinesterase activity of the neural tube cells in early chick embryos. *Acta Histochem. Cytochem.* **18**, 85–96.

Moscona, A. A. (1952). Cell suspensions from organ rudiments of chick embryos. *Exp. Cell Res.* **3**, 535–539.

Muller, F., Dumez, Y., and Massoulié, J. (1985). Molecular forms and solubility of acetylcholinesterase during the embryonic development of rat and human brain. *Brain Res.* **331**, 295–302.

Narayanan, C. H., and Narayanan, Y. (1980). Neural crest and placodal contribution in the development of the glossopharyngeal-vagal complex in the chick. *Anat. Rec.* **196**, 71–82.

Noden, D. M. (1988). Interactions and fates of avian craniofacial mesenchyme. *Development*, Suppl. **103**, 121–140.

Orts-Llorca, F., and Genis-Galvez, J. M. (1960). Experimental production of retinal septa in the chick embryo. Differentiation of pigment epithelium into neural retina. *Acta Anat. (Basel)* **42**, 31–70.

Ozaki, H. (1974). Localization and multiple forms of acetylcholinesterase in sea urchin embryos. *Dev. Growth Diff.* **16**, 267–279.

Park, C. M., and Hollenberg, M. J. (1989). Basic fibroblast growth factor induces retinal regeneration *in vivo*. *Dev. Biol.* **134**, 201–205.

Price, D. L., Whitehouse, P. J., Struble, R. G., Coyle, J. T., Clark, A. W., Delong, M. R., Cork, L. C., and Hedreen, J. C. (1982). Alzheimer's disease and Down's syndrome. *Ann. N. Y. Acad. Sci.* **396**, 145–164.

Prody, C. A., Zevin-Sonkin, D., Gnatt, A., Goldberg, O., and Soreq, H. (1987). Isolation and characterisation of full-length cDNA clones coding for cholinesterase from foetal human tissues. *Proc. Natl. Acad. Sci. U. S. A.* **84**, 3555–3559.

Rager, G. (1976). Morphogenesis and physiogenesis of the retino-tectal connection in the chicken. II. The retino-tectal synapses. *Proc. R. Soc. London B* **192**, 353–370.

Rakonczay, Z. (1988). Cholinesterase and its molecular forms in pathological states. *Prog. Neurobiol.* **31**, 311–330.

Rao, R. V., and Balasubramanian, A. S. (1990). Localization of the peptidase activity of human serum butyrylcholinesterase in an approximately 50 kDa fragment obtained by limited alpha-chymotrypsin digestion. *Eur. J. Biochem.* **188**, 637–643.

Rathjen, F. G. (1991). Neural cell contact and axonal growth. *Curr. Opin. Cell. Biol.* **3**, 992–1000.

Rathjen, F. G., Wolff, J. M., Frank, R., Bonhoeffer, F., and Rutishauser, U. (1987). Membrane glycoproteins involved in neurite fasciculation. *J. Cell Biol.* **104**, 343–353.

Rickmann, M., Fawcett, J. W., and Keynes, R. J. (1985). The migration of neural crest cells and the growth of motor axons through the rostral half of the chick somite. *J. Embryol. Exp. Morphol.* **90**, 437–455.

Riopelle, R. J., McGarry, R. C., and Roder, J. C. (1986). Adhesion properties of a neuronal epitope recognized by the monoclonal antibody HNK-1. *Brain Res.* **367**, 20–25.

Robertson, R. T. (1987). A morphogenetic role for transiently expressed acetylcholinesterase in developing thalamocortical systems? *Neurosci. Lett.* **75**, 259–264.

Robertson, R. T. (1991). Transiently expressed acetylcholinesterase activity in developing thalamocortical projection neurons. *In* "Cholinesterases: Structure, Function, Mechanism, Genetics, and Cell Biology" (J. Massoulié, F. Bacou, E. Barnard, A. Chatonnet, B. Doctor, and D. M. Quinn, eds.), pp. 358–365. ACS-Books, Conference Proceeding Series, American Chemical Society, Washington, DC.

Robertson, R. T., Hanes, M. A., and Yu, J. (1988). Investigations of the origins of transient acetylcholinesterase activity in developing rat visual cortex. *Dev. Brain Res.* **41**, 1–23.

Rotundo, R. L. (1984). Purification and properties of the membrane-bound form of acetylcholinesterase from chicken brain. *J. Biol. Chem.* **259**, 13186–13194.

Rotundo, R. L., and Fambrough, D. M. (1980). Synthesis, transport and fate of acetylcholinesterase in cultured chick embryo muscle cells. *Cell* **22**, 583–594.

Rutishauser, U., Watanabe, M., Silver, J., Troy, F. A., and Vimr, E. R. (1985). Specific alteration of NCAM-mediated cell adhesion by an endoneuraminidase. *J. Cell Biol.* **101**, 1842–1849.

Salceda, R., and Martinez, M. T. (1992). Characterization of acetylcholinesterase and butyrylcholinesterase activities in retinal chick pigment epithelium during development. *Exp. Eye Res.* **54**, 17–22.

Sauer, F. C. (1935). Mitosis in the neural tube. *J. Comp. Neurol.* **62**, 377–406.

Schumacher, M., Camp, S., Maulet, Y., Newton, M., MacPhee-Quigley, K., Taylor, S. S., Friedman, T., and Taylor, P. (1986). Primary structure of *Torpedo californica* acetylcholinesterase deduced from its cDNA sequence. *Nature (London)* **319**, 407–409.

Sethi, J. S., and Tewari, H. B. (1976). Histoenzymological mapping of acetylcholinesterase and butyrylcholinesterase in the diencephalon and mesencephalon of *Uromastix hardwickii*. *J. Hirnforsch.* **17**, 335–349.

Sharon, N., and Lis, H. (1993). Carbohydrates in cell recognition. *Sci. Am.* **268**, 74–81.

Sheffield, J. B., and Moscona, A. A. (1969). Early stages in the reaggregation of embryonic chick neural retina cells. *Exp. Cell Res.* **57**, 462–466.

Sikorav, J. L., Krejci, E., and Massoulié, J. (1987). cDNA sequences of *Torpedo marmorata* acetylcholinesterase: Primary structure of precursor of a catalytic subunit; existence of multiple 5'-untranslated regions. *EMBO J.* **6**, 1865–1873.

Sikorav, J. L., Duval, N., Anselmet, A., Bon, S., Krejci, E., Legay, C., Osterlund, M., Reimund, B., and Massoulié, J. (1988). Complex alternative splicing of acetylcholinesterase transcripts in *Torpedo* electric organ: Primary structure of the precursor of the glycolipid-anchored dimeric form. *EMBO J.* **7**, 2983–2993.

Silman, J., and Futerman, A. H. (1987). Modes of attachment of acetylcholinesterase to the surface membran. *Eur. J. Biochem.* **170**, 11–22.

Silver, A. (1974). "The Biology of Cholinesterases." North-Holland, Amsterdam.

Sine, J. P., Ferrand, R., Cloarec, D., Lehur, P. A., and Colas, B. (1991). Human intestine epithelial cell acetyl- and butyrylcholinesterase. *Mol. Cell. Biochem.* **108**, 145–149.

Small, D. H., Ismael, Z., and Chubb, J. W. (1986). Acetylcholinesterase hydrolyses chromogranin A to yield low molecular weight peptides. *Neuroscience* **19**, 289–295.

Smith, A. D., Wald, N. J., Chuckle, H. S., Stirrat, G. M., Bobrow, M., and Lagercrantz, H.

(1979). Amniotic fluid acetylcholinesterase as a possible diagnostic test for neural tube defects in early pregnancy. *Lancet* **1**, 685–688.

Soreq, H., and Prody, C. A. (1989). Sequence similarities between human acetylcholinesterase and related proteins: Putative implications for therapy of anticholinesterase intoxication. *Prog. Clin. Biol. Res.* **289**, 347–359.

Soreq, H., Zevin-Sonkin, D., and Razon, N. (1984). Expression of cholinesterase gene(s) in human brain tissues: Translational evidence for multiple mRNA species. *EMBO J.* **3**, 1371–1375.

Soreq, H., Zamir, R., Zevin-Sonkin, D., and Zakut, H. (1987). Human cholinesterase genes localized by hybridization to chromosomes 3 and 16. *Hum. Genet.* **77**, 325–328.

Soreq, H., Seidman, P. A., Dreyfus, D., Zevin-Sonkin, D., and Zakut, H. (1989). Expression and tissue specific assembly of cloned human butyrylcholinesterase in microinjected *Xenopus laevis* oocytes. *J. Biol. Chem.* **264**, 10608–10613.

Stefanovic, V., Mandel, P., and Rosenberg, A. (1975). Activation of acetyl- and butyrylcholinesterase by enzymatic removal of sialic acid from intact neuroblastoma and astroblastoma cells in culture. *Biochemistry* **14**, 5257–5260.

Stern, C. D., Sisodiya, S. M., and Keynes, R. J. (1986). Interaction between neurites and somite cells: Inhibition and stimulation of nerve growth in the chick embryo. *J. Embryol. Exp. Morphol.* **91**, 209–226.

Stroeva, O. G. (1960). Experimental analysis of the eye morphogenesis in mammals. *J. Embryol. Exp. Morphol.* **8**, 349–368.

Stroeva, O. G., and Mitashov, V. I. (1983). Retinal pigment epithelium: Proliferation and differentiation during development and regeneration. *Int. Rev. Cytol.* **83**, 221–293.

Sun, Y. A., and Poo, M. M. (1987). Evoked release of acetylcholine from the growing embryonic neuron. *Proc. Natl. Acad. Sci. U. S. A.* **84**, 2540–2544.

Topilko, A., and Caillou, B. (1988). Acetylcholinesterase and butyrylcholinesterase activities in human thyroid-cancer cells. *Cancer* **61**, 491–499.

Toutant, J. P., Massoulié, J., and Bon, S. (1985). Polymorphism of pseudocholinesterase in *Torpedo marmorata* tissues: Comparative study of the catalytic and molecular properties of this enzyme with acetylcholinesterase. *J. Neurochem.* **44**, 580–592.

Treskatis, S. (1990). Die Butyrylcholinesterase des Hühnchens. Isolierung und biochemische Charakterisierung. Diplomarbeit der Fakultät für Physiologische Chemie und Biochemie der Eberhard-Karls-Universität Tübingen, Tübingen, Germany.

Treskatis, S., Ebert, C., and Layer, P. G. (1992). Butyrylcholinesterase from chicken brain is smaller than that from serum: Its purification, glycosylation, and membrane association. *J. Neurochem.* **58**, 2236–2247.

Tsim, K. W. K., Randall, W. R., and Barnard, E. A. (1988a). An asymmetric form of muscle acetycholinesterase contains three subunits and two enzymic activities in one molecule. *Proc. Natl. Acad. Sci. U. S. A.* **85**, 1262–1266.

Tsim, K. W. K., Randall, W. R., and Barnard, E. A. (1988b). Synaptic acetylcholinesterase of chicken muscle changes during development from a hybrid to a homogeneous enzyme. *EMBO J.* **7**, 2451–2456.

Tucker, G. C., Delarue, M., Zada, S., Boucaut, J. C., and Thiery, J. P. (1988). Expression of HNK-1/NC-1 epitope in early vertebrate neurogenesis. *Cell Tissue Res.* **251**, 457–465.

Vellom, D. C., Radic, Z., Li, Y., Pickering, N. A., Camp, S., and Taylor, P. (1993). Amino-acid-residues controlling acetylcholinesterase and butylcholinesterase specificity. *Biochemistry* **32**, 12–17.

Vigny, H., Gisiger, V., and Massoulié, J. (1978). Nonspecific cholinesterase and acetylcholinesterase in rat tissues: Molecular forms, structural and catalytic properties, and significance of the two enzyme systems. *Proc. Natl. Acad. Sci. U. S. A.* **75**, 2588–2592.

Villafruela, M. J., Barat, A., Manrique, E., Villa, S., and Ramirez, G. (1981). Molecular forms of acetylcholinesterase in the developing chick visual system. *Dev. Neurosci.* **4,** 25–36.

Vollmer, G., and Layer, P. G. (1986a). Reaggregation of chick retina and mixtures of retina and pigment epithelial cells: The degree of laminar organization is dependent on age. *Neurosci. Lett.* **63,** 91–95.

Vollmer, G., and Layer, P. G. (1986b). An *in vitro* model of proliferation and differentiation of the chick retina: Coaggregates of retinal and pigment epithelial cells. *J. Neurosci.* **6,** 1885–1896.

Vollmer, G., and Layer, P. G. (1987). Cholinesterases and cell proliferation in "nonstratified" and "stratified" cell aggregates from chicken retina and tectum. *Cell Tissue Res.* **250,** 481–487.

Vollmer, G., Layer, P. G., and Gierer, A. (1984). Reaggregation of embryonic chick retina cells: Pigment epithelial cells induce a high order of stratification. *Neurosci. Lett.* **48,** 191–196.

Walter, J., Kern-Veits, B., Huf, J., Stolze, B., and Bonhoeffer, F. (1987). Recognition of position-specific properties of tectal cell membranes by retinal axons *in vitro*. *Development* **101,** 685–696.

Weikert, T. (1993). Butyrylcholinesterasen und Peanut Agglutinin bindende Proteine als positive und negative Regulatoren des Neuritenwachstums im Hühnerembryo. Immunologische, biochemische und histochemische Untersuchungen. Dissertation der Fakultät für Biologie der Eberhard-Karls-Universität Tübingen, Tübingen, Germany.

Weikert, T., Rathjen, F. G., and Layer, P. G. (1990). Developmental maps of acetylcholinesterase and G4-antigen of the early chicken brain: Long distance tracts originate from AChE-producing cell bodies. *J. Neurobiol.* **21,** 482–498.

Willbold, E. (1991). Die Regeneration der embryonalen Hühner-Retina aus Einzelzellen in der Schüttelkultur. Modellsysteme zur Entstehung neuronaler Netzwerke. Dissertation der Fakultät für Biologie der Eberhard-Karls-Universität Tübingen, Tübingen, Germany.

Willbold, E., and Layer, P. G. (1992a). Formation of neuroblastic layers in chicken retinospheroids: The fibre layer of Chievitz secludes AChE-positive cells from mitotic cells. *Cell Tissue Res.* **268,** 401–408.

Willbold, E., and Layer, P. G. (1992b). A hidden retinal regenerative capacity from the chick ciliary margin is reactivated *in vitro*, that is accompanied by down-regulation of butyrylcholinesterase. *Eur. J. Neurosci.* **4,** 210–220.

Wilson, S. W., Ross, L. S., Parrett, T., and Easter, S. S. (1990). The development of a simple scaffold of axon tracts in the brain of the embryonic zebrafish, *Brachydanio rerio. Development* **108,** 121–145.

Wolburg, H., Willbold, E., and Layer, P. G. (1991). Müller glia endfeet, a basal lamina and the polarity of retinal layers form properly *in vitro* only in the presence of marginal pigmented epithelium. *Cell Tissue Res.* **264,** 437–451.

Wolff, J. M., Brümmendorf, T., and Rathjen, F. G. (1989). Neural cell recognition molecule F11: Membrane interaction by covalently attached phosphatydylinositol. *Biochem. Biophys. Res. Comm.* **161,** 931–938.

Wolfgang, W. J., and Forte, M. A. (1989). Expression of acetylcholinesterase during visual system development in *Drosophila. Dev. Biol.* **131,** 321–330.

Wright, C. I., Geula, C., and Mesulam, M. M. (1993). Protease inhibitors and indolamines selectively inhibit cholinesterases in the histopathologic structures of Alzheimer disease. *Proc. Natl. Acad. Sci. U. S. A.* **90,** 683–686.

Zacks, S. I. (1952). Esterases in the early chick embryo. *Anat. Rec.* **112,** 509–537.

Zakut, H., Matzel, A., Schejter, E., Avni, A., and Soreq, H. (1985). Polymorphism of acetylcholinesterase in discrete regions of the developing human fetal brain. *J. Neurochem.* **45,** 382–389.

Zakut, H., Zamir, R., Sindel, L., and Soreq, H. (1989). Gene mapping on chorionic chromosomes by hybridization in situ: Localization of cholinesterase cDNA binding sites to chromosomes 3q21, 3q26-ter and 16q21. *Hum. Reprod.* **4,** 941–946.

Zimmermann, H. (1992). 5'-Nucleotidase: Molecular structure and functional aspects. *Biochem. J.* **285,** 345–365.

Role of Nuclear Trafficking in Regulating Cellular Activity

Carl M. Feldherr and Debra Akin
Department of Anatomy and Cell Biology, University of Florida, College of
Medicine, Gainesville, Florida 32610

I. Introduction

DNA synthesis, transcription, and RNA processing are dependent on a variety of proteins (e.g., transcription factors, enzymes, and structural components of the nuclear matrix) that are produced in the cytoplasm. The products of transcription, in turn, are exported to the cytoplasm, and, with the exception of specific small nuclear ribonucleoprotein (snRNP) particles, are retained in this compartment where they function in polypeptide synthesis. Since both transcription and translation are interdependent and rely on interactions between the nucleoplasm and cytoplasm, it should be possible to modulate these processes by altering the intracellular trafficking patterns of the essential macromolecules. In this chapter we focus on the specific and nonspecific mechanisms used by cells to regulate the nucleocytoplasmic distribution of proteins and RNA. Theoretically, regulation at this level could be achieved either by variations in the transport machinery or the properties of the permeant molecules.

Central to any discussion of nuclear import and export is the nuclear envelope, which separates the nucleoplasm from the remainder of the cell. The envelope is a double membrane structure that contains the nuclear pores, circular spaces that are 70–80 nm in diameter, which are formed by the fusion of the inner and outer membranes. The pores are not simply openings that permit free communication between the nucleoplasm and cytoplasm, but contain highly organized supramolecular protein structures that regulate the movement of macromolecules. The pores, along with their structural elements, are referred to as pore complexes. Since these organelles are the primary exchange sites, it will be useful to review their structure and composition, as well as the properties of other elements of the transport machinery, before considering the mechanisms regulating nuclear transport.

183

II. The Pore Complex

A. Structure

The pores were first identified by Callan and Tomlin (1950) in air-dried nuclear envelopes obtained from amebas and examined with the electron microscope. This initial observation was quickly followed by a number of studies that employed different electron microscope (EM) techniques, including thin sectioning, negative staining, and freeze-etching. The different structural models of the pore complex resulting from these investigations were described in detail in several early reviews (Stevens and Andre, 1969; Feldherr, 1972; Franke, 1974; Maul, 1977; Kessel, 1988; Scheer *et al.*, 1988). The model that received the most support was proposed by Franke and Scheer (1970, 1974). It consisted of two annular structures, located at the nuclear and cytoplasmic margins of the pore. Both annuli had an outer diameter of approximately 110 nm and were made up of eight spherical granules that ranged in diameter from 10 to 25 nm. Fibrils extended outward from the granular elements of the annuli and away from the pore opening; those projecting from the inner granules (located at the nuclear face of the pores) were longer and frequently in contact with RNP particles. Eight wedge-shaped spokes were positioned between the two annuli. Their bases were attached to the pore margin (region of membrane fusion) and their apices extended toward the center of the pore. A central granule of variable size and shape was frequently but not always observed in the pores.

In recent studies, two basic variations in experimental approaches have resulted in a more detailed understanding of the structure of the pore complex. First, owing to improvements in preparative procedures for EM, more emphasis has been placed on the examination of intact and partially dissociated pore complexes. Second, computer imaging procedures have been employed to analyze and model the specimens. Nuclear envelopes obtained from amphibian oocytes are currently the preferred starting material in these investigations. Not only is there an abundance of pore complexes in these cells (the pores comprise approximately 20% of the nuclear surface), but the envelopes can be manually isolated, thus minimizing structural damage to these organelles. Comparable procedures are not yet available for other cell types. For reviews of the current literature on pore structure see Aebi *et al.* (1990) and Akey (1992).

Based on three-dimensional reconstructions of negatively stained oocyte pore complexes, Unwin and Milligan (1982) modified the tripartite model (a cytoplasmic annulus, central spoke elements, and a nuclear annulus) suggested by Franke and Sheer (1970). In the model proposed by Unwin and Milligan (1982), the granular annuli were replaced by inner and outer rings; the central spoke elements were retained, as was the central granule (plug). The peripheral granules were considered to be variable components. Further evidence supporting the triple ring concept was obtained using unfixed, frozen, hydrated preparations (Akey, 1989); dissociated pores visualized by heavy-metal shadowing (Stewart et al., 1990); unstained, freeze-dried pore complexes (Reichelt et al., 1990); and air-dried, negatively stained specimens (Hinshaw et al., 1992).

Electron-dense areas located within the perinuclear space just adjacent to the margins of the pores were observed by Hoeijmakers et al. (1974) in Triton-treated whole-mount preparations of nuclei isolated from cultured liver cells. Similar structures, termed radial arms, were also seen in computer-enhanced images of oocyte envelopes by Akey (1990), and in sectioned material (Pante et al.,1992). Fibrous elements extending from the cytoplasmic ring, and a nucleocytoplasmic basket associated with the nuclear ring, were visualized using freeze-dried or critical-point dried specimens and low voltage scanning electron microscopy (Ris, 1991). Evidence that the central plug functions as a transporter was obtained by Akey and Goldfarb (1989) and Akey (1990). Originally, it was suggested that the transporter was organized as a double iris structure, with each iris containing eight hinged subunits. This would permit variable dilation, accommodating different-sized permeant substances (discussed later), and at the same time prevent passive diffusion of cytoplasmic materials during transport. More recent studies, performed on frozen envelopes, indicate that rather than being spherical or disc shaped, the transporter is a cylinder 62.5 nm in length and of variable width (Akey and Radermacher, 1993).

Figure 1 is a computer-generated three-dimensional model of the pores proposed by Akey and Radermacher (1993) that incorporates the major structural elements described above. It should be pointed out that models proposed by other investigators using different techniques vary in detail but contain the same basic structures.

B. Composition

The mass of the pore complex has been estimated to be approximately 125×10^6 Da (Reichelt et al., 1990), a size that could accommodate several hundred polypeptides. Of this theoretical number, only a few pore proteins, which will be referred to as nucleoporins, have been identified and characterized; however, the rate of progress in this area can be expected to increase rapidly as new methods for isolating pore components are developed.

The first nucleoporin to be identified is a transmembrane, mannose-rich glycoprotein that is located at the margins of the pores (Gerace et al., 1982). This protein has been sequenced by Wozniak et al. (1989); it has a

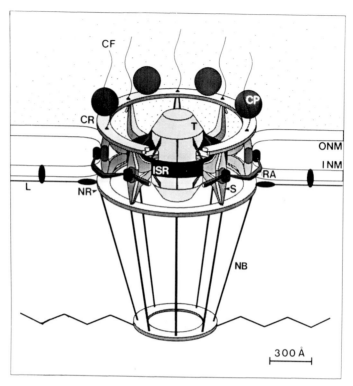

FIG. 1   A three-dimensional, computer-generated diagram of the nuclear pore complex. This model incorporates data from several laboratories and reflects our current understanding of the architecture of the complex. ONM, outer nuclear membrane; INM, inner nuclear membrane; CR, cytoplasmic ring; NR, nuclear ring; ISR, inner spoke ring; S, spokes; T, transporter; CF, thin filaments; CP, cytoplasmic particles; RA, radial arms; NB, nuclear basket; L, nuclear lamina. (After Akey and Radermacher, 1993.)

total mass (including the oligosaccharides) of approximately 210 kDa and is referred to as gp210. The luminal domain (located in the perinuclear space), the hydrophobic transmembrane region, and the cytoplasmic tail contain 1783, 19, and 58 amino acids, respectively (Greber *et al.,*1990). Gp210 is thought to be a component of the radial arms, which could function as attachment sites for the structural components of the pore complexes. There is also evidence that gp210 might be involved in regulating signal-mediated transport (discussed later).

Several investigators subsequently identified a family of O-linked pore complex glycoproteins in vertebrate cells that are not membrane associated. These proteins, which have been studied most extensively in liver cells and *Xenopus* oocytes, contain O-linked *N*-acetylglucosamine (Davis and Blobel, 1986; Finlay *et al.*, 1987; Park *et al.*, 1987; Snow *et al.*, 1987). Using wheat germ agglutinin (WGA) and monoclonal antibodies raised against pore-lamina fractions, eight such proteins have been detected in hepatocytes. They range in molecular mass from 45 to 210 kDa (Snow *et al.*, 1987; Finlay *et al.*, 1987); the most abundant is 62 kDa (p62).

Similar results have been obtained for isolated *Xenopus* oocyte envelopes that were fractionated by gel electrophoresis, transferred to nitrocellulose, and stained with WGA (Scheer *et al.*, 1988; Finlay and Forbes, 1990). In *Xenopus*, the major O-linked pore glycoprotein has a molecular mass of 62–68 kDa. This polypeptide is probably homologous to p62, since both react with RL1, a monoclonal antibody raised against liver nuclear envelopes (Featherstone *et al.*, 1988). When localized with polyspecific immunogold or WGA-ferritin conjugates, the O-linked glycoproteins are found along both the nuclear and cytoplasmic faces of the pores (Finlay *et al.*, 1987; Park *et al.*, 1987; Snow *et al.*, 1987; Scheer *et al.*, 1988). However, there is evidence, obtained with more specific antibodies, that individual nucleoporins preferentially accumulate at a specific pore face (Snow *et al.*, 1987). There is also evidence that the O-linked glycoproteins exist as functional complexes. Thus, Finlay *et al.* (1991) found that three *Xenopus* nucleoporins, p62, p58, and p54, form a stable complex that has a molecular mass of 550–600 kDa, and is required for nuclear transport.

Vertebrate p62 has been cloned and is highly conserved in mouse (Cordes *et al.*, 1991), rat (D'Onofrio *et al.*, 1988; Starr and Hanover, 1990; Starr *et al.*, 1990), human (Carmo-Fonseca *et al.*, 1991), and *Xenopus* (Cordes *et al.*, 1991). The amino-terminal domain includes 15 consensus GFSFG repeats, and is most likely arranged in β-sheets. This domain is followed by a threonine-rich region consisting of a sequence of 25 threonine and serine residues. The carboxy-terminus contains a series of hydrophobic heptad repeats that are characteristic of fibrous proteins. This suggests the presence of α-helices, which could be involved in the formation of structural complexes with other nucleoporins. The O-linked GlcNAc is added to the amino-terminus as described by Haltiwanger *et al.* (1990) and Starr and Hanover (1990).

A second rat liver nucleoporin, designated nup153, was sequenced by Sukegawa and Blobel (1993). This protein has a molecular mass of approximately 153 kDa, and contains a number of XFXFG repeats, a feature it shares with p62. In addition, nup153 contains four zinc finger motifs. The latter finding, plus the fact that nup153 was localized to the nucleoplasmic face of the pores by immunoelectron microscopy, led the authors to suggest that this nucleoporin might be involved in organizing the chromatin and, perhaps, directing specific transcripts to the pores (Blobel, 1985). Recently, a 155-kDa nucleoporin (nup155) has been cloned and sequenced by Radu *et al.* (1993). This protein, also obtained from rat liver nuclei, does not contain repetitive sequences, nor does it bind WGA. Using immunoelectron microscopy, it has been localized to both faces of the pore complex, but its function remains unknown.

Considerable progress has been made in identifying and characterizing yeast nucleoporins. Three such proteins—NSP1 (86 kDa; Hurt, 1988; Nehrbass *et al.*, 1990), NUP1 (130 kDa; Davis and Fink, 1990), and NUP2 (95 kDa; Loeb *et al.*, 1993)—have features indicating that they are homologs of p62. All of these polypeptides contain central domains made up of repeats similar to those present in p62, which suggests a secondary structure composed of β-sheets. These similarities could account for the cross-reactivity of antibodies raised against rat and yeast nucleoporins. The amino- and carboxy-termini of NSP1, NUP1, and NUP2 are not homologous; however, the carboxy region of NSP1 does resemble that of p62 (Carmo-Fonseca *et al.*, 1991). It has a series of heptad repeats that are likely to be involved in targeting and anchoring to the pore complex (Hurt, 1990). NUP1 and NSP1, but not NUP2, are required for cell viability. Interestingly, Nehrbass *et al.* (1990), who investigated the function of different domains of NSP1, found that only the carboxy-terminus is essential for cell growth. Perhaps other nucleoporins with conserved repeats can compensate for the loss of the central domain in NSP1, whereas the carboxy region could have a unique function. In fact, by studying combinations of mutants, Loeb *et al.* (1993) obtained evidence that those nucleoporins are functionally interactive.

Wente *et al.* (1992) identified a second subfamily of yeast nucleoporins that contains at least three proteins, NUP49, NUP100, and NUP116. A common feature of this subfamily is the presence of multiple GLFG repeats in the amino-termini. A high degree of homology also exists between the carboxy domains of NUP100 and NUP116. Figure 2 compares the yeast nucleoporin subfamilies. Wimmer *et al.* (1992) independently identified two yeast pore proteins, NSP49 and NSP116, that are identical to NUP49 and NUP116. The experimental approach (synthetic lethality) employed by the latter investigators also demonstrated that the functions of NUP49 and NUP116 overlap those of NSP1.

Considering that ATP is required for both the efflux of RNA from the nucleus and the import of proteins (see later discussion), there has been considerable interest in identifying the nuceloside triphosphatase [NTPase(s)] associated with the nuclear envelope. The NTPase involved in mRNA export has been extensively investigated (see reviews in Schroder *et al.*, 1987; Agutter, 1988; Riedel and Fasold, 1992); it has a molecular mass of approximately 43–47 kDa and appears to localize along the inner surface of the envelope. The initial evidence that NTPase activity is also present within the pores themselves was based on histochemical studies (Klein and Afzelius, 1966; Yasuzumi *et al.*, 1968). A possible

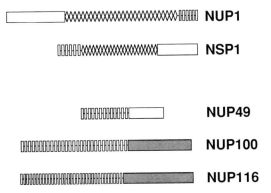

FIG. 2   A comparison of the two known subfamilies of yeast nucleoporins. One family, which includes NUP1, NSP1, and NUP2 (not shown in this diagram), is characterized by the presence of repeated KPAFSFG motifs. The second family (NUP49, NUP100, and NUP116) contains GLFG repeat domains. The number of diamonds and rectangles indicates the number of repeating sequences. The open rectangles indicate unique domains, the shaded rectangles are conserved domains. (After Wente *et al.*, 1992.)

candidate for such activity is a 188-kDa myosin-like adenosine triphosphatase (ATPase); the evidence is based largely on immunogold localization studies and has been reviewed in detail by Berrios (1992).

III.   Macromolecular Transport across the Nuclear Envelope

A.   Evidence, Pathways, and Energy Requirements

Macromolecular exchanges through the pores can occur either by passive diffusion or signal-mediated transport. Evidence for diffusion has been obtained by microinjecting radiolabeled or fluorescent-labeled reporters (either exogenous proteins or synthetic polymers) into the cytoplasm and measuring subsequent uptake into the nucleoplasm. The uptake rates obtained for these reporters are inversely related to molecular size, and are consistent with passive diffusion through a channel 9–10 nm in diameter (Paine *et al.*, 1975; Peters, 1984). The diffusion channels were found in the central regions of the pore complexes by tracing the migration of polyvinylpyrrolidone (PVP)-coated gold particles across the nuclear envelope in amebas (Feldherr, 1962). Based on morphological studies, Hinshaw *et al.* (1992) proposed that diffusion might also occur at the periphery of the pore complex, but direct evidence for peripheral pathways is not available. The diffusion studies have been reviewed by Paine and Horowitz (1980), Peters (1986), and Paine (1992).

Bonner (1975) and Feldherr (1975) found that large karyophilic proteins (molecular mass 60 kDa and above) enter the nucleus much more rapidly than can be accounted for by passive diffusion. This is consistent with, but does not prove, that some form of mediated transport is required for the nuclear import of specific proteins. De Robertis *et al.* (1978) initially suggested that large karyophilic proteins might contain signals that are required for nuclear targeting. Dingwall *et al.* (1982) obtained direct evidence supporting this proposal. They studied the intracellular distribution of native and protease-digested nucleoplasmin, a 110-kDa pentameric protein that is present in the nuclei of *Xenopus* oocytes, and found that the tail region of each of the five monomeric subunits contains a domain necessary for nuclear import.

The pathway for signal-mediated transport into the nucleus was determined by microinjecting nucleoplasmin-coated gold particles into the cytoplasm of *Xenopus* oocytes and locating particles in transit through the pores by transmission electron microscopy (TEM) (Feldherr *et al.*, 1984; Dworetzky *et al.*, 1988). The gold particles acquire the properties of the coating agent and, in this instance, behave as karyophilic proteins. Shortly after injection, the gold had accumulated along the cytoplasmic faces of the pore complexes, presumably bound in these regions, and also extended through the centers of the pores. Particles of approximately 25 nm (including the 3-nm-thick protein coat) were present in both the nucleus and along the central axis of the pores. It was concluded from these studies that mediated transport occurs through the same region of the pore complex as does passive diffusion, except that the functional diameter of the transport channel is approximately 2.5 times greater. It was proposed (see also Feldherr and Dworetzky, 1989) that the central, 9–10 nm-diameter diffusion canal can dilate when activated by appropriate targeting signals, and accommodate molecular complexes up to 25 nm in diameter. In the pore model (see Fig. 1), this channel would be represented by the central transporter element. In fact, Akey (1990) identified four morphological forms of the transporter and suggested that they might reflect different steps in the transport process.

With regard to the sites of RNA efflux, Anderson and Beams (1956), and Franke and Scheer (1970) detected electron-opaque material extending through the centers of the pores in oocytes and suggested that this material might represent RNP in the process of exiting the nucleus. Stevens and Swift (1966), and Daneholt and co-workers (see references in Mehlin *et al.*, 1992) followed the formation and export of a well-characterized messenger RNP particle that encodes a secretory protein of about 1000 kDa in *Chironomus* salivary gland cells. These RNP particles are formed in association with the Balbiani rings as spherical granules 50 nm in diameter. They then migrate to the nuclear surface of the pores, uncoil, and pass through the centers of the pores (5' end first) as rod-like structures with a length of approximately 135 nm and a width of 25 nm.

Dworetzky and Feldherr (1988) identified the export channels in *Xenopus* oocytes by microinjecting gold particles, coated with either RNA or polynucleotides, into the nucleoplasm and following their transit to the cytoplasm. Consistent with results obtained for the *Chironomus* message, it was found that particles approximately 25 nm in diameter exited the nucleus by penetrating the centers of the pores. The latter investigators also demonstrated, by performing double-labeling experiments (different sizes of gold particles were coated with RNA and nucleoplasmin and injected into the nucleus and cytoplasm, respectively, of the same cell), that RNA efflux and protein import can take place through the same pores. The above results indicate that all classes of macromolecules use a common pathway when crossing the nuclear envelope, and that transport in both directions is subject to the same size restrictions.

It was demonstrated by Richardson *et al.* (1988), using Vero cells, and Newmeyer and Forbes (1988), using an *in vitro Xenopus* egg extract system, that signal-mediated protein import through the pores is a two-step process involving initial binding to the pore complex, followed by ATP-dependent translocation. Although it has been difficult to obtain definitive evidence that binding of RNP to the pores is an initial step in export, there are ample data demonstrating that translocation to the cytoplasm is energy dependent (reviewed in Agutter, 1991; see also Bataille *et al.*, 1990; Dargemont and Kuhn, 1992).

B. Transport Signals

1. Protein Import

Specific amino acid sequences that are able to direct proteins to the nucleus (referred to as nuclear localization signals or NLS) were first identified in the yeast transcription factor MATα2 and SV40 large T antigen. The MATα2 signal is contained within the first 13 amino-terminal amino acids (Met–Asn–Lys–Ile–Pro–Ile–Lys–Asp–Leu–Leu–Asn–Pro–Gln), as demonstrated by the fact that fusion of this region of MATα2 to β-galactosidase (a cytoplasmic enzyme) resulted in the nuclear localization of the hybrid protein (Hall *et al.*, 1984). Lanford and Butel (1984) found that a single amino acid substitution at position 128 (Asn for Lys) prevented the nuclear accumulation of SV40 large T antigen. A detailed analysis of this region by Kalderon *et al.* (1984a,b) revealed that the minimal sequence that was both necessary and sufficient for nuclear targeting of SV40 large T extended from amino acid 126–132, and contained

Pro–Lys–Lys–Lys–Arg–Lys–Val. Since these initial studies, NLS have been identified in well over 30 proteins. These data have recently been the subject of comprehensive reviews by Garcia-Bustos *et al.* (1991) and by Richter and Standiford (1992), and will not be considered in detail here; however, the main features of the NLS will be summarized since they are important in understanding regulatory processes.

Although there is no consensus NLS, the different targeting sequences are characterized by an abundance of basic amino acids, and can be divided into two general classes, simple and bipartite. The SV40 large T NLS is considered to be the model for simple targeting signals. The simple sequences are generally 5–10 amino acids in length, and can be located in any region of the protein (unlike the signal found in secretory proteins, which is preferentially located at the amino-terminus).

Just as the SV40 large T NLS is a model for simple signals, nucleoplasmin is the prototype for bipartite signals. Although nucleoplasmin contains four short putative NLS (two resemble the MATα2 NLS, and two the large T NLS), none are sufficient by themselves to target cytoplasmic proteins to the nucleus (Burglin and De Robertis, 1987; Dingwall *et al.*, 1988). Using deletion analysis, Dingwall *et al.* (1988) determined that the functional NLS in nucleoplasmin is 16 amino acids in length (Lys^{155}Arg–Pro–Ala–Ala–Thr–Lys–Lys–Ala–Gly–Gln–Ala–Lys–Lys–Lys–Lys^{170}), much larger than expected. Robbins *et al.* (1991) then showed that the two peripheral basic sequences (Lys–Arg and Lys–Lys–Lys–Lys) are essential and interdependent, and the ten intervening amino acids act as spacers. Increasing the spacer domain by amino acid insertions can be tolerated without loss of function; however, deletions blocked transport. Analysis of other known NLS suggests that the bipartite signal is not unique to nucleoplasmin and in fact similar motifs appear in approximately half of the karyophilic proteins that have been sequenced, including steroid receptors and p53 (see Robbins *et al.*, 1991, for a listing of these proteins).

It is also of interest that a number of karyophilic proteins normally contain two or more different NLS, whereas others contain multiple copies of a single signal. The possible role of multiple signals in regulating the nuclear import rates of proteins under different physiological conditions will be considered below.

NLSs remain functional even when removed from their normal context within proteins; thus, short synthetic peptides that contain targeting sequences can direct nuclear localization when they are conjugated with exogenous proteins such as bovine serum albumin (BSA) or immunoglobulin G (IgG) (Goldfarb *et al.*, 1986; Lanford *et al.*, 1986). However, there are instances where signal activity can be affected by its location within a given polypeptide. For example, Roberts *et al.* (1987) introduced the large T NLS into different regions of pyruvate kinase (a cytoplasmic protein), and found that the signal was inactive when inserted into a putative hydrophobic domain. The effect in this case could simply be due to masking of the NLS, since the signal was functional in other locations where it was presumably exposed.

The results obtained by Rihs and Peters (1989) suggest that normal flanking regions can also modulate the level of signal activity. These investigators constructed hybrid proteins in which short, signal-containing regions of SV40 large T were fused to β-galactosidase. They then used photobleaching procedures to measure nuclear transport kinetics and found, as expected, that the NLS alone was sufficient for nuclear accumulation; however, when the upstream flanking region containing large T residues 111–125 was also present, the rate of import increased significantly, even though the final nucleocytoplasmic concentration ratio remained the same. It was subsequently determined (Rihs et al., 1991) that the increased transport rate was due to phosphorylation at either position 111 or 112 by casein kinase II. Furthermore, phosphorylation of threonine-124 by cdc2 kinase caused a decrease in the level but not the rate of accumulation (Jans et al., 1991). Although the mechanism(s) by which phosphorylation modulates transport remains to be determined, it is clear that the activity of the NLS is subject to regulation by flanking domains.

Transport signals are not only found in large karyophilic proteins, but are also present in small polypeptides (below 40 kDa) that can theoretically diffuse across the envelope. Examples include yeast histone 2B (Moreland et al., 1987), histone H1 (Breeuwer and Goldfarb, 1990), and the ribosomal proteins L3 (Moreland et al., 1985) and L29 (Underwood and Fried, 1990). The possibility that small proteins enter the nucleus as larger, nondiffusive aggregates would explain the need for NLS. This hypothesis is supported by Moreland et al. (1987), who obtained evidence that histones 2B and 2A are imported into the nucleus as a heterodimer.

2. RNA Efflux

In general, the signals necessary for RNA export are not as well understood as those required for protein import, but recent progress has been made, especially with regard to polymerase II transcripts, which include mRNA and most of the U snRNAs. The newly transcribed polymerase II U snRNAs are transported to the cytoplasm where they are processed and assembled into RNP particles. They then return to the nucleus and function in the processing of pre-rRNA and pre-mRNA (for details see the review in Anderson and Zieve, 1991). The polymerase II transcripts are characterized by the presence of a 5' monomethyl guanosine cap, and it has been shown that this cap structure is essential for the export of both mRNA and U snRNAs (Hamm and Mattaj, 1990; Dargemont and Kuhn,

1992). In addition to the 5' cap, there is evidence that the efflux of mRNA is influenced by its poly(A)-tail (reviews in Schroder *et al.*, 1987; Agutter, 1988). In the case of nonpolyadenylated mRNA, normal processing of the 3' end appears to be necessary for export. Thus, Eckner *et al.* (1991) found that prokaryotic and histone mRNA were exported more efficiently when the 3' end was processed through the natural pathway rather than artificially by *cis*-acting ribozyme.

The nuclear efflux of tRNA (a polymerase III transcript) has been investigated by Zasloff (1983) and found to be a saturable, temperature-dependent process. Furthermore, a single nucleotide substitution at position 57 caused a significant reduction in efflux, suggesting the presence of a signal domain within this region. However, in a later investigation by Tobian *et al.* (1985), 30 different mutations of tRNA were analyzed, and in all cases there was a decrease in transport. These results argue against a specific signal and indicate that in this case molecular shape could be a primary recognition factor.

Even in instances where signals have been identified, it is not clear whether they play a primary or secondary role in transport. The RNA signals could represent binding sites for specific proteins, which in turn contain the actual transport domains. The involvement of protein is suggested by a number of studies. For example, Izaurralde *et al.* (1992) identified a cap binding protein that might mediate the transport of polymerase II transcripts. These investigtors found that the export of U1 snRNA can be blocked by the addition of 5' monomethyl guanosine cap analogs; furthermore, a direct correlation existed between the inhibitory effect of particular analogs and their affinity for a specific 80-kDa cap binding protein. Guddat *et al.* (1990) reported that the nuclear export of 5S RNA (a polymerase III transcript) in *Xenopus* oocytes is dependent on its interaction with either L5 (a ribosomal protein) or transcription factor IIIA.

Singh and Green (1993) obtained evidence for sequence-specific binding of glyceraldehyde 3-phosphate dehydrogenase (GAPDH) to tRNA; in addition, the intracellular distribution of GAPDH is consistent with that of a shuttling protein. These results, combined with the fact that GAPDH does not bind as effectively to tRNA mutants that have reduced transport capabilities, led to the suggestion that GAPDH might be involved in tRNA transport. Several nucleolar proteins, which are known to shuttle between the nucleus and cytoplasm, have been implicated in the export of ribosomal subunits. These include C23 and B24 (Borer *et al.*, 1989), and Nopp140 (Meier and Blobel, 1992). The transport of ribosomal subunits, including possible signaling mechanisms, has been reviewed in detail by Fried (1992).

Pinol-Roma and Dreyfuss (1992) found that a number of mRNA-binding proteins shuttle across the envelope, and proposed that they might also be involved in transport. Martin and Helenius (1991) determined that the bidirectional nucleocytoplasmic exchange of influenza viral (v) RNP is controlled by the viral matrix protein M1. In order for vRNP to enter the nucleus following infection, it must first dissociate from M1. However, newly synthesized vRNP can leave the nucleus only if it is bound to M1.

RNA efflux studies are further complicated by the likelihood that RNA processing and, perhaps, transport to the pores occur in close association with elements of the nuclear matrix. This solid-state mechanism was first proposed by Agutter (1985) and is supported by the work of Lawrence *et al.* (1989) and Xing *et al.* (1993), who observed that mRNA is not randomly distributed throughout the nucleoplasm, but is restricted to tracks, which presumably direct the message to the nuclear surface. Furthermore, these tracks remain intact following extraction procedures designed to preserve elements of the nuclear matrix (Xing and Lawrence, 1991). Similar tracks could also function in the export of proteins from the nucleus (Meier and Blobel, 1992). Given the possibility that migration along the matrix is necessary for export, it becomes difficult to distinguish at what level (intranuclear transport or translocation through the pores) specific signals function. Detailed discussions of RNA efflux, including signal requirements, can be found in reviews by Riedel and Fasold (1992), Fried (1992), and Izaurralde and Mattaj (1992).

3. U snRNA Import

Following transcription, the polymerase II U snRNAs enter the cytoplasm where they are assembled into mature snRNP particles. The maturation process involves (1) the addition of proteins, including a set of eight common core polypeptides (U3 snRNA is an exception, in that it complexes with a unique set of core proteins) plus specific proteins that associate with individual U snRNAs, and (2) hypermethylation of the 5' end to form a trimethylguanosine cap. U6 snRNA is a polymerase III transcript; as such, it contains a γ-monomethyl phosphate at its 5' end, and, in *Xenopus* oocytes, is normally retained in the nucleus. However, when microinjected into the cytoplasm, U6 can reenter the nucleoplasm.

It was originally found that the import of U snRNAs is prevented by mutations that block binding of the common core proteins (Mattaj and De Robertis, 1985). However, since the bound core proteins are also necessary for hypermethylation of the 5' end, it remained unclear whether the 5' cap or the core proteins were primarily involved in import. This question was resolved by Fischer and Luhrmann (1990) and Hamm *et al.* (1990), who determined that the transport of U1 and U2 snRNPs require both hypermethylation and protein binding. The particles were retained in the cytoplasm by either altering the cap structure or by modifications that prevented protein assembly. Both signals (the 5' cap and the core proteins) are also necessary for the nuclear import of U4 and U5; however, the requirement for the cap structure is less stringent (Fischer *et al.*, 1991). Thus, alterations in the cap structure which block the import of U1 and U2 have only a limited effect on U4 and U5. Fischer *et al.* (1993) have obtained evidence that the differential requirement for the trimethylguanosine cap is related to the structure of the specific U RNAs.

In contrast to the results obtained for polymerase II U snRNPs, the 5' (γ-monomethyl phosphate) cap is not necessary for the nuclear uptake of U6 snRNP; however, U6 import can be prevented by mutations in the region containing nucleotides 21–26 (Hamm and Mattaj, 1989). This region is thought to be the binding site for a protein that contains an NLS essential for transport.

Interestingly, the polymerase II and III U snRNPs use different kinetic pathways to enter the nucleus. This was initially demonstrated in competition studies performed on *Xenopus* oocytes by Michaud and Goldfarb (1991). In this investigation it was found that saturating amounts of BSA conjugated with synthetic peptides containing the SV40 large T NLS reduced the nuclear import of karyophilic proteins (nucleoplasmin and BSA containing the large T NLS) and U6 snRNP, but not U2 snRNP. Further evidence for multiple pathways was obtained by Fischer *et al.* (1991), who observed that a dinucleotide analog of the trimethylguanosine cap blocked the transport of polymerase II U snRNPs, but not U6 snRNP or lamin L1 (a karyophilic protein). It has also been suggested that a third pathway might exist for the import of U3 snRNA (Michaud and Goldfarb, 1992).

C. Nuclear Localization Signal Receptors

The involvement of receptors in facilitating nuclear protein import is implied from the fact that specific signals are required to initiate the transport process. Supporting data were first reported by Goldfarb *et al.* (1986), who found that the nuclear uptake of proteins conjugated with synthetic peptides containing the SV40 large T targeting sequence is saturable, and can be competed by adding an excess of synthetic NLS. Newmeyer and Forbes (1988) and Richardson *et al.* (1988) subsequently demonstrated that binding to the surface of the pore complex is a prerequisite for mediated translocation into the nucleoplasm.

With this background, a number of laboratories initiated searches for receptor molecules. The underlying approach was to identify proteins that specifically bound peptides containing NLS. Binding was studied by using photoactivatable cross-linkers, affinity columns, and blotting procedures. The results of these experiments are summarized in Table I. Although the ability to bind NLS is clearly an essential feature of a receptor, it should be pointed out that binding in experimental situations does not necessarily prove that these proteins are actually involved in transport (potential problems that can arise in these studies are discussed in Forbes, 1992). Thus, unless functional data are also available, these molecules are best described as candidate receptors. The fact that candidate receptors have been found in nuclear, cytoplasmic, and envelope fractions (see Table I) implies that NLS binding proteins are not localized exclusively to the nuclear surface, and raises the possibility that targeted proteins first bind to factors within the cytoplasm and are then transferred to secondary pore complex receptors. Furthermore, the intracellular distributions are consistent with the hypothesis that the transport factors shuttle between the cytoplasm and the nucleus.

Direct evidence for the existence of cytoplasmic transport factors was first obtained by Newmeyer and Forbes (1990), and Adam et al. (1990). Newmeyer and Forbes (1990) studied transport in isolated liver nuclei that were added to Xenopus egg extracts. The extracts are able to maintain the integrity of the envelopes, and can also support signal-mediated import of proteins through the pores. In this system, transport was prevented when the extracts were treated with the sulfhydryl alkylating agent N-ethylmaleimide (NEM). Analysis of fractions of untreated extracts that were able to overcome the NEM block revealed that at least two cytoplasmic factors are involved in transport, NIF1 and NIF2. The composition of these factors has not been established, but it is known that NIF1 facilitates binding of targeted proteins to the pores.

Adam and co-workers (1990) found that the nuclei of cells made permeable with digitonin could import karyophilic proteins if they were supplemented with a 100,000 g cytosolic fraction and ATP. In a later study, Adam and Gerace (1991) presented evidence that the cytosolic fraction contains three transport factors, two of which are NEM-sensitive. In addition, they purified two NEM-sensitive, NLS binding proteins of 54 and 56 kDa. These proteins alone caused a 2–3-fold increase in transport when added to permeabilized cells.

Stochaj and Silver (1992b) isolated a conserved group of 70-kDa phosphoproteins from Drosophila, HeLa cells and Zea mays that bind NLS. These proteins are found in both the nucleus and cytoplasm (as would be expected of shuttling proteins) and, as demonstrated using permeabilized cells, can mediate nuclear import. One interesting feature of these proteins is that they must be phosphorylated in order to bind NLS.

TABLE I
Proteins Binding Nuclear Transport Signals[a]

Species	Molecular weight[b]	Location[c]	Interacting signal	Signal method	Reference
Rat	70K, 60K	N + C, N + C	SV40-T	Cross-linking	Adam et al. (1989)
	140K, 100K, 70K, 55K	N, C, C, N	SV40-T, nucleoplasmin, E1A, MATα2	Cross-linking	Yamasaki et al. (1989)
	76K, 67K, 59K, 58K	N, N, N, N	SV40-T	Cross-linking	Benditt et al. (1989)
	140K, 55K	NO, N	SV40-T	Blotting	Meier and Blobel (1990)
	69K	N	SV40-T, nucleoplasmin	Affinity chromatography	Yoneda et al. (1988) Imamoto-Sonobe et al. (1990)
	60KD	N	SV40-T, nucleoplasmin	Blotting, affinity chromatography	Haino et al. (1993)
Human	38K (B23)	NO	SV40-T	Affinity chromatography	Goldfarb (1988)
	66K	N	SV40-T, protein A	Cross-linking	Li and Thomas (1989)
Yeast	140K, 95K, 70K, 59K	N, N, N, N	SV40-T, H2B, GAL4	Blotting	Silver et al. (1989)
	70K (NBP70)	N + C	SV40-T, nucleoplasmin	Blotting, affinity chromatography	Stochaj and Silver (1992a) Lee and Melese (1989)
	67K (NSR1)	NO	SV40-T, H2B	Blotting, affinity chromatography	Lee et al. (1991)

[a] Adapted from Yamasaki and Lanford (1992, Table III, p. 152).
[b] Molecular weight refers to the apparent weight, including any modifying signal peptide.
[c] Location: N, in nuclei; C, in cytoplasm; N + C, in nuclei and cytoplasm; NO, in nucleoli.

Moore and Blobel (1992) identified two transport factors in *Xenopus* oocytes; one (fraction A) facilitates the accumulation of NLS-containing proteins at the nuclear surface (similar to NIF1), and the second (fraction B) is necessary for translocation of bound material through the pores. The involvement of two heat shock proteins (hsp70 and hsc70) in signal-mediated transport has been suggested by Shi and Thomas (1992), and Imamoto *et al.* (1992).

The relationships among the different NLS binding proteins and cytoplasmic factors that have been identified in various laboratories remain to be worked out (some differences are most likely due to the diversity of the experimental approaches); however, there is general agreement that protein import involves the activity of at least two cytoplasmic factors.

D. Role of Pore Proteins in Transport

Two general procedures have been used to investigate the function of the pore proteins in signal-mediated nuclear transport. The first method involves treatment with agents that specifically bind to, and interfere with the activity of the nucleoporins. This would include WGA, which reacts with the O-linked GlcNAc residues of the glycoproteins, and antibodies. In the second approach, the composition of the pores is altered either by deleting or modifying the component proteins.

The effects of WGA, or antipore antibodies on nuclear protein import in vertebrate cells has been studied by microinjection or by treating nuclei that were reconstituted in *Xenopus* egg extracts. In all of these investigations (Finlay *et al.*, 1987; Yoneda *et al.*, 1987; Dabauvalle *et al.*, 1988a,b; Featherstone *et al.*, 1988; Wolff *et al.*, 1988; Newmeyer and Forbes, 1988), mediated protein transport was inhibited by the binding reactions, but passive diffusion of small macromolecules was not affected, demonstrating that the pores were not sterically blocked by the bound proteins. Newmeyer and Forbes (1988) concluded that the pore glycoproteins function in the translocation step of protein import since binding of signal-containing proteins occurred even in the presence of WGA. Sterne-Marr *et al.* (1992) reported that the O-linked pore glycoproteins interact with a cytoplasmic factor in reticulocyte lysates that is necessary for nuclear transport in permeabilized cells. This factor has not yet been identified, but it could be related to fraction B identified by Moore and Blobel (1992).

In addition to protein import, RNA efflux is also inhibited by the appropriate antibodies against pore glycoproteins, or WGA. This was demonstrated in *Xenopus* oocytes by Featherstone *et al.* (1988) for 5S ribosomal RNA and tRNA. The antibody used in these experiments (RL1; Snow *et al.*, 1987) was more effective when injected into the nucleus than into the cytoplasm, which could reflect an asymmetric distribution of the antigen between the two surfaces of the pores. WGA was found to inhibit the efflux of rRNA in oocytes (Bataille *et al.,* 1990); however, the lectin was only effective when injected into the nucleus.

Two laboratories reconstituted nuclear envelopes in *Xenopus* egg extracts that had been depleted of O-linked glycoproteins and obtained somewhat different results. Finlay and Forbes (1990) found that extracts previously treated with WGA could support the formation of both nuclear envelopes and pores upon the addition of sperm chromatin. Although the pores appeared normal when visualized with the EM, they were unable to transport or bind signal-containing proteins. Dabauvalle *et al.* (1990) used λ DNA as a substrate for nuclear envelope formation in extracts pretreated with either WGA or antibody against the glycoprotein p68. These investigators found that the inner and outer nuclear membranes re-formed, but not the pores. The reason for the disparity in these results is not known, but differences in experimental conditions are very likely involved.

Greber and Gerace (1992) showed that the transmembrane glycoprotein gp210 might also function in translocation. In this study mRNA (obtained from hybridoma cells) coding for a monoclonal IgG against an epitope located in the luminal domain of gp210 was microinjected into rat kidney cells. Following its expression and assembly in the rough endoplasmic reticulum, the antibody was able to migrate through the reticulum to the lumen of the nuclear envelope and bind gp210. The result was a significant decrease in both the signal-mediated transport of nucleoplasmin-coated gold particles, and the passive diffusion of 10-kDa dextran. Whether gp210 itself, or in conjunction with other regulatory substances, actually functions to control transport *in vivo* is not known, but it clearly has the capacity to do so.

The function of the yeast NSP1 pore protein has been investigated by Mutvei *et al.* (1992). These investigators placed the NSP1 gene under the control of the GAL10 promoter; thus, the expression of NSP1 could be repressed by substituting glucose for galactose in the medium. In the repressed state, the nuclei were unable to accumulate chimeric NLS-containing reporter proteins, suggesting a role for NSP1 in mediated transport. NSP1 is also likely to be involved in pore formation, as indicated by the fact that repression was accompanied by a marked decrease in pore density (pores/μm2).

IV. Regulation of Macromolecular Exchanges

In this section we consider the different mechanisms of regulating transport. Consistent with our current understanding of signal-mediated nuclear transport, control of macromolecular exchanges could depend on differences in the properties of the permeant molecules, or variations in the transport process itself. The characteristics of the permeant molecules that are most likely to have a major influence on nucleocytoplasmic exchanges relate to (1) signal composition, which is encoded in the primary structure of the protein, (2) the accessibility of the signal to the receptor(s), which could be affected by either binding reactions that anchor the permeant molecule to cytoplasmic elements or masking of the signal, and (3) post-translational changes, such as phosphorylation, that interfere with signal activity.

Broadly defined, the transport process can be thought of as consisting of three fundamental steps: interactions between the NLS and cytoplasmic factors, binding to pore receptors, and translocation through the pores, including subsequent release into the targeted compartment. Regulation could occur at any level in the process. Although less specific in its effects than alterations in the properties of the permeant substances, regulation by modulation of components of the transport machinery could significantly alter a number of fundamental metabolic events that underlie, for example, cell division and transformation.

A. Variations in the Properties of the Permeant Molecules

1. Signal Composition

By using both recombinant DNA procedures and carrier proteins conjugated with synthetic peptides containing different NLSs, it has been possible to evaluate the effects of the number and amino acid sequence of NLSs on nuclear transport. The initial evidence that there is a direct relationship between signal number and the rate of nuclear import was obtained by Dingwall *et al.* (1982), who analyzed the transport of the pentameric oocyte protein nucleoplasmin. Pentamers containing 0 to a maximum of 5 NLS domains were obtained by mild digestion with proteolytic enzymes, and could be distinguished by polyacrylamide gel electrophoresis (PAGE) owing to different migration rates. It was found, following cytoplasmic injection into oocytes, that the rate of nuclear import, but not the equilibrium distribution, was dependent on the number of NLSs. The nuclear uptake of synthetic NLS-protein conjugates injected into cultured cells is also dependent on the number of signal peptides. Thus, increasing the number of large T NLSs bound to ovalbumin from two to six reduced the import time 8-fold (Lanford *et al.*, 1986). Furthermore, mutant forms of the large T signal, in which either Asn or Thr was substituted for Lys at position 128, are inactive when present in low copy numbers, but have a limited capacity to support transport when the number is increased (Goldfarb *et al.*, 1986; Lanford *et al.*, 1990). Roberts *et al.* (1987) genetically introduced defective large T NLSs into pyruvate kinase (normally a cytoplasmic protein) and found that whereas a single copy of the signal was unable to induce nuclear import, multiple copies did cause a shift in distribution to the nucleoplasm, again demonstrating that signals have an additive effect.

Dworetzky *et al.* (1988) reported that signal number not only affected the rate of nuclear import but also the functional size of the transport channel. In these experiments, colloidal gold particles of various sizes were coated with BSA that was conjugated with different numbers of synthetic peptides containing the large T NLS. The coated gold was injected into *Xenopus* oocytes and the number and size of the particles that entered the nucleoplasm was determined by electron microscopy. A direct correlation was found between the the size of the particles that entered the nucleus and the number of NLS. The exclusion limits of the transport channels for BSA containing 3, 5, and 8 signals were approximately 11, 21, and 26 nm, respectively.

Yamasaki and Lanford (1992) suggested that increased signal number could enhance transport rates by (1) interacting with higher affinity to cytoplasmic receptors (assuming that the receptors contain multiple binding sites), (2) reaching threshold binding levels at the pore surface more rapidly, or (3) facilitating transfer from cytoplasmic to pore receptors. The results obtained by Dworetzky *et al.* (1988) further suggest that the degree of dilation of the central pore channel is variable and depends on the number of binding events at the pore surface.

Chelsky *et al.* (1989) and Lanford *et al.* (1990) conjugated chicken serum albumin and mouse IgG, respectively, with synthetic peptides that contained known NLSs or putative signals. The conjugates were microinjected into cultured cells, and nuclear import was assayed at specific times after injection by immunofluorescence. The results demonstrated that not all signals were equally effective in mediating nuclear transport. Consistent with these findings, Dworetzky *et al.* (1988) showed that nucleoplasmin-gold was transported at a significantly greater rate than particles coated with BSA–large T NLS conjugates, even though the large T signals were more numerous. In interpreting these results, however, it is important to keep in mind that the activity of the NLS tested could be differentially affected by the experimental manipulations. As demonstrated by Rihs and Peters (1989), the capacity of signals to facilitate transport can be modulated by their flanking sequences, which were not included in the carrier–NLS conjugates. Moreover, most of the data comparing signal activity is qualitative. Further quantitative studies might also reveal differences in uptake rates.

Overall, there are indications that signal content might be involved in regulating nucleocytoplasmic exchanges. This suggests a functional role for the observed variations in signal sequences, as well as the presence of multiple signals in some proteins.

2. Accessibility of the Signal

a. Anchoring to Cytoplasmic Elements Type II cAMP-dependent protein kinase (PKA), which is involved in transcriptional control, is a prime example of anchorage-dependent regulation. The inactive form of PKA consists of two catalytic and two cAMP-binding regulatory subunits that are associated with the Golgi apparatus. The localization of these subunits following the treatment of cultured cells with forskolin (an activator of adenylate cyclase) has been analyzed by Nigg *et al.* (1985) and Nigg (1990). Immunofluorescence studies using antibodies against each of the subunits revealed that in response to increased levels of cAMP, the catalytic subunits were released and migrated from the Golgi region to the nucleus; however, the regulatory subunits did not redistribute. When treated cells were incubated in forskolin-free medium, the catalytic subunits left the nucleus and reassociated with the regulatory components in the Golgi region.

b. Masking the Nuclear Localization Signal: NRD Transcription Factors Numerous transcription factors are regulated at the posttranslational, rather than the translational level. This allows cells to respond more rapidly to external signals, but requires that the factors be prevented from binding with their target DNA sequences in the absence of the appropriate stimulus. The activity of one group of transcription factors, which includes NF-κB, the products of the *rel* oncogene family, and the *Drosophila* regulatory factor dorsal, are posttranslationally controlled by cytoplasmic inhibitory subunits that restrict nuclear import. All of these factors share a highly conserved 300-amino acid, amino-terminal domain, referred to as the NRD (*NF-κB*/Rel/*dorsal) motif. This motif contains DNA-binding and dimerization sequences.

NF-κB is a heterodimer composed of 50 (p50) and 65 (p65) kDa subunits. Both subunits contain the NRD domain, but only p65 has a transactivating region, which is located separately in the carboxy-terminus of the protein. Putative NLSs for p65 and p50 (Arg–Lys–Arg–Gln–Lys and Lys–Arg–Lys–Arg, respectively) are also present in the NRD regions (Grimm and Baeuerle, 1993). Variants of p50 and p65 have been identified, including *rel,* which is closely related to p65 (reviewed in Blank *et al.,* 1992). In addition, four inhibitor proteins in the IκB family have been purified (reviewed in Schmitz *et al.,* 1991). These NRD motif and inhibitor homologs are, to a large extent, interchangeable and able to form functional regulatory complexes. Since p65 and p50 have been extensively studied, we will limit the discussion of nuclear transport regulation to these sub-units.

NF-κB was originally identified as an enhancer element for the κ light chain in B-lymphocytes. It is now known to be functional in most cell types and is likely to be involved in the activation of over 20 genes (see reviews in Schmitz *et al.,* 1991; Grimm and Baeuerle, 1993). Initial indications that nuclear transport of NF-κB is regulated was obtained from fractionation studies which showed that NF-κB DNA binding activity was not present in either the cytoplasmic or nuclear fractions of nonstimulated cells, but was detected following treatment of the cytosol with sodium deoxycholate. This activity was identical to that found in the nuclei of stimulated cells; in addition, an inhibitory factor, referred to as IκB, was released by deoxy-cholate (Baeuerle and Baltimore, 1988, 1989). The IκB-like proteins all contain ankyrin-like repeats (33 amino acid domains that include Thr–Pro–Leu–His–Leu–Ala sequences) that are thought to be directly in-volved in controlling the activity of NF-κB (Blank *et al.,* 1992). The fractionation studies provided indirect evidence that an inactive form of NF-κB is retained in the cytoplasm and enters the nucleus after disso-ciating from the inhibitor subunit.

More direct evidence for inducible transport has recently been obtained by Zabel *et al.* (1993). In these studies it was shown that p65, p50, and IκB all entered the nucleus when they were overexpressed individually in Vero cells. The nuclear import of p65 and p50 is dependent on functional NLSs; the uptake of IκB, which has a mass of 35 kDa, is most likely the result of passive diffusion. When p65 or p50 were overexpressed along with IκB, nuclear transport was blocked. Since it could be shown that this was a specific effect, these findings indicate that IκB can regulate the intracellu-lar distribution of the NF-κB subunits. Furthermore, in the presence of IκB, antibodies against epitopes overlapping the p65 and p50 NLS were unable to bind. The latter results favor the view that the inhibitor acts by masking the NLSs (this could be a steric and/or allosteric effect) rather than by anchoring NF-κB to cytoplasmic elements. The NF-κB inhibitor complex can be dissociated *in vitro* by phosphorylation of IκB; however, *in vivo,* oxygen radicals might also be involved (Grimm and Baeuerle, 1993). The process regulating the intracellular distribution of NF-κB is illustrated in Fig. 3.

Prior to its incorporation into NF-κB, p50 is retained in the cytoplasm as a 110-kDa precursor (p110). p50 is located in the N-terminal half of the precursor, whereas the C-terminus is homologous to IκB. The mechanism by which the p50 NLS is suppressed in the precursor has been investigated by Henkel *et al.* (1992). It was found that specific antibodies to the p50 NLS were unable to react with p110; however, antibody binding, as well as nuclear import, was obtained if 191 amino acids were deleted from the

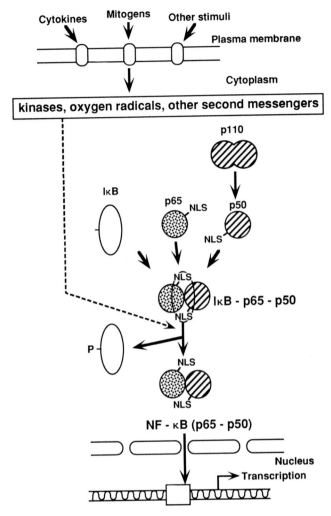

FIG. 3   A model of the pathways for the activation and nuclear transport of NF-κB. Stimulation by extracellular signals leads to the phosphorylation of IκB, its release from NF-κB, and the unmasking of the p50 and p65 NLS. p110 is the precursor form of p50. (After Blank *et al.,* 1992.)

carboxy-terminus, or if p110 was denatured. Based on these results, and additional data indicating that p110 is not tightly associated with other cytoplasmic components, Henkel and co-workers proposed a novel intramolecular mechanism in which the hinged carboxy-terminus is able to bend back and mask the NLS.

The dorsal protein was cloned and sequenced by Steward (1987). It contains an NRD motif and functions in the formation of dorsoventral polarity in the *Drosophila* embryo (reviewed in Govind and Steward, 1991). Dorsal can act either as an activator or repressor of a number of genes, depending on its concentration in the nucleus. Consistent with its regulatory role, it is asymmetrically distributed within the blastoderm. A nucleocytoplasmic gradient exists between the ventral pole, where dorsal is primarily nuclear, and the dorsal pole, where it is localized in the cytoplasm (Rushlow *et al.*, 1989; Steward, 1989; Roth *et al.*, 1989). Establishing the necessary gradient is a complex process involving several different dorsal group genes (Govind and Steward, 1991); however, the protein cactus seems to play a central role in regulating selective nuclear transport. In mutants lacking the cactus gene, dorsal is nuclear in both the dorsal and ventral regions (Roth *et al.*, 1989). There is also evidence that cactus, like IκB, contains ankyrin repeats (Schmitz *et al.*, 1991). These findings support the model that dorsal complexes with cactus (analogous to the interaction between NF-κB and IκB) and, as a result, is retained in the cytoplasm. The complex is dissociated by the action of two additional proteins, pelle and tube.

c. Masking the Nuclear Localization Signal: Steroid Receptors   In the absence of hormone, the glucocorticoid receptor is localized in the cytoplasm; however, upon the addition of steroid it is rapidly translocated to the nucleus. Picard and Yamamoto (1987) identified two NLS (NL1 and NL2) within the receptor. NL1 is homologous to the SV40 large T NLS, and when this signal region is fused to β-galactosidase, the chimeric protein enters the nucleus in a hormone-independent fashion. NL2 is located in the steroid binding domain and, when fused with β-galactosidase, its activity was found to be hormone dependent. Furthermore, NL2 appears to be the dominant signal since fusion proteins containing NL2 alone are imported into the nucleus at rates similar to those observed for the native receptor.

One explanation for these results is that hormone binding facilitates nuclear transport by unmasking the NLS; however, interpretation of the data is complicated by the fact that within the cytoplasm the unligated receptor is bound to at least two molecules of hsp90 and one 59-kDa protein (Muller and Renkawitz, 1991), forming a 9S complex. It is possible that this complex is bound to cytoskeletal elements (Pratt, 1992) and that the steroid acts by dissociating the proteins, thereby releasing the receptor from its cytoplasmic anchorage site. Figure 4 illustrates the possible events leading to nuclear import.

The progesterone receptor also contains two NLSs, one of which is constitutive and the second hormone dependent. Since the unligated re-

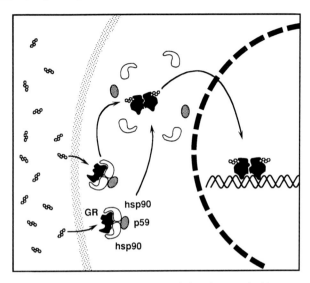

FIG. 4   Requirements for the nuclear transport of the glucocorticoid receptor. Upon the addition of hormone, the receptor–hsp90–p59 complex dissociates and the receptor is rapidly transported to the nucleus. Hormone binding facilitates the transport process by unmasking the NLS and/or releasing the receptor from a cytoplasmic anchorage site. (After Muller and Renkawitz, 1991.)

ceptor is present in the nucleus, it appears that the dominant signal (homologous to the large T NLS) acts in the absence of hormone. When this signal is deleted, the receptor localizes in the cytoplasm but can reenter the nucleus if the second NLS is activated by the addition of steroid (Guiochon-Mantel *et al.*, 1989, 1992). Only one NLS has been identified in the estrogen receptor, and it is not regulated by steroids (Picard *et al.*, 1990).

3. Post-translational Modifications

The finding (discussed earlier) that the activity level of the SV40 large T NLS can be experimentally altered by phosphorylation of specific sites in the adjacent flanking region suggests that this could be an effective strategy to control transport *in vivo*. In fact this appears to be the mechanism that is employed in yeast to regulate the cell cycle-dependent nuclear transport of the transcription factor SWI5. This factor is required for the transcription of the HO gene, which codes for an endonuclease that causes mating-type switching, and is expressed only during G_1 in the mother cells. Consistent with its role in transcription, SWI5 is incorporated in the nucleus when the cells enter G_1, but is retained in the cytoplasm during the S, G_2, and M phases of the cycle (Nasmyth *et al.*, 1990). The intracellular distribution of SWI5 is dependent on the phosphorylation state of three serine residues located within or adjacent to the bipartate (nucleoplasmin-like) NLS. When these residues are phosphorylated, presumably by cdc28 kinase, SWI5 is cytoplasmic. Dephosphorylation prior to G_1 is a prerequisite for nuclear import (Moll *et al.*, 1991).

Recently, Hennekes *et al.* (1993) investigated the effect of phosphorylation on the transport of chicken lamin B2 into the nuclei of permeabilized HeLa cells. They determined that phosphorylation of serines 410 and 411 (the NLS occupies positions 414–420) by protein kinase C (PKC) blocks import into the nucleus, and suggested that PKC, which is under hormonal control, might mediate the nucleocytoplasmic exchange of a variety of other proteins as well.

B. Variations in the Transport Machinery

1. Passive Diffusion

Early evidence that the properties of the nuclear envelope can vary during the life of a cell was obtained by analyzing the nuclear uptake of PVP-coated gold particles in amebas (Feldherr, 1966, 1968). These investigations revealed that for the first 2 hr after the completion of mitosis, gold particles were incorporated into the nucleus at a significantly greater rate than at later stages of the cell cycle. The fact that PVP lacks an NLS implies that nuclear import occurred by passive diffusion. Since the nuclear membranes had re-formed prior to the first experimental time point, as demonstrated by EM, the results indicate that there are differences in the properties of the pores.

Feldherr and Akin (1990) microinjected BSA-coated gold particles into HeLa cells at different times in the cell cycle and found that the relative rates of diffusion into the nucleus, expressed as the nuclear to cytoplasmic (N/C) gold ratio, peaked during the 1st and 5th hours after division. Although changes in the import rates were observed, the exclusion limit for the gold particles (approximately 12 nm) remained unchanged. The diffusion peaks were closely correlated with the maximum rates of pore formation in HeLa cells, which occur at 0–1 and 4–5 hr after division (Maul *et al.*, 1972). This relationship lead Feldherr and Akin to suggest that newly forming pore complexes are more permeable (i.e., have larger passive diffusion channels) than mature pores, and that the differences in BSA-gold import reflected the proportion of the two populations. Whether variations in diffusion are physiologically significant, or simply a consequence of increasing pore number is not known.

2. Signal-Mediated Transport in Dividing and Quiescent Cells

Mediated nuclear transport has been compared during different phases of cellular activity by analyzing the intracellular distribution of nucleoplasmin-coated gold particles following cytoplasmic injection (Feldherr and Akin, 1990, 1991, 1993). Relative rates of nuclear import were obtained by establishing the N/C gold ratios; in addition, the functional size of the transport channel was determined by injecting different-sized gold fractions and measuring the dimensions of the particles that are able to enter the nucleus.

Using this procedure, Feldherr and Akin (1990) investigated signal-mediated nuclear transport during the cell cycle in HeLa cells. Particles 11–27 nm in diameter (including the protein coat) were injected 30 min (early G_1), 4.5 hr (G_1), 8.5 hr (S phase), or 17–20 hr (G_2) after division, and the cells were fixed for electron microscopy 15 min later. At all time intervals, the exclusion limit for mediated transport was 23–25 nm and the N/C gold ratios ranged from 2.5 to 3.5. The differences in ratios were not statistically significant, indicating that transport is maintained at the same level throughout the cell cycle. However, since relatively few time points were studied, transient changes, which could be physiologically significant, might have gone undetected. The results also show that mediated transport capacity is reestablished within 30 min after division. Rapid recovery of this function had previously been demonstrated by Benavente et al. (1989). Variations in passive diffusion rates during the cell cycle (see earlier discussion) would not have affected the import of nucleoplasmin-gold, since essentially all of the injected nucleoplasmin particles were larger than the diffusion channels.

Although no differences in transport were observed during the cell cycle, Feldherr and Akin (1990) did detect highly significant decreases during quiescence. In these experiments, nucleoplasmin-gold import was compared in proliferating and confluent (G_0) 3T3-L1 cells. The N/C ratios at 30 min were 0.67 and 0.09, respectively, and there was an accompanying decrease of approximately 4 nm in the maximum size of the transport channel (from about 24 to 20 nm in diameter) in confluent cells. The distribution of particle sizes in nucleus and cytoplasm is shown in Fig. 5. When the confluent cells were induced to differentiate, transport capacity returned to the proliferating level. These differences in transport were not due to changes in pore number.

A variation of these results was obtained for proliferating and G_0 BALB/c 3T3 cells (Feldherr and Akin, 1991). As BALB/c cells became confluent, the N/C gold ratio decreased as much as 78-fold, but, in contrast to the 3T3-L1 data, there was no accompanying change in the size distribution of the particles located in the nucleoplasm, suggesting that a small proportion of the pores remained fully competent. To determine whether the remain-

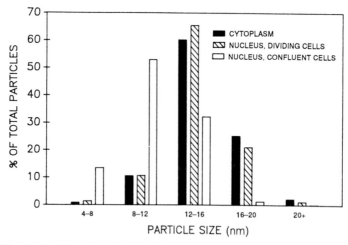

FIG. 5   The size distribution of nucleoplasmin-gold particles in the nuclei of proliferating and confluent 3T3-L1 cells 30 min after microinjection. The dimensions of the particles in the cytoplasm that are available for transport are also included in the histogram. (From Feldherr and Akin, 1991.)

der of the pores became nonfunctional, or functioned at a reduced level, small nucleoplasmin-gold particles (total diameter 5–8 nm) were injected. These particles were incorporated into the nuclei at equivalent rates in quiescent and proliferating populations, confirming the view that all or most of the pores remained active in quiescent cells, but that dilation of the transport channel was restricted. Analysis of nuclear import using intermediate-sized gold fractions (total diameter 5–15 nm) revealed that the functional pore size decreased by as much as 10 nm in the G_0 cells. Similar results were obtained when the G_0 state was induced by serum depletion.

The available evidence (discussed earlier) indicates that macromolecular exchanges both into and out of the nucleus occur through the same central channel within the pores, and are subject to the same size restrictions. Assuming, therefore, that the changes in transport capacity during G_0, which result in reduced import, have a similar effect on nuclear export, then the endogenous substances that are most likely to be affected are ribosomal subunits and mRNP particles, since these are the largest materials that routinely cross the nuclear envelope. Control of the export of these substances could also help explain the overall decrease in metabolic activity in quiescent cells (Levine *et al.*, 1965). Both the large ribosomal subunit, which is approximately 22 nm wide, and mRNP particles, which differ in size but can be as large as 25 × 135 nm (e.g., BR RNP in *Chironomus;* Mehlin *et al.*, 1992), are approaching the exclusion limits of the pores in proliferating cells; a reduction in functional pore size, to the extent observed in G_0 cells, could significantly restrict export. The intracellular distributions of rRNA and mRNA in proliferating and serum-deprived cells are consistent with this proposal.

A comparison of proliferating and quiescent 3T3 cells by Johnson et al. (1974) revealed that the amounts of rRNA and mRNA were significantly greater (2.8- and 4.0-fold, respectively) in the proliferating cultures. Johnson et al. (1975, 1976) have shown that mRNA is not regulated at the transcriptional level (heterogeneous RNA—hnRNA—is synthesized at the same rates during both functional states); instead, processing and export to the cytoplasm appear to be more efficient in proliferating cells. Regulation of rRNA occurs at several levels, including transcription (Mauck and Green, 1973), turnover (Abelson et al., 1974), and processing (Johnson et al., 1976). However, it also appears that rRNA, and particularly the large subunit, is retained in the nuclei of quiescent cells for a longer time. Although it is clearly a possibility, it should be emphasized that there is no definitive evidence showing that translocation through the pores is the rate-limiting step for RNP efflux.

C. Regulation of the Transport Machinery

1. The Pore Complex and Cytoplasmic Factors

The decrease in transport capacity in quiescent cells could be caused by differences in the availability of cytoplasmic factors, specifically ATP or NLS receptors, or by changes in the properties of the pores. To distinguish between these possibilities, Feldherr and Akin (1993) mechanically fused proliferating and serum-depleted quiescent BALB/c cells and compared the import of nucleoplasmin-gold into the two nuclei after allowing sufficient time for soluble factors to be distributed throughout the heterokaryon. Significant differences in transport capacity between the proliferating and quiescent nuclei were maintained even though they shared the same cytoplasm. This implies that the properties of the pore complexes, rather than the availability of cytoplasmic factors, were primarily involved in regulating nuclear transport in the two functional states. Furthermore, the number of gold particles located at the cytoplasmic faces of the pores was significantly greater in proliferating nuclei, suggesting that the differences in transport capacity might be related to the concentration or activity of the pore complex receptors.

The macro- and micronuclei of Tetrahymena also exhibit different transport properties that can be attributed to the functional state of the pores (White et al., 1989). Following microinjection, histone H1 and BSA–SV40 large T NLS conjugates were incorporated only into the macronuclei, regardless of the replicating state of the cells. Histone H4, on the other hand, was taken up only by the macronuclei in nondividing cells, whereas in dividing cells import was restricted to the micronuclei. Thus, highly specialized transport mechanisms appear to be operational in this organism.

Li *et al.* (1992) produced antibodies to a 66-kDa protein that binds the SV40 large T NLS, and determined its intracellular distribution in HeLa and monkey BSC-1 cells by immunofluorescence. In proliferating cultures, the NLS binding protein was cytoplasmic, except for a weak punctate fluorescence at the nuclear surface. In confluent monkey cells, the punctate staining remained, but the cytoplasmic fluorescence was greatly reduced. These results raise the possibility that, in addition to the pores, cytoplasmic receptors might also be involved in controlling nuclear import. The degree of regulation at this level could vary according to the cell type or the specific sequence of the targeting signal.

One possible regulatory role for NLS receptors was suggested by the experiments of Slavicek *et al.* (1989) and Standiford and Richter (1992). These investigators studied the effects of two E1a NLS on mediated nuclear transport in *Xenopus* oocytes and embryos. One signal (Lys–Arg–Pro–Arg–Pro), originally identified by Lyons *et al.* (1987), is located at the carboxy-terminus. It is similar to the large T NLS, and can direct nuclear import in both oocytes and embryos. The second signal, located between amino acids 140–185, is developmentally regulated (drNLS) and functions in oocytes and embryos up to, but not beyond, the early neurula stage. The drNLS is complex and unique in three respects: (1) it has two spacer regions, (2) it is hydrophobic, and (3) it contains no basic amino acids. The two E1a signals compete for nuclear import, demonstrating that they use one or more common elements of the transport machinery. Although several methods of modulating drNLS activity have been proposed by Standiford and Richter (1992) (e.g., binding to cytoplasmic anchoring sites or piggybacking on a second developmentally regulated protein), one intriguing possibility involves the production (or activation) of an isoform of a signal receptor that preferentially binds the drNLS. This mechanism could also explain the competition between the signals.

Although there is no direct evidence that the binding characteristics of cytoplasmic receptors vary in relation to cellular activity, this could serve as an effective means of regulating the nuclear import of specific proteins, and should be investigated further.

2. Effect of Cell Shape

A number of cellular functions, including growth and differentiation, that require DNA synthesis and/or specific gene expression are dependent on cell shape (see review in Ingber and Folkman, 1989). The relationship between cell shape and nuclear activity is illustrated by the studies of Ingber *et al.* (1987). In this investigation, DNA synthesis was measured in cultured endothelial cells that were grown on different substrates. A direct relationship existed between the degree of cell spreading (a function of the substrate) and the level of DNA synthesis. Furthermore, as the projected area of the cells increased, there was a corresponding increase in the area of the nucleus, suggesting a structural interaction between the cell and nuclear surfaces. As proposed by Ingber and Folkman (1989), the shape-related increase in DNA synthesis could be caused by mechanical forces exerted on the nuclear matrix, which in turn, influence the organization of the associated DNA. Alternatively, expansion of the nucleus could affect the permeability of the pore complexes. Evidence supporting the latter possibility was obtained by Jiang and Schindler (1988), who detected increases in the passive diffusion of dextran into the nuclei of cells that were induced to spread by treatment with growth factors.

To determine if signal-mediated transport through the pores is also shape dependent, Feldherr and Akin (1993) compared the intracellular distribution of nucleoplasmin-gold in rounded and extended BALB/c cells that were prepared as described by Ireland *et al.* (1987). In this procedure, different sizes of palladium domains are deposited through a copper mask on a coverslip previously coated with nonadhesive, poly-2-hydroxyethyl methacrylate. The cells plated on this surface adhere only to the palladium, and their shapes are dependent on the dimensions of the domains that they occupy. Cells attached to small domains remained rounded, whereas those located on large areas of palladium were flattened. Analysis of the gold distribution in injected cells showed that the functional size of the pore complexes, as well as the N/C ratios obtained using 11–27 nm particles, were significantly greater in the extended cells. No changes were detected in either the number or diameter (membrane to membrane distance) of the pores in rounded versus extended cells.

One mechanism (reviewed in Hanson and Ingber, 1992) that could account for the effect of cell shape on nuclear transport involves the transmission of tension from the cell surface to the nuclear envelope by elements of the cytoskeleton. Ingber and Folkman (1989) proposed a model in which actin filaments, microtubules, and intermediate filaments form an integrated system that acts to adjust cell and nuclear shape in response to external forces applied at the cell attachment sites (see also Wang *et al.*, 1993). It is likely that the intermediate filaments play a direct role in force transmission since these structures are known to interact with both the plasma membrane and nuclear surface (Goldman *et al.*, 1986). Tension applied to the envelope or perhaps to the pores themselves (Carmo-Fonseca *et al.*, 1987) could modulate the structure and functional capacity of the pore complexes.

Elements of cytoskeleton, particularly actin filaments, could also mediate nuclear transport by acting as binding sites for a variety of kinases, phosphatases, and transcription factors (Ben-Ze'ev, 1991). Release of these regulatory substances as a result of cytoskeletal reorganization could affect the activity of the transport machinery by introducing posttranslational modifications of existing proteins or by altering the protein composition.

3. Effect of Growth Factors

Serum-depleted BALB/c cells can be stimulated to reenter the cell cycle either by increasing the serum concentration (from 0.5 to 10%) or by adding physiological concentrations of platelet-derived growth factor (PDGF), epidermal growth factor (EGF), and insulin-like growth factor-1 (IGF-1). These growth factors act in a coordinated fashion as described by Pardee (1989); PDGF is necessary for the initial transition from G_0 to G_1 (Scher *et al.*, 1979; Ross *et al.*, 1987), and both EGF and IGF-1 are required for subsequent DNA synthesis (Leof *et al.*, 1982; Yang and Pardee, 1986). Feldherr and Akin (1993) studied the effect of these growth factors on the recovery of nuclear transport capacity in quiescent cells that were maintained in 0.5% serum for 4 days. Growth factors were added either individually or in combination, and nuclear transport was assayed 6 12, or 18 hr later using nucleoplasmin-gold. For comparison, normal recovery levels were obtained by incubating starved cells for similar periods in medium containing 10% serum.

The main results of these experiments can be summarized as follows. (1) Nuclear transport capacity in quiescent cultures returned to the proliferating cell level approximately 18 hr after the addition of 10% serum. (2) At physiological concentrations, all of the growth factors, when applied individually, caused a partial recovery of transport. That is, nuclear import was significantly greater than that observed in serum-depleted cells, but failed to reach the rates obtained in 10% serum. (3) Treatment of quiescent cells in a mixture of EGF and IGF-1 for 18 hr resulted in recovery of nuclear transport to the normal proliferating level. (4) PDGF acted as an antagonist when it was added to the EGF/IGF-1 mixture. The last effect could be due to downregulation of the EGF receptors (Wharton *et al.*, 1982).

Considering the wide-ranging effects of growth factors on cell function (reviews in Haley, 1990; Ullrich and Schlessinger, 1990; Aaronson, 1991; Roth *et al.*, 1992), modulation of signal-mediated nuclear transport could be caused by changes in ionic composition, gene expression, or the degree of phosphorylation of specific components of the transport machinery. Furthermore, growth factors could indirectly affect nuclear import by disrupting cytoskeletal elements (Herman and Pledger, 1985). Owing to the limited amount of data, it is not possible to determine the relative importance of these processes in transport regulation. It is interesting, however, that phosphorylation of the nucleoporins during mitosis appears to be required for dissociation of the pore complexes (Macauly and Forbes, 1991). Although this finding is not directly related to the control of transport, it does demonstrate that the molecular organization of the pores is sensitive to posttranslational modifications. Under different physiological conditions (e.g., in the presence of the appropriate kinases), such changes could conceivably alter transport capacity.

Establishing the mechanism by which growth factors regulate transport is further complicated by the fact that independent signaling systems are present at both the cell and nuclear surfaces (reviewed in Dingwall and Laskey, 1992). For example, there is evidence that protein kinase C is located at the nuclear surface (Buchner *et al.*, 1992) along with the components of the signaling system that are necessary for its stimulation. The latter substances include phosphoinositidase C (Martelli *et al.*, 1992), which cleaves phosphatidylinositol-4,5-biphosphate (PIP_2), diacylglycerol (Divecha *et al.*, 1991), a product of PIP_2 cleavage that activates PKC, and guanosine triphosphate (GTP)-binding proteins (Rubins *et al.*, 1990), which regulate phosphoinositidase C. Considering its proximity to the pores, it is possible that the nuclear signaling system is primarily involved in modulating transport, but supporting evidence is not available.

D. Nuclear Transport in SV40-Transformed Cells

Transformation by simian virus requires the expression of the early region of the SV40 genome, and involves an overall increase in nuclear activity, including the stimulation of host cell DNA and RNA synthesis (Tjian *et al.*, 1978; Soprano *et al.*, 1983). To determine if these metabolic changes are accompanied by an increase in nuclear transport capacity, Feldherr *et al.* (1992) investigated the effect of transformation on the nuclear import of 11–32 nm nucleoplasmin-gold particles.

Initially, nuclear uptake was compared in proliferating and SV40-transformed (SV-T2) BALB/c cells. As can be seen in Table II,A, both the relative rate of import (N/C gold ratios) and functional diameter of the transport channels (the percentage of particles above 20 nm in diameter present in the nuclei) were significantly greater in the SV-T2 transformed cell line. It was estimated that the exclusion limit for particles entering the nuclei of transformed cells was approximately 28 nm, about 4 nm greater than in nontransformed cells. Increases in nuclear transport were also observed in BALB/c cells that were transfected with a plasmid (pSV3neo) that encodes both the large T and small t antigens. These data are given in Table II,B.

The effect of large T antigen alone on transport capacity was investigated by injecting purified antigen into BALB/c cells and analyzing nucleoplasmin-gold import either 1 or 6 hr later. Proliferating BALB/c cells and cells injected with BSA 6 hr prior to transport analysis served as controls. It was found (see Table IIC) that 6 hr after the injection of large T, nuclear transport had increased significantly and reached the level observed in transformed cultures; however, there was no significant increase after 1 hr, demonstrating that the increase is time dependent. The site of action of large T was determined by investigating the effect of a mutant, (cT)-3, that is deficient in nuclear transport. The NLS in this mutant is inactive owing to a substitution of asparagine for lysine at position 128. A significant increase in transport occurred when (cT)-3 was injected directly into the nucleus, but not when it was injected into the cytoplasm (Table II,C), demonstrating that large T acts within the nucleoplasm. This finding,

TABLE II

N/C Ratios and the Size Distribution of Particles Transported into the Nucleus[a]

Experiment	No. of cells measured	% of particles in nucleus[b]		N/C ± S.E.	Significance[c]	
		<20 nm	>20 nm		Size	N/C
A. BALB/c Controls	10	95.7	4.3	1.00 ± 0.07	—	—
SV-T2	12	65.7	34.3	1.80 ± 0.15	s	s
Size of injected gold[d]		63.3	36.7			
B. BALB/c Controls	22	94.6	5.4	0.99 ± 0.16	—	—
pSV3 neo transformed	25	82.8	17.2	2.11 ± 0.15	s	s
Size of injected gold		76.9	23.1			
C. BALB/c Controls	16	95.8	4.2	0.71 ± 0.12	—	—
BSA injection 6 hr	16	96.4	3.6	0.63 ± 0.07	ns	ns
TAg injection 6 hr	19	76.6	23.4	2.03 ± 0.11	s	s
TAg injection 1 hr	16	91.0	9.0	0.72 ± 0.07	ns	ns
(cT)-3 cytoplasmic injection 6 hr	16	94.5	5.5	1.10 ± 0.11	ns	ns
(cT)-3 nuclear injection 6 hr	18	85.2	14.8	1.52 ± 0.13	s	s
Size of injected gold		69.8	30.2			

[a] With the exception of (cT)-3 nuclear injection data, table adapted from Feldherr et al. (1992, Table I, p. 11003).

[b] Particle measurements do not include the 3-nm protein coat.

[c] Statistical analyses indicate whether the results are (s), or are not (ns) significantly different than the controls. Probability values of 0.01 or below were considered statistically significant.

[d] Size distribution of injected particles is included because different gold fractions were used in each experiment.

together with the observed time delay, suggests that the increase in nuclear transport mediated by large T requires the expression of one or more effector proteins.

The results obtained by studying transformed cells provide further evidence that nuclear transport capacity can fluctuate in response to changes in cellular activity. The increase in nucleocytoplasmic interactions following the expression of large T in infected cells would theoretically facilitate virion formation by enhancing the production of host cell factors that are necessary for the synthesis of viral DNA and late stage mRNA. Clever *et al.* (1991) and Yamada and Kasamatsu (1993) reported that SV40 virions themselves are transported through the pores following cytoplasmic injection into permissive monkey kidney cells. Since the virion particles (approximately 50 nm in diameter) are considerably larger than the transport channels in normal proliferating cells, import would require either partial dissociation of the particles or an increase in functional pore size. It is not yet possible to distinguish between these mechanisms; however, if an increase in pore diameter does occur, it could not involve large T since this antigen is not yet present at this stage of the infectious cycle.

V. Conclusions

Maintaining the proper nucleocytoplasmic distribution of specific proteins and RNAs in relation to changing cellular activities is an essential aspect of cellular regulation. Two strategies are available for controlling macromolecular exchanges. The first is based on the properties of the permeant molecules, including (1) the number and effectiveness of the NLS, (2) flanking sequences that can modulate signal activity, and (3) binding domains for cytoplasmic inhibitors that can mask the NLS or act as anchoring sites. Since the regulatory components are incorporated into the structure of the molecules, this control mechanism is highly specific and has been found to function in the nuclear import of transcription factors.

The second strategy is focused on the transport machinery, which consists of the pore complexes, the primary sites for protein and RNA trafficking across the nuclear envelope, and cytoplasmic receptors for the NLS. Signal-mediated transport of targeted molecules involves (1) binding to cytoplasmic (and perhaps nuclear) receptors, (2) transfer and binding to the surfaces of the pore complexes, and (3) translocation through the pores and subsequent targeting within the nucleus. There are indications that regulation can occur at any of the three steps. Although this mechanism of control does not appear to be highly specific, alterations in the transport machinery can affect the rates of exchange across the envelope and also limit exchanges on the basis of molecular size. The latter process could be especially important in regulating the levels of RNP efflux from the nucleoplasm. Furthermore, changes in the binding properties of the cytoplasmic receptors could differentially affect the transport of molecules, depending on the characteristics of their targeting signals.

Considering the effect of growth factors on nuclear import, it is likely that variations in transport capacity are accompanied by posttranslational modifications of the nucleoporins or changes in pore composition. Other mechanisms involving mechanical forces exerted on the pores by cytoskeletal elements have also been suggested. However, development of a detailed molecular model will require a better understanding of the composition and organization of the pore complex.

In view of the results obtained with SV40 transformed cells, it is interesting to speculate that factors such as large T antigen might override the normal cellular mechanisms that control the translocation step. This could be accomplished by blocking endogenous transport suppressors or mimicking the the action of enhancers. Large T and other well-characterized transforming factors could serve as useful probes for studying these and other fundamental questions relating to transport control.

Acknowledgment

We thank Dr. Robert Cohen for critically reviewing the manuscript.

References

Aaronson, S. (1991). Growth factors. *Science* **254**, 1146–1153.

Abelson, H., Johnson, L., Penman, S., and Green, H. (1974). Changes in RNA in relation to growth of the fibroblast. II. The lifetime of mRNA, rRNA, and tRNA in resting and growing cells. *Cell* **1**, 161–165.

Adam, S., and Gerace, L. (1991). Cytosolic proteins that specifically bind nuclear location signals are receptors for nuclear import. *Cell* **66**, 837–847.

Adam, S., Lobl, T., Mitchell, M., and Gerace, L. (1989). Identification of specific binding proteins for a nuclear location sequence. *Nature (London)* **337**, 276–279.

Adam, S., Marr, R., and Gerace, L. (1990). Nuclear protein import in permeabilized mammalian cells requires soluble cytoplasmic factors. *J. Cell Biol.* **111**, 807–816.

Aebi, U., Jarnik, M., Reichelt, R., and Engel, A. (1990). Structural analysis of the nuclear pore complex by conventional and scanning electron microscopy. *EMSA Bull.* **20**, 69–76.

Agutter, P. (1985). RNA processing, RNA transport and nuclear structure. *In* "The Nuclear Envelope and RNA Maturation" (G. Clawson and E. Smuckler, eds.), Vol. 26, pp. 539–560. Alan R. Liss, New York.

Agutter, P. (1988). Nucleocytoplasmic transport of mRNA: Its relationship to RNA metabolism, subcellular structures and other nucleocytoplasmic exchanges. *Prog. Mol. Subcell. Biol.* **10**, 15–96.

Agutter, P. (1991). "Between Nucleus and Cytoplasm." Chapman & Hall, London.

Akey, C. (1989). Interactions and structure of the nuclear pore complex revealed by cryo-electron microscopy. *J. Cell Biol.* **109**, 955–970.

Akey, C. (1990). Visualization of transport-related configurations of the nuclear pore transporter. *Biophys. J.* **58**, 341–355.

Akey, C. (1992). The nuclear pore complex: a macromolecular transporter. *In* "Nuclear Trafficking" (C. Feldherr, ed.), pp. 31–70. Academic Press, San Diego.

Akey, C., and Goldfarb, D. (1989). Protein import through the nuclear pore complex is a multistep process. *J. Cell Biol.* **109**, 971–982.

Akey, C., and Radermacher, M. (1993). Architecture of the *Xenopus* nuclear pore complex revealed by three dimensional cryo-electron microscopy. *J. Cell Biol.* **122**, 1–19.

Anderson, E., and Beams, H. (1956). Evidence from electron micrographs for the passage of material through the pores of the nuclear membrane. *J. Biophys. Biochem. Cytol.* **2S**, 439–443.

Anderson, J., and Zieve, G. (1991). Assembly and intracellular transport of snRNP particles. *BioEssays* **13**, 57–64.

Baeuerle, P., and Baltimore, D. (1988). IκB: A specific inhibitor of the NF-κB transcription factor. *Science* **242**, 540–546.

Baeuerle, P., and Baltimore, D. (1989). A 65kD subunit of active NF-κB is required for inhibition of NF-κB by IκB. *Genes Dev.* **3**, 1689–1698.

Bataille, N., Helser, T., and Fried, H. (1990). Cytoplasmic transport of ribosomal subunits microinjected into the *Xenopus laevis* oocyte nucleus: a generalized, facilitated process. *J. Cell Biol.* **111**, 1571–1582.

Benavente, R., Dabauvalle, M., Scheer, U., and Chaly, N. (1989). Functional role of newly formed pore complexes in postmitotic nuclear reorganization. *Chromosoma* **98**, 233–241.

Benditt, J., Meyer, C., Fasold, F., Barnard, F., and Reidel, N. (1989). Interaction of a nuclear localization signal with isolated nuclear envelopes and identification of signal binding proteins by photoaffinity labeling. *Proc. Natl. Acad. Sci. U.S.A.* **86**, 9327–9331.

Ben-Ze'ev, A. (1991). Animal cell shape changes and gene expression. *BioEssays* **13**, 207–212.

Berrios, M. (1992). Nuclear pore complex-associated ATPase. *In* "Nuclear Trafficking" (C. Feldherr, ed.), pp. 203–230. Academic Press, San Diego.

Blank, V., Kourilsky, P., and Israel, A. (1992). NF-κB and related proteins: Rel/dorsal homologies meet ankyrin-like repeats. *Trends Biochem. Sci.* **17**, 135–140.

Blobel, G. (1985). Gene gating: a hypothesis. *Proc. Natl. Acad. Sci. U.S.A.* **82**, 8527–8529.

Bonner, W. (1975). Protein migration into nuclei. II. Frog oocyte nuclei accumulate a class of microinjected oocyte nuclear proteins and exclude a class of microinjected oocyte cytoplasmic proteins. *J. Cell Biol.* **64**, 431–437.

Borer, R., Lehner, C., Eppenberger, H., and Nigg, E. (1989). Major nucleolar proteins shuttle between nucleus and cytoplasm. *Cell* **56**, 379–390.

Breeuwer, M., and Goldfarb, D. (1990). Facilitated nuclear transport of histone H1 and other small nucleophilic proteins. *Cell* **60**, 999–1008.

Buchner, K., Otto, H., Hilbert, R., Lindschau, C., Haller, H., and Hucho, F. (1992). Properties of protein kinase C associated with nuclear membranes. *Biochem. J.* **286**, 369–375.

Burglin, T., and De Robertis, E. (1987). The nuclear migration signal of *Xenopus laevis* nucleoplasmin. *EMBO J.* **6**, 2617–2625.

Callan, H., and Tomlin, S. (1950). Experimental studies on amphibian oocyte nuclei. I. Investigation of the structure of the nuclear membrane by means of the electron microscope. *Proc. R. Soc. London, Ser. B* **137**, 367–378.

Carmo-Fonseca, M., Cidadao, A., and David-Ferreira, J. (1987). Filamentous cross-bridges link intermediate filaments to the nuclear pore complexes. *Eur. J. Cell Biol.* **45**, 282–290.

Carmo-Fonseca, M., Kern, H., and Hurt, E. (1991). Human nucleoporin p62 and the essential yeast nuclear pore protein NSP1 show sequence homology and a similar domain organization. *Eur. J. Cell Biol.* **55**, 17–30.

Chelsky, D., Ralph, R., and Jonak, G. (1989). Sequence requirements for synthetic peptide-mediated translocation to the nucleus. *Mol. Cell. Biol.* **9**, 2487–2492.

Clever, J., Yamada, M., and Kasamatsu, H. (1991). Import of simian virus 40 virions through nuclear pore complexes. *Proc. Natl. Acad. Sci. U.S.A.* **88**, 7333–7337.

Cordes, V., Waizennegger, I., and Krohne, G. (1991). Nuclear pore complex glycoprotein p62 of *Xenopus laevis* and mouse: cDNA cloning and identification of its glycosylated region. *Eur. J. Cell Biol.* **55**, 31–47.

Dabauvalle, M., Benavente, R., and Chaly, N. (1988a). Monoclonal antibodies to a Mr 68000 pore complex glycoprotein interfere with nuclear protein uptake in *Xenopus* oocytes. *Chromosoma* **97**, 193–197.

Dabauvalle, M., Schulz, B., Scheer, U., and Peters, R. (1988b). Inhibition of nuclear accumulation of karyophilic proteins in living cells by microinjection of the lectin wheat germ agglutinin. *Exp. Cell Res.* **179**, 291–296.

Dabauvalle, M., Loos, K., and Scheer, U. (1990). Identification of a soluble precursor complex essential for nuclear pore assembly *in vitro*. *Chromosoma* **100**, 56–66.

Dargemont, C., and Kuhn, L. (1992). Export of mRNA from microinjected nuclei of *Xenopus laevis* oocytes. *J. Cell Biol.* **118**, 1–9.

Davis, L., and Blobel, G. (1986). Identification and characterization of a nuclear pore complex protein. *Cell* **45**, 699–709.

Davis, L., and Fink, G. (1990). The NUP1 gene encodes an essential component of the yeast nuclear pore complex. *Cell* **61**, 965–978.

De Robertis, E., Longthorne, R., and Gurdon, J. (1978). Intracellular migration of nuclear proteins in *Xenopus* oocytes. *Nature (London)* **272**, 254–256.

Dingwall, C., and Laskey, R. (1992). The nuclear membrane. *Science* **258**, 942–947.

Dingwall, C., Sharnick, S., and Laskey, R. (1982). A polypeptide domain that specifies migration of nucleoplasmin into the nucleus. *Cell* **30**, 449–458.

Dingwall, C., Robbins, J., Dilworth, S., Roberts, B., and Richardson, W. (1988). The nucleoplasmin nuclear location sequence is larger and more complex than that of SV-40 large T antigen. *J. Cell Biol.* **107**, 841–849.

Divecha, N., Banfic, H., and Irvine, R. (1991). The polyphosphoinositide cycle exists in the nuclei of Swiss 3T3 cells under the control of a receptor (for IGF-1) in the plasma membrane, and stimulation of the cycle increases nuclear diacylglycerol and apparently induces translocation of protein kinase C to the nucleus. *EMBO J.* **10**, 3207–3214.

D'Onofrio, M., Starr, C., Park, M., Holt, G., Haltiwanger, R., Hart, G., and Hanover, J. (1988). Partial cDNA sequence encoding a nuclear pore protein modified by O-linked N-acetylglucosamine. *Proc. Natl. Acad. Sci. U.S.A.* **85**, 9595–9599.

Dworetzky, S., and Feldherr, C. (1988). Translocation of RNA-coated gold particles through the nuclear pores of oocytes. *J. Cell Biol.* **106**, 575–584.

Dworetzky, S., Lanford, R., and Feldherr, C. (1988). The effects of variations in the number and sequence of targeting signals on nuclear uptake. *J. Cell Biol.* **107**, 1279–1287.

Eckner, R., Ellmeier, W., and Birnstiel, M. (1991). Mature mRNA 3' end formation stimulates RNA export from the nucleus. *EMBO J.* **10**, 3513–3522.

Featherstone, C., Darby, M., and Gerace, L. (1988). A monoclonal antibody against the nuclear pore complex inhibits nucleocytoplasmic transport of protein and RNA *in vivo*. *J. Cell Biol.* **107**, 1289–1297.

Feldherr, C. (1962). The nuclear annuli as pathways for nucleocytoplasmic exchanges. *J. Cell Biol.* **14**, 65–72.

Feldherr, C. (1966). Nucleocytoplasmic exchanges during cell division. *J. Cell Biol.* **31**, 199–203.

Feldherr, C. (1968). Nucleocytoplasmic exchanges during early interphase. *J. Cell Biol.* **39**, 49–54.

Feldherr, C. (1972). Structure and function of the nuclear envelope. *Adv. Cell Mol. Biol.* **2**, 273–307.

Feldherr, C. (1975). The uptake of endogenous proteins by oocyte nuclei. *Exp. Cell Res.* **93**, 411–419.

Feldherr, C., and Akin, D. (1990). The permeability of the nuclear envelope in dividing and nondividing cell cultures. *J. Cell Biol.* **111**, 1–8.

Feldherr, C., and Akin, D. (1991). Signal-mediated nuclear transport in proliferating and growth-arrested BALB/c 3T3 cells. *J. Cell Biol.* **115**, 933–939.

Feldherr, C., and Akin, D. (1993). Regulation of nuclear transport in proliferating and quiescent cells. *Exp. Cell Res.* **205**, 179–186.

Feldherr, C., and Dworetzky, S. (1989). Transport of macromolecules through the nuclear pores. *In* "Immuno-Gold Labeling in Cell Biology" (A. Verkleij and J. Leunissen, eds.), pp. 305–315. CRC Press, Boca Raton, Florida.

Feldherr, C., Kallenbach, E., and Schultz, N. (1984). Movement of a karyophilic protein though the nuclear pores of oocytes. *J. Cell Biol.* **99**, 2216–2222.

Feldherr, C., Lanford, R., and Akin, D. (1992). Signal mediated nuclear transport in simian virus 40 transformed cells is regulated by large tumor antigen. *Proc. Natl. Acad. Sci. U.S.A.* **89**, 11002–11005.

Finlay, D., and Forbes, D. (1990). Reconstitution of biochemically altered nuclear pores: transport can be eliminated and restored. *Cell* **60**, 17–29.

Finlay, D., Newmeyer, D., Price, T., and Forbes, D. (1987). Inhibition of *in vitro* nuclear transport by a lectin that binds to nuclear pores. *J. Cell Biol.* **104**, 189–200.

Finlay, D., Meier, E., Bradley, P., Horecka, J., and Forbes, D. (1991). A complex of nuclear pore proteins required for pore function. *J. Cell Biol.* **114**, 169–183.

Fischer, U., and Luhrmann, R. (1990). An essential signalling role for the m3G cap in the transport of U1 snRNA to the nucleus. *Science* **249**, 786–790.

Fischer, U., Darzynkiewicz, E., Tahara, S., Dathan, N., Luhrmann, R., and Mattaj, I. (1991). Diversity in the signals required for nuclear accumulation of U snRNPs and variety in the pathways of nuclear transport. *J. Cell Biol.* **113**, 705–714.

Fischer, U., Sumpter, V., Sekine, M., Satoh, T., and Luhrmann, R. (1993). Nucleocytoplasmic transport of U snRNPs: defination of a nuclear location signal in the Sm core domain that binds a transport receptor independently of the m₃G cap. *EMBO J.* **12**, 573–583.

Forbes, D. (1992). Structure and function of the nuclear pore complex. *Annu. Rev. Cell Biol.* **8**, 495–527.

Franke, W. (1974). Structure, biochemistry and functions of the nuclear envelope. *Int. Rev. Cytol., Suppl.* **4**, 71–236.

Franke, W., and Scheer, U. (1970). The ultrastructure of the nuclear envelope of amphibian oocytes: a reinvestigation. I. The mature oocyte. *J. Ultrastruct. Res.* **30**, 288–316.

Franke, W., and Scheer, U. (1974). Pathways of nucleocytoplasmic translocation of ribonucleoproteins. *Symp. Soc. Exp. Biol.* **28**, 249–282.

Fried, H. (1992). Transport of ribosomal proteins and rRNA, tRNA, and snRNA. *In* "Nuclear Trafficking" (C. Feldherr, ed.), pp. 291–333. Academic Press, San Diego.

Garcia-Bustos, J., Heitman, J., and Hall, M. (1991). Nuclear protein localization. *Biochim. Biophys. Acta* **1071**, 83–101.

Gerace, L., Ottaviano, Y., and Konder-Koch, C. (1982). Identification of a major polypeptide of the nuclear pore complex. *J. Cell Biol.* **95**, 826–837.

Goldfarb, D. (1988). Karyophilic peptides: applications to the study of nuclear transport. *Cell Biol. Int. Rep.* **12**, 809–832.

Goldfarb, D., Gariepy, J., Schoolnik, G., and Kornberg, R. (1986). Synthetic peptides as nuclear localization signals. *Nature (London)* **322**, 641–644.

Goldman, R., Goldman, K., Green, K., Jones, J., Jones, S., and Yang, H. (1986). Intermediate filament networks: organization and possible functions of a diverse group of cytoskeletal elements. *J. Cell Sci.* **5**, 69–97.

Govind, S., and Steward, R. (1991). Dorsoventral pattern formation in *Drosophila:* signal transduction and nuclear targeting. *Trends Genet.* **7**, 119–125.

Greber, U., and Gerace, L. (1992). Nuclear protein import is inhibited by an antibody to a lumenal epitope of a nuclear pore complex glycoprotein. *J. Cell Biol.* **116**, 15–30.

Greber, U., Senior, A., and Gerace, L. (1990). A major glycoprotein of the nuclear pore complex is a membrane-spanning polypeptide with a large lumenal domain and a small cytoplasmic tail. *EMBO J.* **9**, 1495–1502.

Grimm, S., and Baeuerle, P. (1993). The inducible transcription factor NF-kB: structure-function relationship of its protein subunits. *Biochem. J.* **290**, 297–308.

Guddat, U., Bakken, A., and Pieler, T. (1990). Protein-mediated nuclear export of RNA: 5S rRNA containing small RNPs in *Xenopus* oocytes. *Cell* **60**, 619–628.

Guiochon-Mantel, A., Loosfelt, H., Lescop, P., Sae, S., Atger, M., Perrot-Applanat, M., and Milgrom, E. (1989). Mechanisms of the nuclear localization of the progesterone receptor: evidence for interaction between monomers. *Cell* **57**, 1147–1154.

Guiochon-Mantel, A., Loosfelt, H., Lescop, P., Christin-Maitre, S., Perrot-Applanat, M., and Milgrom, E. (1992). Mechanisms of nuclear localization of the progesterone receptor. *J. Steroid Biochem.* **41**, 209–215.

Haino, M., Kawahire, S., Omata, S., and Horigome, T. (1993). Purification of a 60kDa nuclear localization signal binding protein in rat liver nuclear envelopes and characterization of its properties. *J. Biochem. (Tokyo)* **113**, 308–313.

Haley, J. (1990). Regulation of epidermal growth factor receptor expression and activation: a brief review. *Symp. Soc. Exp. Biol.* **44**, 21–37.

Hall, M., Hereford, L., and Herskowitz, I. (1984). Targeting of *E. coli* β-galactosidase to the nucleus in yeast. *Cell* **36**, 1057–1065.

Haltiwanger, R., Holt, G., and Hart, G. (1990). Enzymatic addition of O-GlcNAc to nuclear and cytoplasmic proteins. *J. Biol. Chem.* **265**, 2563–2568.

Hamm, J., and Mattaj, I. (1989). An abundant U6 snRNP found in germ cells and embryos of *Xenopus laevis*. *EMBO J.* **8**, 4179–4187.

Hamm, J., and Mattaj, I. (1990). Monomethylated cap structures facilitate RNA export from the nucleus. *Cell* **63**, 109–118.

Hamm, J., Darzynkiewicz, E., Tahara, S., and Mattaj, I. (1990). The trimethylguanosine cap structure of U1 snRNA is a component of a bipartite nuclear targeting signal. *Cell* **62**, 569–577.

Hanson, L., and Ingber, D. (1992). Regulation of nucleocytoplasmic transport by mechanical forces transmitted through the cytoskeleton. *In* "Nuclear Trafficking" (C. Feldherr, ed.), pp. 71–89. Academic Press, San Diego.

Henkel, T., Zabel, U., Van Zee, K., Muller, J., Fanning, E., and Baeuerle, P. (1992). Intramolecular masking of the nuclear location signal and dimerization domain in the precursor for the p50 NF-κB subunit. *Cell* **68**, 1121–1133.

Hennekes, H., Peter, M., Weber, K., and Nigg, E. (1993). Phosphorylation on protein kinase C sites inhibits nuclear import of lamin B$_2$. *J. Cell Biol.* **120**, 1293–1304.

Herman, B., and Pledger, W. (1985). Platelet-derived growth factor induced alterations in vinculin and actin distribution in BALB/c-3T3 cells. *J. Cell Biol.* **100**, 1031–1040.

Hinshaw, J., Carragher, B., and Milligan, R. (1992). Architecture and design of the nuclear pore complex. *Cell* **69**, 1133–1141.

Hoeijmakers, J., Schel, J., and Wanka, F. (1974). Structure of the nuclear pore complex in mammalian cells. *Exp. Cell Res.* **87**, 195–206.

Hurt, E. (1988). A novel nucleoskeletal-like protein located at the nuclear periphery is required for the life cycle of *Saccharomyces cerevisiae*. *EMBO J.* **7**, 4323–34.

Hurt, E. (1990). Targeting of a cytosolic protein to the nuclear periphery. *J. Cell Biol.* **111,** 2829–2837.

Imamoto, N., Matsuoka, Y., Kurihara, T., Kohno, K., Miyagi, M., Sakiyama, F., Okada, Y., Tsunasawa, S., and Yoneda, Y. (1992). Antibodies against 70kD heat shock cognate protein inhibit mediated nuclear import of karyophilic proteins. *J. Cell Biol.***119,** 1047–1061.

Imamoto-Sonobe, N., Matsuoka, Y., Semba, T., Okada, Y., Uchida, T., and Yoneda, Y. (1990). A protein recognized by antibodies to asp-asp-asp-glu-asp shows specific binding activity to heterogenous nuclear transport signals. *J. Biol. Chem.* **265,** 16504–16508.

Ingber, D., and Folkman, J. (1989). Tension and compression as basic determinants of cell form and function: utilization of a cellular tensegrity mechanism. *In* "Cell Shape: Determinents, Regulation, and Regulatory Role" (W. Stein and F. Bronner, eds.), pp. 3–31. Academic Press, San Diego.

Ingber, D., Madri, J., and Folkman, J. (1987). Endothelial growth factors and extracellular matrix regulate DNA synthesis through modulation of cell and nuclear expansion. *In Vitro Cell. Dev. Biol.* **23,** 387–394.

Ireland, G., Dopping-Hepensal, P., Jordan, P., and O'Neill, C. (1987). Effect of patterned surfaces of adhesive islands on the shape, cytoskeleton, adhesion and behavior of Swiss mouse 3T3 fibroblasts. *J. Cell Sci.* **8,** 19–33.

Izaurralde, E., and Mattaj, I. (1992). Transport of RNA between nucleus and cytoplasm. *Semin. Cell Biol.* **3,** 279–288.

Izaurralde, E., Stepinski, J., Darzynkiewicz, E., and Mattaj, I. (1992). A cap binding protein that may mediate nuclear export of RNA polymerase II-transcribed RNAs. *J. Cell Biol.* **118,** 1287–1295.

Jans, D., Ackermann, M., Bischoff, J., Beach, D., and Peters, R. (1991). p34^{cdc2}-Mediated phosphorylation at T124 inhibits nuclear import of SV-40 T antigen proteins. *J. Cell Biol.* **115,** 1203–1212.

Jiang, L., and Schindler, M. (1988). Nuclear transport in 3T3 fibroblasts: effects of growth factors, transformation and cell shape. *J. Cell Biol.* **106,** 13–19.

Johnson, L., Abelson, H., Green, H., and Penman, S. (1974). Changes in RNA in relation to growth of the fibroblast. I. Amounts of mRNA, rRNA, and tRNA in resting and growing cells. *Cell* **1,** 95–100.

Johnson, L., Williams, J., Abelson, H., Green, H., and Penman, S. (1975). Changes in RNA in relation to growth of the fibroblast. III. Posttranslational regulation of mRNA formation in resting and growing cells. *Cell* **4,** 69–75.

Johnson, L., Levis, R., Abelson, H., Green, H., and Penman, S. (1976). Changes in RNA in relation to growth of the fibroblast. IV. Alterations in the production and processing of mRNA and rRNA in resting and growing cells. *J. Cell Biol.* **71,** 933–938.

Kalderon, D., Richardson, W., Markham, A., and Smith, A. (1984a). Sequence requirements for nuclear location of simian virus 40 large T antigen. *Nature (London)* **311,** 33–38.

Kalderon, D., Roberts, B., Richardson, W., and Smith, A. (1984b). A short amino acid sequence able to specify nuclear location. *Cell* **39,** 499–509.

Kessel, R. (1988). The contribution of the nuclear envelope to eukaryotic cell complexity: architecture and functional roles. *Crit. Rev. Anat. Cell Biol.* **1,** 327–423.

Klein, R., and Afzelius, B. (1966). Nuclear membrane hydrolysis of adenosine triphosphate. *Nature (London)* **212,** 609.

Lanford, R., and Butel, J. (1984). Construction and characterization of an SV40 mutant defective in nuclear transport of T antigen. *Cell* **37,** 801–813.

Lanford, R., Kanda, P., and Kennedy, R. (1986). Induction of nuclear transport with a synthetic peptide homologous to the SV40 T antigen transport signal. *Cell* **46,** 575–582.

Lanford, R., Feldherr, C., White, R., Dunham, R., and Kanda, P. (1990). Comparison of diverse transport signals in synthetic peptide induced nuclear transport. *Exp. Cell Res.* **186**, 32–38.

Lawrence, J., Singer, R., and Marselle, L. (1989). Highly localized tracks of specific transcripts within interphase nuclei visualized by in situ hybridization. *Cell* **57**, 493–502.

Lee, W., and Melese, T. (1989). Identification and characterization of a nuclear localization sequence binding protein in yeast. *Proc. Natl. Acad. Sci. U.S.A.* **86**, 8808–8812.

Lee, W., Xue, Z., and Melese, T. (1991). The NSR1 gene encodes a protein that specifically binds nuclear localization sequences and has two RNA-recognition motifs. *J. Cell Biol.* **113**, 1–12.

Leof, E., Wharton, W., Van Wyk, J., and Pledger, W. (1982). Epidermal growth factor (EGF) and somatomedin C regulate G1 progression in competent BALB/c 3T3 cells. *Exp. Cell Res.* **141**, 107–115.

Levine, E., Becker, Y., Boone, C., and Eagle, H. (1965). Contact inhibition, macromolecular synthesis and polyribosomes in cultured human diploid fibroblasts. *Proc. Natl. Acad. Sci. U.S.A.* **53**, 350–356.

Li, R., and Thomas, J. (1989). Identification of a human protein that interacts with nuclear localization signals. *J. Cell Biol.* **109**, 2623–2632.

Li, R., Shi, Y., and Thomas, J. (1992). Intracellular distribution of a nuclear localization signal binding protein. *Exp. Cell Res.* **202**, 355–365.

Loeb, J., Davis, L., and Fink, G. (1993). NUP2, a novel yeast nucleoporin, has functional overlap with other proteins of the nuclear pore. *Mol. Biol. Cell* **4**, 209–222.

Lyons, R., Ferguson, B., and Rosenberg, M. (1987). Pentapeptide nuclear localization signal in adenovirus E1a. *Mol. Cell. Biol.* **7**, 2451–2456.

Macauly, C., and Forbes, D. (1991). Cell cycle dependent phosphorylation of nuclear pore glycoproteins. *J. Cell Biol.* **115**, 373a.

Martelli, A., Gilmour, R., Bertagnola, V., Neri, L., Manzoli, L., and Cocco, L. (1992). Nuclear localization and signalling activity of phosphoinositidase C in Swiss 3T3 cells. *Nature (London)* **358**, 242–245.

Martin, K., and Helenius, A. (1991). Nuclear transport of influenza virus ribonucleoproteins: the viral matrix protein (M1) promotes export and inhibits import. *Cell* **67**, 117–130.

Mattaj, I., and De Robertis, E. (1985). Nuclear segregation of U2 snRNA requires binding of specific snRNP proteins. *Cell* **40**, 111–118.

Mauck, J., and Green, H. (1973). Regulation of RNA synthesis in fibroblasts during transition from resting to growing state. *Proc. Natl. Acad. Sci. U.S.A.* **70**, 2819–2822.

Maul, G. (1977). The nuclear and the cytoplasmic pore complex: structure, dynamics, distribution, and evolution. *Int. Rev. Cytol.* **6**, 75–186.

Maul, G., Maul, H., Scogna, J., Lieberman, M., Stein, G., Hsu, B., and Borun, T. (1972). Time sequence of nuclear pore formation in phytohemagglutinin-stimulated lymphocytes and in HeLa cells during the cell cycle. *J. Cell Biol.* **55**, 433–447.

Mehlin, H., Daneholt, B., and Skogland, U. (1992). Translocation of a specific premessenger ribonucleoprotein particle through the nuclear pore studied with electron microscope tomography. *Cell* **69**, 605–613.

Meier, U., and Blobel, G. (1990). A nuclear localization signal binding protein in the nucleolus. *J. Cell Biol.* **111**, 2235–2245.

Meier, U., and Blobel, G. (1992). Nopp140 shuttles on tracks between nucleolus and cytoplasm. *Cell* **70**, 127–138.

Michaud, N., and Goldfarb, D. (1991). Multiple pathways in nuclear transport: the import of U2 sn RNP occurs by a novel kinetic pathway. *J. Cell Biol.* **112**, 215–223.

Michaud, N., and Goldfarb, D. (1992). Microinjected U snRNAs are imported to oocyte nuclei via the nuclear pore complex by three distinquishable targeting pathways. *J. Cell Biol.* **116**, 851–861.

Moll, T., Tebb, G., Surana, U., Robitsch, H., and Nasmyth, K. (1991). The role of phosphorylation and the CDC28 protein kinase in cell cycle-regulated nuclear import of the *S. cerevisiae* transcription factor SWI5. *Cell* **66**, 743–58.

Moore, M., and Blobel, G. (1992). The two steps of nuclear transport, targeting to the nuclear envelope and translocation through the nuclear pore, require different cytosolic factors. *Cell* **69**, 939–950.

Moreland, R., Nam, H., Hereford, L., and Fried, H. (1985). Identification of a nuclear localization signal of a yeast ribosomal protein. *Proc. Natl. Acad. Sci. U.S.A.* **82**, 6561–6565.

Moreland, R., Langevin, G., Singer, R., Garcea, R., and Hereford, L. (1987). Amino acid sequences that determine the nuclear localization of yeast histone 2B. *Mol. Cell. Biol.* **7**, 4048–4057.

Muller, M., and Renkawitz, R. (1991). The glucocorticoid receptor. *Biochim. Biophys. Acta* **1088**, 171–182.

Mutvei, A., Dihlmann, S., Herth, W., and Hurt, E. (1992). NSP1 depletion in yeast affects nuclear pore formation and nuclear accumulation. *Eur. J. Cell Biol.* **59**, 280–295.

Nasmyth, K., Adolf, G., Lydall, D., and Seddon, A. (1990). The identification of a second cell cycle control of the HO promoter in yeast: cell cycle regulation of SWI5 nuclear entry. *Cell* **62**, 631–647.

Nehrbass, U., Kern, H., Mutvei, A., Horstmann, H., Marshallsay, B., and Hurt, E. (1990). NSP1: a yeast nuclear envelope protein localized at the nuclear pores exerts its essential function by its carboxy-terminal domain. *Cell* **61**, 979–989.

Newmeyer, D., and Forbes, D. (1988). Nuclear import can be separated into distinct steps *in vitro:* nuclear pore binding and translocation. *Cell* **52**, 641–653.

Newmeyer, D., and Forbes, D. (1990). An N-ethylmaleimide sensitive cytosolic factor necessary for nuclear protein import: requirement in signal mediated binding to the nuclear pore. *J. Cell Biol.* **110**, 547–557.

Nigg, E. (1990). Mechanisms of signal transduction to the cell nucleus. *Adv. Cancer Res.* **55**, 271–310.

Nigg, E., Hilz, H., Eppenberger, H., and Dutley, F. (1985). Rapid and reversible translocation of the catalytic subunit of cAMP-dependent protein kinase type II from the golgi complex to the nucleus. *EMBO J.* **4**, 2801–2806.

Paine, P. (1992). Diffusion between nucleus and cytoplasm. *In* "Nuclear Trafficking" (C. Feldherr, ed.), pp. 3–14. Academic Press, San Diego.

Paine, P., and Horowitz, S. (1980). The movement of material between nucleus and cytoplasm. *In* "Cell Biology: A Comprehensive Treatise" (D. Prescott and L. Goldstein, eds.), Vol. 4, pp. 299–338. Academic Press, New York.

Paine, P., Moore, L., and Horowitz, S. (1975). Nuclear envelope permeability. *Nature (London)* **254**, 109–114.

Pante, N., Jarnik, M., Heitlinger, E., and Aebi, U. (1992). Structure, assembly and interactions of the nuclear lamina and the nuclear pore complex. *Electron Microsc. Am. Proc.* pp. 490–491.

Pardee, A. (1989). G1 events and regulation of cell proliferation. *Science* **246**, 603–608.

Park, M., D'Onofrio, M., Willingham, M., and Hanover, J. (1987). A monoclonal antibody against a family of nuclear pore proteins (nucleoporins): O-linked N-acetylglucosamine is part of the immunodeterminant. *Proc. Natl. Acad. Sci. U.S.A.* **84**, 6462–6466.

Peters, R. (1984). Nucleocytoplasmic flux and intracellular mobility in single hepatocytes measured by fluorescence microphotolysis. *EMBO J.* **3**, 1831–1836.

Peters, R. (1986). Fluorescence microphotolysis to measure nucleocytoplasmic transport and intracellular mobility. *Biochim. Biophys. Acta* **864**, 305–359.

Picard, D., and Yamamoto, K. (1987). Two signals mediate hormone-dependent nuclear localization of the glucocorticoid receptor. *EMBO J.* **6**, 3333–3340.

Picard, D., Kumar, V., Chambon, P., and Yamamoto, K. (1990). Signal transduction by steroid hormones: nuclear localization is differentially regulated in estrogen and glucocorticoid receptors. *Cell Regul.* **1**, 291–299.

Pinol-Roma, S., and Dreyfuss, G. (1992). Shuttling of pre-mRNA binding proteins between nucleus and cytoplasm. *Nature (London)* **355**, 730–732.

Pratt, W. (1992). Control of steroid receptor function and cytoplasmic-nuclear transport by heat shock proteins. *BioEssays* **14**, 1–8.

Radu, A., Blobel, G., and Wozniak, R. (1993). Nup155 is a novel nuclear pore complex protein that contains neither repetitive sequence motifs nor reacts with WGA. *J. Cell Biol.* **121**, 1–9.

Reichelt, R., Holzenburg, A., Buhle, E., Jarnik, M., Engel, A., and Aebi, U. (1990). Correlation between structure and mass distribution of the nuclear pore complex and distinct pore complex components. *J. Cell Biol.* **110**, 883–894.

Richardson, W., Mills, A., Dilworth, S., Laskey, R., and Dingwall, C. (1988). Nuclear protein migration involves two steps: rapid binding at the nuclear envelope followed by slower translocation through nuclear pores. *Cell* **52**, 655–664.

Richter, J., and Standiford, D. (1992). Structure and regulation of nuclear localization signals. *In* "Nuclear Trafficking" (C. Feldherr, ed.), pp. 90–121. Academic Press, San Diego.

Riedel, N., and Fasold, H. (1992). Role of mRNA transport in postranscriptional control of gene expression. *In* "Nuclear Trafficking" (C. Feldherr, ed.), pp. 231–290. Academic Press, San Diego.

Rihs, H., and Peters, R. (1989). Nuclear transport kinetics depend on phosphorylation-site-containing sequences flanking the karyophilic signal of the simian virus 40 T antigen. *EMBO J.* **8**, 1479–1484.

Rihs, H., Jans, D., Fan, H., and Peters, R. (1991). The rate of nuclear cytoplasmic protein transport is determined by the casein kinase II site flanking the nuclear localization sequence of the SV40 T-antigen. *EMBO J.* **10**, 633–639.

Ris, H. (1991). The three-dimensional structure of the nuclear pore complex as seen by high voltage electron microscopy and high resolution low voltage scanning electron microscopy. *EMSA Bull.* **21**, 54–56.

Robbins, J., Dilworth, S., Laskey, R., and Dingwall, C. (1991). Two interdependent basic domains in nucleoplasmin nuclear targeting sequence: identification of a class of bipartite nuclear targeting sequences. *Cell* **64**, 615–623.

Roberts, B., Richardson, W., and Smith, A. (1987). The effect of protein content on nuclear location signal function. *Cell* **50**, 465–475.

Ross, R., Raines, E., and Bowen-Pope, D. (1987). The biology of platelet derived growth factor. *Cell* **46**, 155–169.

Roth, R., Zhang, B., Chin, J., and Kovacina, K. (1992). Substrates and signalling complexes: the tortured path to insulin action. *J. Cell. Biochem.* **48**, 12–18.

Roth, S., Stein, D., and Nusslein-Volhard, C. (1989). A gradient of nuclear localization of the dorsal protein determines dorsoventral pattern in *Drosophila* embryo. *Cell* **59**, 1189–1202.

Rubins, J., Benditt, J., Dickey, B., and Riedel, N. (1990). GTP-binding proteins in rat liver nuclear envelopes. *Proc. Natl. Acad. Sci. U.S.A.* **87**, 7080–7084.

Ruslow, C., Han, K., Manley, J., and Levine, M. (1989). The graded distribution of the dorsal morphogen is initiated by selective nuclear transport in *Drosophila*. *Cell* **59**, 1165–1177.

Scheer, U., Dabauvalle, M., Merkert, H., and Benavente, R. (1988). The nuclear envelope and the organization of the pore complexes. *Cell Biol. Int. Rep.* **12**, 669–689.

Scher, C., Shepard, R., Antoniades, H., and Stiles, C. (1979). Platelet derived growth factor and the regulation of the mammalian fibroblast cell cycle. *Biochim. Biophys. Acta* **560**, 217–241.

Schmitz, M., Henkel, T., and Baeuerle, P. (1991). Proteins controlling the nuclear uptake of NF-kB, Rel and dorsal. *Trends Cell Biol.* **1**, 130–137.

Schroder, H., Bachmann, M., Diehl-Seifert, B., and Muller, W. (1987). Transport of mRNA from nucleus to cytoplasm. *Prog. Nucleic Acid Res. Mol. Biol.* **34**, 89–142.

Shi, Y., and Thomas, J. (1992). The transport of proteins into the nucleus requires the 70-kilodalton heat shock protein or its cytosolic cognate. *Mol. Cell. Biol.* **12**, 2186–2192.

Silver, P., Sadler, I., and Osborne, M. (1989). Yeast proteins that recognize nuclear localization signals. *J. Cell Biol.* **109**, 983–989.

Singh, R., and Green, M. (1993). Sequence-specific binding of transfer RNA by glyceraldehyde-3-phosphate dehydrogenase. *Science* **259**, 365–368.

Slavicek, J., Jones, N., and Richter, J. (1989). A karyophilic signal sequence in adenovirus type 5 E1a is functional in *Xenopus* oocytes but not in somatic cells. *J. Virol.* **63**, 4047–4050.

Snow, C., Senior, A., and Gerace, L. (1987). Monoclonal antibodies identify a group of nuclear pore complex glycoproteins. *J. Cell Biol.* **104**, 1143–1156.

Soprano, K., Galanti, N., Jonak, G., McKercher, S., Pipas, J., Peden, K., and Baserga, R. (1983). Mutational analysis of simian virus 40 T antigen: stimulation of cellular DNA synthesis and activation of ribosomal RNA genes by mutants with deletions in the T antigen gene. *Mol. Cell. Biol.* **3**, 214–219.

Standiford, D., and Richter, J. (1992). Analysis of a developmentally regulated nuclear localization signal in *Xenopus*. *J. Cell Biol.* **118**, 991–1002.

Starr, C., and Hanover, J. (1990). Glycosylation of nuclear pore protein p62. *J. Biol. Chem.* **265**, 6868–6873.

Starr, C., D'Onofrio, M., Park, M., and Hanover, J. (1990). Primary sequence and heterologous expression of nuclear pore glycoprotein p62. *J. Cell Biol.* **110**, 1861–1871.

Sterne-Marr, R., Blevitt, J., and Gerace, L. (1992). O-linked glycoproteins of the nuclear pore complex interact with a cytosolic factor required for nuclear protein import. *J. Cell Biol.* **116**, 271–280.

Stevens, B., and Andre, J. (1969). The nuclear envelope. *In* "Handbook of Molecular Cytology" (A. Lima de Faria, ed.), pp. 837–871. North-Holland Publ., Amsterdam.

Stevens, B., and Swift, H. (1966). RNA transport from nucleus to cytoplasm in *Chironomus* salivary glands. *J. Cell Biol.* **31**, 55–77.

Steward, R. (1987). Dorsal, an embryonic polarity gene in *Drosophila*, is homologous to the vertebrate proto-oncogene c-rel. *Science* **238**, 692–694.

Steward, R. (1989). Relocalization of the dorsal protein from the cytoplasm to the nucleus correlates with its function. *Cell* **59**, 1179–1188.

Stewart, M., Whytock, S., and Mills, A. (1990). Association of gold-labeled nucleoplasmin with the centres of ring components of *Xenopus* oocyte nuclear pore complexes. *J. Mol. Biol.* **213**, 575–582.

Stochaj, U., and Silver, P. (1992a). A conserved phosphoprotein that specifically binds nuclear localization sequences is involved in nuclear import. *J. Cell Biol.* **117**, 473–482.

Stochaj, U., and Silver, P. (1992b). Nucleocytoplasmic traffic of proteins. *Eur. J. Cell Biol.* **59**, 1–11.

Sukegawa, J., and Blobel, G. (1993). A nuclear pore complex protein that contains zinc finger motifs, binds DNA, and faces the nucleoplasm. *Cell* **72**, 29–38.

Tjian, R., Fey, G., and Graessmann, A. (1978). Biological activity of purified simian virus 40 T antigen proteins. *Proc. Natl. Acad. Sci. U.S.A.* **75**, 1279–1283.

Tobian, J., Drinkard, L., and Zasloff, M. (1985). tRNA nuclear transport: defining the critical regions of human tRNA^{met} by point mutagenesis. *Cell* **43**, 415–422.

Ullrich, A., and Schlessinger, J. (1990). Signal transduction by receptors with tyrosine kinase activity. *Cell* **61**, 203–212.

Underwood, M., and Fried, H. (1990). Characterization of nuclear localizing sequences derived from yeast ribosomal protein L29. *EMBO J.* **9**, 91–99.

Unwin, P., and Milligan, R. (1982). A large particle associated with the perimeter of the nuclear pore complex. *J. Cell Biol.* **93**, 63–75.

Wang, N., Butler, J., and Ingber, D. (1993). Mechanotransduction across the cell surface and through the cytoskeleton. *Science* **260**, 1124–1127.

Wente, S., Rout, M., and Blobel, G. (1992). A new family of yeast nuclear pore complex proteins. *J. Cell Biol.* **119**, 705–723.

Wharton, W., Leof, E., Pledger, W., and O'Keefe, E. (1982). Modulation of the epidermal growth factor receptor by platelet derived growth factor and choleragen: effects on mitogenesis. *Proc. Natl. Acad. Sci. U.S.A.* **79**, 5567–5571.

White, E., Allis, C., Goldfarb, D., Srivastva, A., Weir, J., and Gorovsky, M. (1989). Nucleus specific and temporally restricted localization of proteins in *Tetrahymena* macronuclei and micronuclei. *J. Cell Biol.* **109**, 1983–1992.

Wimmer, C., Doye, V., Grandi, P., Nehrbass, U., and Hurt, E. (1992). A new subclass of nucleoporins that functionally interact with nuclear pore protein NSP1. *EMBO J.* **11**, 5051–5061.

Wolff, B., Willingham, M., and Hanover, J. (1988). Nuclear protein import: specificity for transport across the nuclear pore. *Exp. Cell Res.* **178**, 318–334.

Wozniak, R., Bartnik, E., and Blobel, G. (1989). Primary structure analysis of an integral membrane glycoprotein of the nuclear pore. *J. Cell Biol.* **108**, 2083–2092.

Xing, Y., and Lawrence, J. (1991). Preservation of specific RNA distribution within the chromatin-depleted nuclear substructure demonstrated by in situ hybridization coupled with biochemical fractionation. *J. Cell Biol.* **112**, 1055–1063.

Xing, Y., Johnson, C., Dobner, P., and Lawrence, J. (1993). Higher level organization of individual gene transcription and RNA splicing. *Science* **259**, 1326–1330.

Yamada, M., and Kasamatsu, H. (1993). Role of nuclear pore complex in simian virus 40 nuclear targeting. *J. Virol.* **67**, 119–130.

Yamasaki, L., and Lanford, R. (1992). Nuclear transport receptors: specificity amid diversity. *In* "Nuclear Trafficking" (C. Feldherr, ed.), pp. 122–174. Academic Press, San Diego.

Yamasaki, L., Kanda, P., and Lanford, R. (1989). Identification of four nuclear transport signal binding proteins that interact with diverse transport signals. *Mol. Cell. Biol.* **9**, 3028–3036.

Yang, H., and Pardee, A. (1986). Insulin-like growth factor regulation of transcription and replicating enzyme induction necessary for DNA synthesis. *J. Cell Physiol.* **127**, 410–416.

Yasuzumi, G., Tsubo, I., Okada, K., Terawaki, A., and Enomoto, Y. (1968). Fine structure of nuclei as revealed by electron microscopy. V. Intracellular inclusion bodies in hepatic parenchymal cells in case of serum hepatitis. *J. Ultrastruct. Res.* **23**, 321.

Yoneda, Y., Imamoto-Sonobe, N., Yamaizumi, M., and Uchida, T. (1987). Reversible inhibition of protein transport into the nucleus by wheat germ agglutinin injected into cultured cells. *Exp. Cell Res.* **173**, 586–595.

Yoneda, Y., Imamoto-Sonobe, N., Matsuoka, Y., Iwamoto, R., Kiho, U., and Uchida, T. (1988). Antibodies to asp-asp-glu-asp can inhibit transport of nuclear proteins into the nucleus. *Science* **242**, 275–278.

Zabel, U., Henkel, T., dos Santos Silva, M., and Baeuerle, P. (1993). Nuclear uptake control of NF-κB by MAD-3, an IκB protein present in the nucleus. *EMBO J.* **12**, 201–211.

Zasloff, M. (1983). tRNA transport from the nucleus in aneukaryotic cell: carrier-mediated translocation process. *Proc. Natl. Acad. Sci. U.S.A.* **80**, 6436–6440.

Phenolic Components of the Plant Cell Wall

Graham Wallace and Stephen C. Fry

Centre for Plant Science, Division of Biological Sciences, University of
Edinburgh, Edinburgh EH9 3JH, United Kingdom

I. Introduction

The role of compounds containing a phenolic group in the structure of
many secondary cell walls has long been recognized. The secondary walls
of mature tracheids and xylem vessel elements, for example, are typically
30–40% lignin, and cork cell walls are very rich in suberin. Phenol (**1**) is the
structural component common to all phenolic compounds. Both lignin and
suberin are phenolic, or phenol-rich, polymers, and their role in the life of
the plant is well established. In contrast, the existence and significance of
phenolic compounds in primary cell walls is less widely acknowledged.
Primary cell walls (i.e., wall layers whose cellulosic microfibrils were
deposited while the cell was still expanding) usually contain much less
phenolic material than do the secondary walls of wood and cork. Nevertheless, we argue that the phenolic components of the primary cell wall
have a qualitative biological importance that belies their quantitative abundance.

II. Lignin

A. Occurrence and Detection *in Situ*

Work on lignin–polysaccharide cross-links has inevitably set an important
precedent in the development of concepts and methodologies that are used
in studies of other phenol–polysaccharide cross-links more typical of
growing primary cell walls. It is therefore convenient to begin our discussion with a consideration of lignins, because knowledge of them is further
advanced than that of the other wall-bound phenolics.

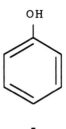

1

Lignins, as conventionally defined (see below), are found in spermato-phytes and pteridophytes, except certain aquatic species (Erickson *et al.*, 1973; Erickson and Miksche, 1974). Related phenolic polymers are also present in some bryophytes (Miksche and Masuda, 1978), although these polymers are devoid of methoxy (—O—CH$_3$) groups and apparently are held only weakly within the cell wall (Wilson *et al.*, 1989). Lignins occur in greatest quantity in the secondary cell walls of certain tissues, especially fibers, xylem vessels, and tracheids—indeed, lignin may be the second most abundant organic compound on earth (after cellulose). However, lignin is often also present in the primary cell walls of fibers, xylem vessels, and tracheids. Smaller amounts of lignin may be deposited in the primary walls of other cell types, especially in response to stresses (Vance *et al.*, 1980; Moerschbacher *et al.*, 1986), and lignins or related polymers occur in the walls of some suspension-cultured cells (Nimz and Ebel, 1975; Mollard and Robert, 1984; Robert *et al.*, 1989).

Lignin can be stained cytochemically by use of a range of relatively specific reagents (Sarkanen and Ludwig, 1971), including phloroglucinol-HCl, which responds to the cinnamaldehyde groups present in lignin but absent in most other phenolics (Clifford, 1974; Lewis and Yamamoto, 1990). However, the number of aldehyde groups per unit mass of lignin is variable from sample to sample, so this test cannot readily be used to investigate lignin quantitatively. Another convenient stain is toluidine blue O, which imparts a green or blue-green color to lignified cell walls (O'Brien *et al.*, 1964); the precise basis of this reaction is still unclear, although it appears to be due to an ability of the dye to "stack" with phenolic compounds and may also be given by phenolics other than lignin (C. Wenzel and S. C. Fry, unpublished). Diazonium compounds are also highly effective and stable stains for phenolics in general, including lignin (Akin *et al.*, 1990).

Phenolic compounds can also be detected *in situ* by their auto-fluorescence under ultraviolet light (Harris and Hartley, 1976). However, different phenolic compounds give very different fluorescence yields, and indeed there is evidence that any given compound can vary in its fluo-rescence yield depending on the precise architectural organization of the cell walls in which it occurs (Willemse and Emons, 1991).

Attempts to make quantitative measurements of lignin in tissue extracts also run into problems of distinguishing lignin from the numerous other phenolic substances present in plants; these problems have been reviewed (Lewis and Yamamoto, 1990).

The precise site of lignin deposition in developing tissues has been studied, after *in vivo* feeding of radioactive precursors, by microautoradiography. For instance, Terashima *et al.* (1988) fed tissue slices of differentiating pine xylem with [^{14}C]coniferin (**2**; a major indirect precursor of lignin in some trees; * = ^{14}C) and demonstrated at least three histological zones of lignification, one of which began at or shortly after the onset of formation of the S$_1$ layer. 4-*O*-β-D-Glucosyl-[*ring*-2-^{3}H]coumaryl alcohol (**3**; * = ^{3}H) was shown to be incorporated into the lignin of the middle lamellae of the vessel and fiber cell walls, i.e., apparently in close association with the pectic polysaccharides (Fukushima and Terashima, 1990).

During development, lignification often begins in the middle lamellae and/or primary cell walls and only later spreads into the secondary wall layers; in other tissues, however, only the secondary walls lignify. For example, within the pea epicotyl, the phloem fibers begin lignification in the middle lamella, whereas the xylem vessels begin in the secondary wall (O'Brien *et al.*, 1964).

B. Composition and Heterogeneity of Lignin

Lignin is a phenolic polymer built up by oxidative coupling of three major C$_6$–C$_3$ (phenylpropanoid) units (**4**), substrates of the last known enzymati-

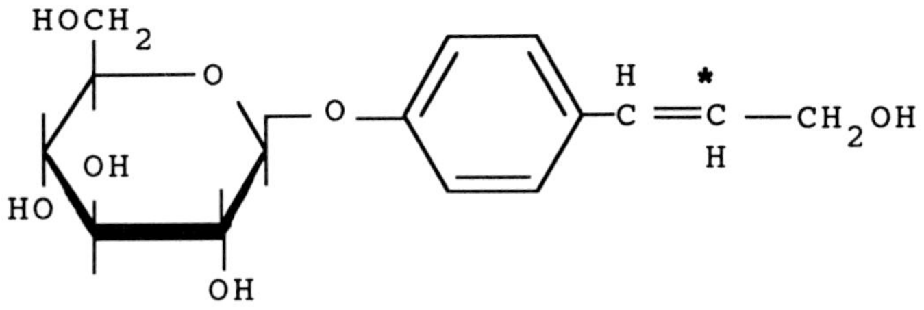

2

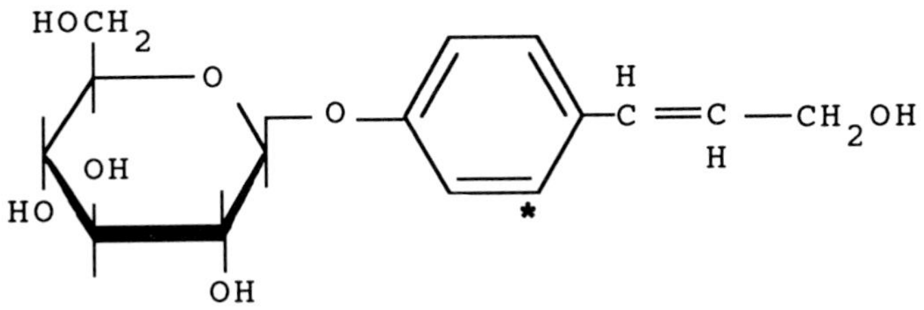

3

cally catalyzed reaction leading to lignin synthesis, namely, p-coumaryl alcohol (H), coniferyl alcohol (G), and syringyl alcohol (S). The ratio of these three major units, and the types of interunit linkages, vary among taxa (Gross, 1979), tissues (Whiting and Göring, 1982), and physiological conditions (see below). The major interunit linkage is an aryl–alkyl ether (Higuchi, 1990). In addition to the phenylpropanoid units, smaller amounts of C_6–C_1 units are found in some lignin samples, especially p-hydroxybenzoic acid (5) units, which may be linked via ester and ether bonds to the rest of the lignin molecule (Terashima et al., 1979).

Qualitative analysis of lignin generally has the aim of defining the H/S/G ratio and the nature of the interunit bonds. The traditional method for this is alkaline nitrobenzene oxidation, the products of which have recently been reinvestigated (Iiyama and Lam, 1990). A new and more successful alternative involves the use of thioacidolysis (treatment with C_2H_5SH/BF_3) (Lapierre et al., 1986, 1991; Aloni et al., 1990; Rolando et al., 1991). Lignin can also be characterized with respect to its monomeric constituents and their cross-links, by means of Curie-point pyrolysis–mass spectrometry (Scheijen and Boon, 1989), a method particularly appropriate for lignin because of its insolubility. A further novel means of probing the heterogeneity of lignin samples, after solubilization in alkali, is by isoelectric focusing electrophoresis: this gives a series of discrete bands with apparent pI values in the range ~3–7, although the basis of the banding was probably pH-dependent precipitation rather than immobility of zwit-

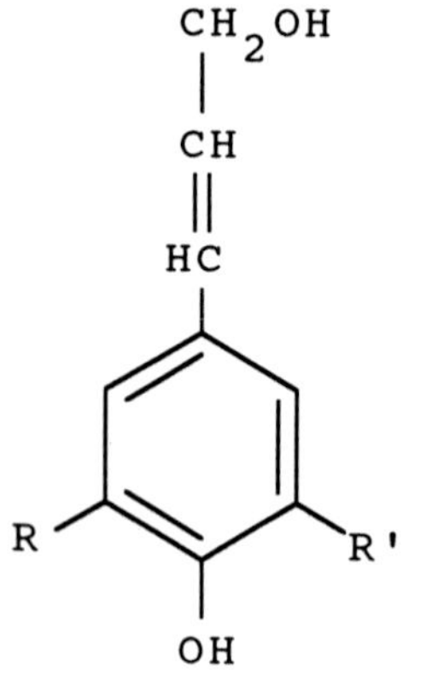

4

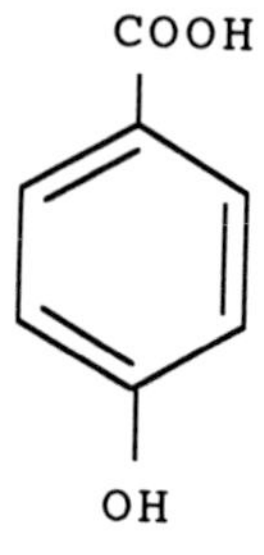

5

p-Coumaryl alcohol (R=R'=H)
Coniferyl alcohol (R=H, R'=O—CH$_3$)
Syringyl alcohol (R=R'=O—CH$_3$)

terions at their p*I* values (Niku-Paavola, 1991). This potentially useful method of electrophoresis deserves further exploration.

C. Biosynthesis of Lignin

The biosynthetic pathway for phenolic acids is well characterized and documented (Gross, 1981). *In vivo* labeling experiments with radioactive precursors have established that the general pathway for incorporation of the phenylpropanoid units into lignin is as follows:

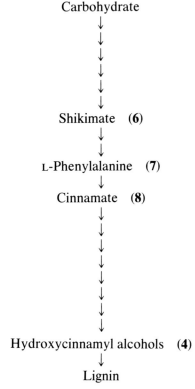

The shikimic acid pathway has been reviewed (Floss, 1979). A key enzyme is L-phenylalanine ammonia-lyase (PAL; EC 4.3.1.5) (Hansen and Havir, 1981). Shikimate (**6**) is a key intermediate in the biosynthesis of many aromatic compounds. L-Phenylalanine (**7**) occurs at a key branch point between primary metabolism (protein synthesis) and secondary metabolism (phenolic synthesis). Cinnamate (*trans*-cinnamic acid; **8**) is a product of the enzyme PAL. The deamination of L-tyrosine directly to *p*-coumaric acid also occurs but would appear rare outside graminaceous species

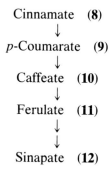

6	**7**	**8**	**9**

(Young *et al.*, 1966; Jangaard, 1974). It is unclear by what precise route the cinnamate is converted to cinnamyl alcohols. The majority of the intermediate compounds involved occur as conjugates, e.g., *O*-feruloyl-β-D-glucopyranoside, feruloyl-*S*-CoA, and *N*-feruloyl putrescine. Thus, although it is clear that the pathway for the aromatic moiety is as shown, it is unclear whether R is H, glucose, CoA, polyamines, or any combination thereof.

$$\text{Cinnamate} \quad (\textbf{8})$$
$$\downarrow$$
$$p\text{-Coumarate} \quad (\textbf{9})$$
$$\downarrow$$
$$\text{Caffeate} \quad (\textbf{10})$$
$$\downarrow$$
$$\text{Ferulate} \quad (\textbf{11})$$
$$\downarrow$$
$$\downarrow$$
$$\text{Sinapate} \quad (\textbf{12})$$

trans-p-Coumaric acid (**9**) is the first-formed phenolic unit, although it is not necessarily formed as the free acid. An enzyme has been found that specifically methoxylates caffeoyl-CoA (forming feruloyl-CoA) rather than *trans*-caffeic acid (**10**) (Schmitt *et al.*, 1991). Whatever the precise compounds involved in aromatic interconversion, the metabolites are chaneled to hydroxycinnamoyl-CoA thioesters, which are then reduced to the corresponding hydroxycinnamyl alcohols by the action of specific hydroxycinnamyl alcohol dehydrogenases (CADs), e.g.,

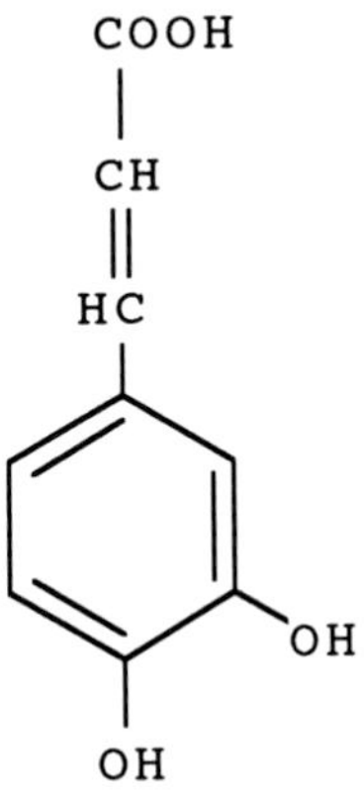

COOH
|
CH
‖
HC

(benzene ring with OH and OH)

10

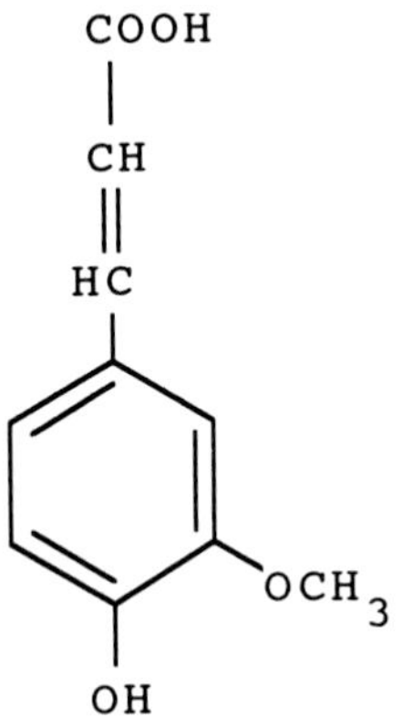

COOH
|
CH
‖
HC

(benzene ring with OCH₃ and OH)

11

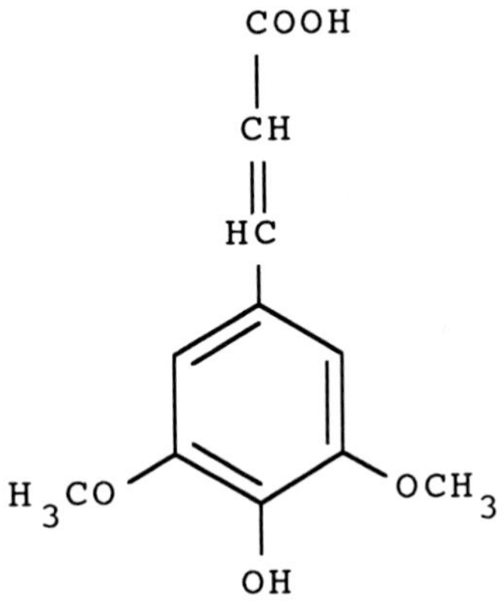

COOH
|
CH
‖
HC

(benzene ring with H₃CO, OCH₃ and OH)

12

CHO
|
CH
‖
HC

(benzene ring with OCH₃ and OH)

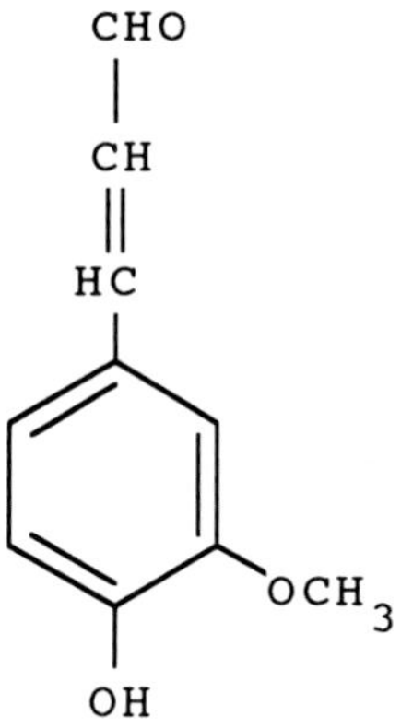

13

Feruloyl-CoA → coniferaldehyde (**13**) → coniferyl alcohol   (**4**)

By use of specific inhibitors of CAD, evidence has been obtained that the reaction catalyzed by this enzyme represents a major rate-limiting step in lignin biosynthesis (Grand *et al.*, 1985). Furthermore, in CAD-deficient ("brown midrib") mutants, lignin levels are decreased and hydroxycinnamyl aldehydes are incorporated into "lignin" in place of the missing hydroxycinnamyl alcohols (Bucholtz *et al.*, 1980; Pillonel *et al.*, 1991).

The hydroxycinnamyl alcohols, generated by CAD under normal circumstances, may be cis/trans isomerized, possibly by means of an enzyme (Yamamoto et al., 1990). The cis- or trans-hydroxycinnamyl alcohols may then be glycosylated, usually but not always (see Harmatha et al., 1978) at the phenolic oxygen, to produce compounds such as coniferin (2). At the site of lignification, in the cell wall, the coniferin is assumed to be enzymatically deglycosylated by β-glucosidases, to yield free hydroxycinnamyl alcohols. However, the turnover of coniferin appears to be insufficient to account for some observed rates of lignification (Marcinowski and Grisebach, 1977), so the direct secretion of hydroxycinnamyl alcohols can also be postulated.

The hydroxycinnamyl alcohols are then thought to be oxidatively coupled to form the lignin polymer—the last known enzymatic step in lignin synthesis (Freudenberg and Neish, 1968). The formation of lignin is very precisely controlled in its subcellular location and in the orientation of the polymer within the wall layers (Terashima, 1990). In *in vitro* systems utilizing peroxidase + H_2O_2, both cis- and trans-coniferyl alcohol can be converted to ligninlike polymers, and the products are virtually indistinguishable (Morelli et al., 1986); this suggests that either of the geometrical isomers can participate *in vivo*.

It is not certain where the H_2O_2—thought to be required for lignification—comes from. One widely discussed proposal is that secreted malate is oxidized to oxaloacetate by a cell wall-localized malate dehydrogenase, and that the (enzyme-bound) NADH produced is used to reduce O_2 to H_2O_2 (Gross et al., 1977). An alternative hypothesis, however, is that wall-bound polyamine oxidases participate, oxidizing apoplastic polyamines and in the process reducing O_2 to H_2O_2:

$$R—CH_2—NH_2 + H_2O + O_2 \rightarrow R—CHO + NH_3 + H_2O_2$$

This idea is supported by a strong positive correlation between the induction of peroxidase and that of polyamine oxidase in response to various treatments that evoke lignification (Angelini et al., 1990; Augeri et al., 1990).

The first products of oxidative coupling of hydroxycinnamyl alcohols, at least in *in vitro* systems utilizing peroxidase + H_2O_2, are dimers known as lignans. Later, these dimers couple further and the polymer lignin is formed. In some plants, however, lignans accumulate to high concentrations, e.g., pinoresinol (14; a dimer that is an oxidative coupling product of coniferyl alcohol) in *Forsythia suspensa*. This pinoresinol is often optically active, occurring very predominantly in the (+) form. Peroxidase + H_2O_2 converts coniferyl alcohol to (±)-pinoresinol, and [^{14}C]coniferyl alcohol applied *in vivo* is also converted to (±)-pinoresinol; however, exogenous

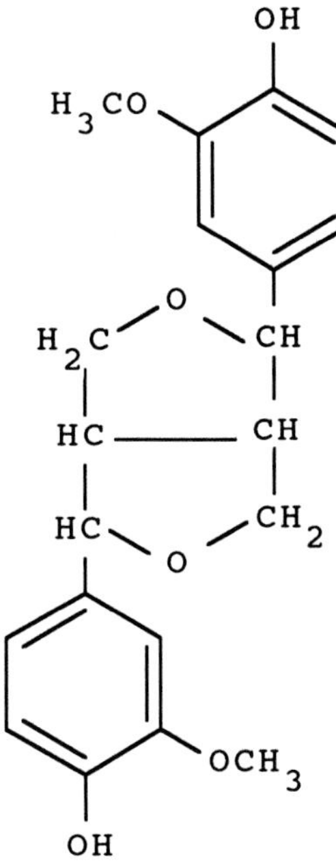

OH

H$_3$CO

H$_2$C

O

CH

HC —— CH

HC

O

CH$_2$

OCH$_3$

OH

14

[^{14}C]phenylalanine is incorporated specifically into (+)-pinoresinol and not into (−)-pinoresinol (Umezawa et al., 1990). This was interpreted as evidence for endogenous coniferyl alcohol, generated via phenylalanine, being "properly" compartmentalized in contrast to exogenous coniferyl alcohol. The picture of oxidative coupling of hydroxycinnamyl alcohols being a haphazard, uncontrolled process may thus be misleading, and the nature of the last step in lignification will certainly be an area of interest in the future.

A different labeling method—involving the *in vivo* feeding of ^{13}C-labeled precursors, e.g., ferulic acid—has also recently been used. Solid-state ^{13}C NMR of the labeled tissue allowed further characterization of some of the structural features of the lignin (and suberin) deposited (Lewis et al., 1987b) and showed that there are some significant differences in bonding patterns between lignins formed *in vivo* and *in vitro*, as well as differences between a grass lignin and a dicot lignin (Lewis et al., 1989). ^{13}C NMR has also been used to demonstrate differences between lignin produced *in vivo* in living cultured rose cells and *in vitro* by the peroxidase isolated from the same cells (Mollard and Robert, 1984; Robert et al., 1989).

The C_6–C_1 (*p*-hydroxybenzoate) groups of lignin also arise from hydroxycinnamates, as shown by the fact that radioactivity is incorporated from L-[3-^{14}C]phenylalanine (**15**; * = ^{14}C) but not L-[1,2-^{14}C]phenylalanine (**16**; * = ^{14}C) (Terashima *et al.*, 1975). This is in contrast to the situation with the C_6–C_1 units of gallotannins, which probably arise as a direct branch from the shikimate pathway without going via C_6–C_3 intermediates (Zenk, 1971).

D. Control of Lignin Biosynthesis

The lignin content of tissues can change quantitatively and qualitatively in response to various stimuli. The formation of lignin in tissue cultures can be induced by various treatments, especially hormonal (Jeffs and Northcote, 1966; Miller and Roberts, 1986), but in these studies it is often difficult to distinguish satisfactorily between the induction of lignification per se and the deposition of lignin as an inevitable consequence of the differentiation of xylem tissue. However, clear-cut evidence for the induction of lignin (by orthovanadate) in suspension-cultured *Petunia* cells was provided by Hagendoorn *et al.* (1990). Furthermore, a more direct piece of evidence for an effect of hormones on lignification per se comes from a study of *Coleus* stems: treatment with gibberellic acid led to the development of phloem fibers with a high lignin content but low S/G ratio; in contrast, auxin treatment gave phloem fibers in which the lignin content

15   **16**

was low but the S/G ratio was high (Aloni *et al.*, 1990). Different responses occurred with respect to lignification in the xylem vessel elements.

Lignin synthesis is also activated during the response of plants to potentially pathogenic microorganisms (for review, see Vance *et al.*, 1980). For instance, wheat leaves rapidly deposit lignin on infection with nonpathogenic fungi (Ride, 1978; Ride and Pearce, 1979a,b), and a similar response can be evoked by chitin and related compounds in the absence of living fungi (Pearce and Ride, 1982). Carnation (*Dianthus caryophyllus*) xylem undergoes a rapid increase in the G/S ratio of its lignin as part of the successful defense response when challenged with the fungus *Fusarium oxysporum* f. sp. *dianthi* (Niemann et al., 1990b). This was determined on minute particles of xylem isolated from the infection zones and analyzed for lignin quality (as well as the lignin/hemicellulose ratio) by pyrolysis—mass spectrometry. The enhanced lignification in carnation stems was not restricted to the xylem vessel elements, but also occurred in xylem parenchyma; an area of suberized tissue also differentiated (Baayen, 1988).

Developmental changes in the composition of lignin are evident during the normal ontogeny of the healthy plant. For example, Terashima *et al.* (1979) have shown that xylem tissue close to the cambium incorporates a much higher proportion of C_6–C_1 units into the nascent lignin compared to xylem 500–1000 μm away from the cambium. Similarly, pyrolysis–mass spectrometry has been applied to the analysis of lignin in very small samples of various tissues, of various ages, microdissected from plant material (Niemann *et al.*, 1990a). The G/S ratio was found to be much higher in the lignin of protoxylem than in that of metaxylem. There were also qualitative differences between the lignin of phloem fibers and that of xylem vessel elements. The lignin content of the shoot of *Arundo donax* increases 10-fold from apex to base, and the lignin from the basal tissue was enriched in S units (Joseleau *et al.*, 1977).

The above studies indicate that lignin is a subtly variable commodity that the plant is able to lay down when and where it is required. Much remains to be learned, however, about the biological advantages of different types of lignin in specific tissues and under particular biological circumstances.

E. Bonds between Lignin and Polysaccharides

At least some of the lignin within the plant cell wall occurs in covalent attachment to the structural polysaccharides (Eriksson *et al.*, 1980; Wallace *et al.*, 1991). This has been demonstrated by extraction of materials, lignin–carbohydrate complexes (LCCs), which, after satisfactorily rigorous purification have proved to contain both lignin and carbohydrate within the same macromolecule. It is necessary to distinguish carefully between lignin–carbohydrate complexes and O-hydroxycinnamoylated polysaccharides (see Section III,B); for example, the "LCC" extracted from *Lolium perenne* in dimethyl sulfoxide should probably be classified as a feruloyl–polysaccharide complex, because the aromatic component was largely converted to ferulic acid by mild alkaline hydrolysis (Morrison, 1974).

After vigorous physical distruption (e.g., "vibroimilling"), lignin and LCCs with a high lignin/carbohydrate ratio can be extracted, e.g., with 80% (aq) dioxane (Watanabe *et al.*, 1989), and the rest of the LCCs are then extractable in cold and hot water and 1 *M* (aq) NaOH. The isolated LCCs, which may form micelles (Yaku *et al.*, 1979), can then be fractionated (Koshijima *et al.*, 1976) and enzymatically "pruned" by hydrolysis with a mixture of glycanases, to leave a core enriched in the lignin–carbohydrate linkage point, which can be purified by adsorption chromatography (Takahashi *et al.*, 1982; Watanabe *et al.*, 1985) and whose structure can be probed. In some cases, the LCC has been pruned naturally, e.g., by digestion in the sheep rumen (Gaillard and Richards, 1975; Conchie *et al.*, 1988). The major linkage point has been shown to be a benzyl ether structure in which a 1'-carbon atom of a phenylpropanoid unit is O-linked to a sugar residue of a polysaccharide (**17**; a simple representative O-benzylglucose ether): this is a linkage that can be oxidatively cleaved by the use of 2,3-dichloro-5,6-dicyano-1,4-benzoquinone (DDQ). In one study (Watanabe *et al.*, 1989), the lignin moiety was attached to O-2 and O-3 of the xylose residues of a glucuronoxylan. Morrison (1974) suggested that, in a grass LCC, the linkage of lignin was mainly to a β-(1 → 4)-D-xylan and to smaller amounts of a β-(1 → 4)-D-glucan. Another potential lignin–carbohydrate linkage is an ester bond between the lignin and the carboxy (C-6) group of a uronic acid residue (**18**; a simple representative benzyl glucuronate ester) (Takahashi and Koshijima, 1988; Watanabe and Koshijima, 1988).

F. Formation of Lignin–Polysaccharide Bonds

The formation of a covalent (probably ether) bond between lignin and a carbohydrate has been mimicked *in vitro*. Thus, if the enzymatic oxidation of coniferyl alcohol is carried out in the presence of a high concentration of sucrose or sorbitol, some LCC is generated in which the artificial lignin is ether linked to the sugar (Freudenberg and Grion, 1959). The mechanism

17

of this reaction is assumed to involve the formation of some quinone methides (**19**) among the range of peroxidase reaction products; these are highly reactive with weak nucleophiles, which are particularly abundant in cell walls. For example, quinone methides can potentially add to water (to form a benzyl alcohol, **20**) (Freudenberg and Schluter, 1955), to an aliphatic alcohol (to form a benzyl–alkyl ether, **21**) (Freudenberg and Grion, 1959; Tanaka *et al.*, 1979; Katayama *et al.*, 1980), to a phenol (to form a benzyl–phenyl ether, **22**) (Freudenberg and Friedmann, 1960; Brunow *et al.*, 1989), or to an acid (to form a benzyl ester, **23**) (Tanaka *et al.*, 1976). Any carbohydrate could contribute an aliphatic alcohol, and uronic acids could contribute a carboxy group—either of which could add nonenzymatically to the quinone methide stuctures formed during lignification and thereby generate a cross-link (Freudenberg and Neish, 1968; Leary, 1980).

In model studies in which a purified quinone methide was incubated with vanillyl alcohol (**24**) in aqueous solution, there were three groups that could have added to the quinone methide: the phenolic —OH, the alkyl

18

—OH, and the H_2O. At low pH (≤ 4), reaction with H_2O was favored, and still occurred to some extent at pH 5; at pH ≥ 5, reaction with the phenolic —OH was favored and no reaction with the alkyl —OH was detected (Sipilä and Brunow, 1991a–c), despite the fact that several workers, starting with Freudenberg and Grion (1959), had demonstrated such a reaction. In the work of Sipilä and Brunow (1991a–c), the reaction mixtures contained 1 mmol quinone methide, 2 mmol vanillyl alcohol, 280 mmol water (5 ml), 0.25 mmol phthalate (added as a buffer, pH 2.5–7.0), and 15 ml dioxane (added as an organic solvent), i.e., the H_2O was present in a large excess. Thus, although the results clearly show that addition of the phenol is preferred at pH ≥ 5, they do not dismiss the possibility of addition of an alkyl alcohol at pH ≤ 5, because the water was present in great excess. It seems likely that under physiological conditions within the primary cell wall, the local concentration of H_2O would be quite low relative to that of the —OH groups of polysaccharides. It also seems conceivable that the relatively unfavorable reaction between quinone methides and specific (alkyl) —OH groups on polysaccharides is catalyzed by a novel enzyme.

19   **20**

Covalent linkages have also been proposed between lignin and other structural polymers of the cell wall, e.g., proteins (Whitmore, 1978; Dill *et al.*, 1984; Ford, 1986).

G. Special Structural Features of Grass Lignin

The lignins of the Gramineae show some major differences from those of the Dicotyledoneae (Scalbert *et al.*, 1986; He and Terashima, 1989a; Monties, 1990; Morrison *et al.*, 1991). For example, in grass tissues, extraction of "milled wood" in 90% dioxane by the method of Björkman (1956) solubilizes a lignin-rich preparation that, on saponification, yields significant amounts of *p*-coumaric acid, suggesting that some *p*-coumarate is esterified to the lignin core (Shimada *et al.*, 1971). Because it has been established that grass cell walls contain p-coumaroylated polysaccharides (see Section III,B), and because Björkman lignin is known to contain some lignin–carbohydrate complexes (Azuma *et al.*, 1981; Lam *et al.*, 1990), it is important to establish whether the *p*-coumarate groups demonstrated by

21

22

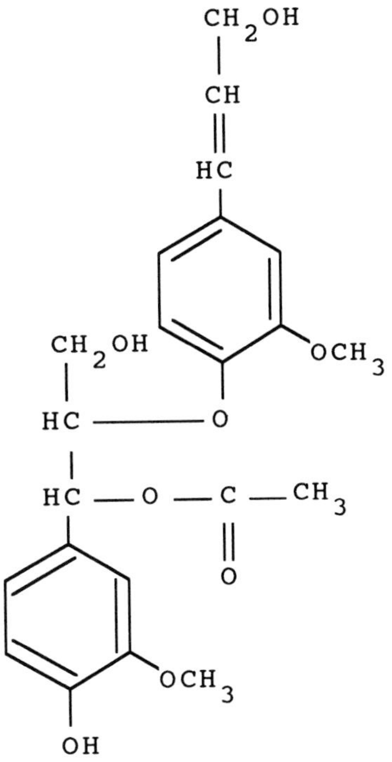

23

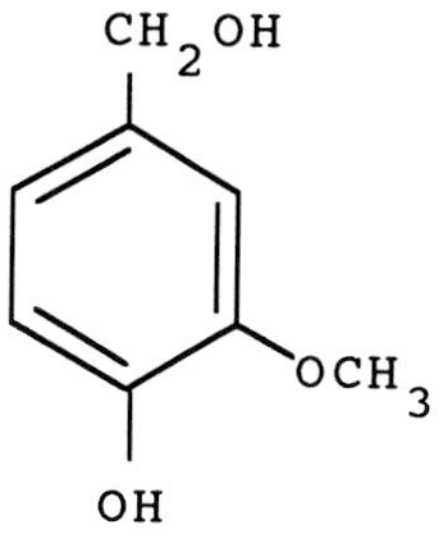

24

Shimada *et al.* (1971) are truly linked to the lignin rather than to accompanying carbohydrate.

Ferulate groups were deposited in rice and sugarcane stem cell walls mainly before lignification, whereas *p*-coumarate groups were incorporated both before and during lignification (He and Terashima, 1989b). Some of the hydroxycinnamate units of grass lignins are ether linked through their phenolic oxygen to lignin, and can be released by methanolysis but not by cold dilute alkali (Scalbert *et al.*, 1985); these units are thought to be incorporated into the lignin by addition to a quinone methide intermediate.

p-Coumaric acid could be released from the grass milled wood lignin by alkaline hydrolysis but not by mild acid-catalyzed methanolysis (0.5% HCl in methanol at room temperature for 2 days) (Shimada *et al.*, 1971), suggesting that the linkage was between the carboxy group of the *p*-coumarate and a terminal —CH_2OH group (rather than the benzylic position) of the C_3 side chains of lignin (Nakamura and Higuchi, 1976). It is unclear how the terminal —CH_2OH groups would become p-coumaroylated *in vivo:* the model compound coniferyl *p*-coumarate (**25**), which already has the terminal ester group, has been synthesized and shown to be incorporated into a

25

ligninlike material by peroxidase + H_2O_2 (Nakamura and Higuchi, 1978); however, there appears to be no evidence for the widespread natural occurrence of coniferyl p-coumarate.

H. Biological Roles of Lignin

There is no doubt that lignin plays a very important rôle in the structure of mature wood: it renders wood very resistant to enzymatic digestion, provides physical strength by enabling stress transfer between neighboring microfibrils, and seals the side walls of xylem vessels and tracheids so that air is not drawin into the tensioned column of water in the transpiration stream. The precise roles of lignin in the primary cell wall, and the reason why lignification often begins in the primary wall, are more controversial.

Akin *et al.* (1987) have shown that chemical treatments, e.g., with $KMnO_4$, that degrade lignin cause loss of tissue coherency, e.g., of the cortex, in bermuda grass stems, suggesting that lignin (or other phenolic material) acts as an intercellular cement.

Much evidence shows that lignin decreases the digestibility of plant cell walls by microbial enzymes (Wallace *et al.*, 1991). Changes in lignin content of maize tissues due to different growth temperatures caused correlative changes in *in vitro* digestibility (Cone and Engels, 1990). Also, the brown midrib mutant of sorghum, which has a modified lignin structure, is more susceptible to digestion than is the wild type (Akin *et al.*, 1986). $KMnO_4$-treated tissues were also much more susceptible to digestion by rumen microorganisms, indicating a role for lignin in resistance to digestion. Of two cultivars of bermuda grass, the one with the greater amount of phenolic material (determined by UV microspectrophotometry at the λ_{max} values of ~245, 291, and 320 nm) in its bundle sheath cell walls was more resistant to digestion (Akin *et al.*, 1990). However, Engels (1989) showed that the most enzyme-resistant zones of the cell walls of grasses were the middle lamellae and primary walls; these withstood enzyme mixtures that were able to digest quite heavily lignified secondary walls, and $KMnO_4$ failed to render the middle lamellae/primary walls more digestible.

Another possible reason why lignification could retard microbial digestion of cell walls is that the low molecular weight degradation products of lignin affect the growth and physiology of microorganisms. *p*-Coumaric acid was particularly inhibitory to ruminal bacteria, although some other phenolics, e.g., *p*-hydroxybenzoic acid, actually promoted bacterial growth and digestion (Borneman *et al.*, 1986).

Although lignin is hydrophobic, insoluble, and generally thought of as a barrier to the enzymatic digestion of polysaccharides, it can, under certain circumstances, be attacked by enzymes (Young and Frazer, 1987; Kivaisi *et al.*, 1990). Peroxidase has been suggested to depolymerize lignin (Dordick *et al.*, 1986), although this could not be confirmed (Lewis *et al.*, 1987a), even with the so-called lignin peroxidase produced by white-rot basidiomycetes (Sarkanen *et al.*, 1991).

Many plants can rapidly switch on lignin synthesis in response to microbial penetration (Sherwood and Vance, 1976; Beardmore *et al.*, 1983; Moerschbacher *et al.*, 1988, 1990a,b). This is apparently induced by certain components of the cell walls of the microorganisms, e.g., chitin oligosaccharides, and also by pectic oligosaccharides liberated by the fungus from the host cell walls (Aldington and Fry, 1992), and has been demonstrated to suppress the colonization of the invading microbes. The lignin synthesized in response to pectic oligosaccharides has been shown to exhibit substantial chemical differences from the constitutive lignin of the same plant (Robertsen and Svalheim, 1990). A role for the induced lignification in disease resistance is strongly suggested by the discovery that chemical inhibition of CAD [e.g., by *N*-(*O*-aminophenyl)sulfinamoyltertiobutyl acetate], which inhibits lignification, was able to break down the resistance mechanism of wheat (Moerschbacher *et al.*, 1990a).

Lignin-related compounds have also been reported to possess antitumor, antiviral, and immunopotentiating activities (Sakagami *et al.*, 1991). Whether this has any bearing on the medical benefits of the consumption of dietary fiber deserves investigation.

III. Phenolic Acids

A. Occurrence and Cytology

Phenolic acids, as their name implies, are aromatic compounds with at least one carboxylic and one phenolic functional group. A large number of such acids and their derivatives are found, within the protoplast, throughout the plant kingdom (Van Sumere, 1989). They are also known to exist as covalently bound cell wall components, especially in angiosperms. Wall-bound phenolic acids are best known in monocotyledonous species, especially the Gramineae. They are derivatives of cinnamic and benzoic acids, predominantly p-coumaric (**9**) and ferulic (**11**) acids with lesser amounts of vanillic (**26**), sinapic (**12**), and p-hydroxybenzoic (**5**) acids (Gordon et al., 1985). Benzaldehyde derivatives [e.g., p-hydroxyben-zaldehyde (**27**) and vanillin (**28**)] have also been found in alkaline hydroly-sates of cell walls of members of the Gramineae, but in minute amounts (<0.026% of the cell wall); the nature of the linkage to a wall polymer is still uncertain (Hartley and Keene, 1984). All the C_6–C_3 derivatives are found mainly in their trans form; there remains some doubt as to whether the small amounts of cis isomer, usually found, are artifacts of the isolation procedure (Morelli et al., 1986) or have a more profound significance (Towers and Abeysekera, 1984; Lewis et al., 1987c).

The bright blue fluorescence of ferulate, following irradiation by UV light, makes it particularly amenable to histological study by UV micro-scopy. This technique has been used to study ferulate distribution in wheat aleurone layers (Fulcher et al., 1972), wheat root epidermis (Smith and O'Brien, 1979), whole organs from a number of graminaceous species (Harris and Hartley, 1976), various monocotyledonous and dicotyle-donous species (Harris and Hartley, 1980; Hartley and Harris, 1981), and tissue cultures of spinach (Fry, 1979). In general, it is found that ferulate would appear to be uniformly distributed across the cell wall and it is particularly abundant in epidermis, xylem vessels, bundle sheaths, and sclerenchyma. In leek roots, fluorescence was concentrated in the walls of the epidermis, hypodermis, and endodermis; it was also more intense in the radial and outer tangential walls of the hypodermis than in the inner tangential walls (Codignola et al., 1989). These details vary taxonomically; thus, for example, in the Ginkgo root the hypodermis fluoresces but the epidermis does not (Codignola et al., 1989). By use of radiolabeled p-coumaric acid administered to rice plants followed by microautoradiogra-phy of sections cut at intervals thereafter (He and Terashima, 1989b), it

26   **27**   **28**

was found that phenolic acids are present in every type of cell wall and that *p*-coumarate levels increase during increased lignification. Large amounts of sinapate were also found (comparable to amounts of ferulate) and appeared to be mostly associated with lignin.

The cell walls of leek roots, on saponification, yield not only phenolic acids (vanillic, syringic, and ferulic), but also the novel compound tyrosol [**29**; 2-(4-hydroxyphenyl)ethanol] (Codignola *et al.*, 1989); the mild alkali used and large quantities of tyrosol obtained suggest that it was not a degradation product of a phenolic acid. Because tyrosol has no carboxy group, the possibility is raised that it is ester linked to the free carboxy groups of another wall polymer—perhaps pectin or glucuronoarabinoxylan.

B. Bonds between Phenolic Acids and Polysaccharides

Phenolic acids have long been known to be alkali-extractable components of cell walls, suggesting ester linkages to cell wall polymers. However, it is only relatively recently that detailed structural information on the nature of such linkages and the polymers involved has come to light.

To date, most studies have concentrated on the relationship of phenolic acids with polysaccharides, especially in the Gramineae (grasses and cereals) and the Chenopodiaceae (spinach, beet, etc.). The general strategy has relied on producing low molecular weight cell wall fragments, usually by enzymatic hydrolysis followed by the isolation and structural characterization of phenolic–carbohydrate complexes. In the Chenopodiaceae, the phenolic acids appear to be associated with pectins. This was demonstrated by the isolation of two major feruloyl conjugates, Fer-Gal$_2$

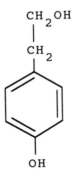

29

[4-O-(6-O-feruloyl-β-D-galactopyranosyl)-D-galactose; **30**] and 3 -O-(3-O-feruloyl-α-L-arabinopyranosyl)-L-arabinose, and the equivalent p-coumaroyl esters, from the walls of spinach cell cultures on digestion with Driselase (Fry, 1982). This finding of a rather small number of precisely defined chemical structures established that ferulate and p-coumarate residues were attached to polysaccharides at specific loci rather than randomly. Similarly, feruloylated pectic polysaccharides have been isolated from sugar beet (Rombouts and Thibault, 1986), but their precise linkage pattern has not been deduced.

In the Gramineae, the phenol–carbohydrate conjugates isolated from wall digests consist, predominantly, of Fer-Ara-Xyl [3-O-(5-O-feruloyl-α-L-arabinofuranosyl)-D-xylose], Fer-Ara-Xyl$_2$ {4-O-[3-O-(5-O-feruloyl-α-L-arabinofuranosyl)-β-D-xylopyranosyl]-D-xylose; **31**}, and the corresponding p-coumaroyl esters (Cou-Ara-Xyl and Cou-Ara-Xyl$_2$) (Smith and Hartley, 1983; Kato and Nevins, 1985; Mueller-Harvey *et al.*, 1986; Kato *et al.*, 1987). This finding strongly suggests that the hydroxycinnamates are predominantly associated with arabinoxylan (the major hemicellullose in the Gramineae) rather than with a pectic polysaccharide. The O-feruloylated oligosaccharides may also be acetylated at the O-2 of the arabinose (Azuma *et al.*, 1990; Ishii, 1991a). Hartley's group showed that in wheat bran ~1 in 150 pentose units is feruloylated whereas in barley straw ~1 in 121 is feruloylated and ~1 in 243 is p-coumaroylated.

More recently, another structure has been discovered in digests of

30

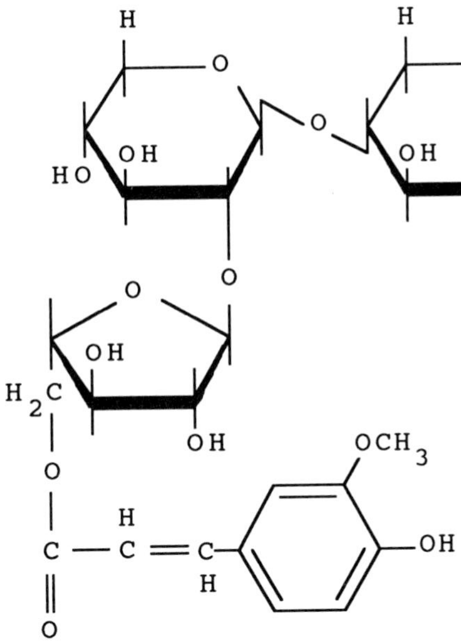

31

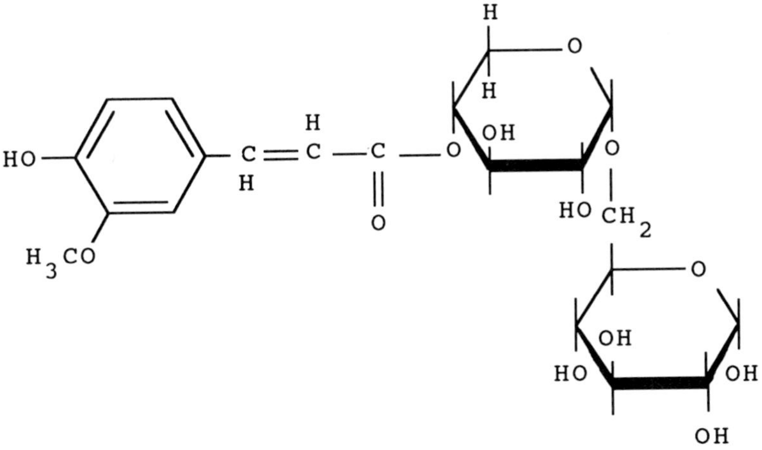

32

bamboo shoots, 6-*O*-(4-*O*-feruloyl-α-D-xylopyranosyl)-D-glucose (**32**), indicating a linkage of ferulate to xyloglucan (Ishii and Hiroi, 1990). It is
interesting that such a structure has been found in a member of the Gramineae, in which xyloglucan is a relatively minor component, rather than
in a dicot, in which there are often much higher concentrations. The

absence of such a structure, to date, in dicots may suggest a different role for xyloglucans in grasses.

C. Biosynthesis of Hydroxycinnamoyl–Polysaccharides

The origin of the hydroxycinnamate moieties has been discussed in Section II. The benzoates are mainly produced by β-oxidation of the appropriately substituted phenylpropanoids (Alibert and Ranjeva, 1971), although they may also be synthesized by aromatization of shikimic and dehydroshikimic acids (Zenk, 1971).

The mechanism whereby ferulate is incorporated into cell wall polymers is a contentious issue. Yamamoto and Towers (1985) found that, in barley coleoptiles, ferulate incorporation continued to increase long after the net deposition of cell wall polymers had stopped, suggesting feruloylation within the cell wall. Similar results were more recently obtained by Nishitani and Nevins (1990) using maize coleoptiles. Such results could, however, merely reflect a higher deposition rate of feruloylated polymers, compared to nonferuloylated, in the later stages of development. Fry (1987) studied the feruloylation of newly synthesized ^{3}H-labeled polysaccharides in cultured spinach cells fed L-[1-^{3}H]arabinose. It was found that feruloylated ^{3}H-labeled polymers could be detected as little as 4 minutes after the administration of [^{3}H]arabinose, yet ^{3}H-labeled polymers did not arrive at the cell wall until after 25 minutes. Similar data have recently been obtained for cultured cells of the grass *Festuca arundinacea* (K.E. Myton and S.C. Fry, unpublished). These observations indicated an intraprotoplasmic feruloylation of nascent polysaccharides, contradicting the previous views.

For net ester formation to occur, the acyl group must be activated. Likely candidates in the case of ferulate are the CoA thioester and the β-D-glucosyl ester, widely found in plants and previously shown to act as feruloyl donors in transacylation reactions (Zenk, 1979; Tkotz and Strack, 1980). More recent work (Meyer *et al.*, 1991) has gone a long way to settling the issue. Studying microsomal fractions from cultured parsley cells, Meyer *et al.* (1991) found a subfraction of the Golgi apparatus that will transfer feruloyl residues from feruloyl-CoA to pectic polysaccharides, thereby strongly supporting the idea of intraprotoplasmic feruloylation using feruloyl-CoA as the donor. *p*-Coumaroyl-CoA is likely to act in a similar reaction.

Although it has been found esterified to hemicellulose, *p*-coumarate is thought to be, quantitatively, more important as an esterified component of lignin (Higuchi *et al.*, 1967; Shimada *et al.*, 1971; He and Terashima, 1989b; Ford, 1990). This would suggest that the bulk of *p*-coumarate is incorporated into the cell wall differently from ferulic acid and probably by mechanisms similar to those for coniferyl and syringyl alcohols for lignin synthesis.

D. Oxidative Coupling of Hydroxycinnamoyl—Polysaccharides

It has been suggested that polysaccharide-bound phenolic acids may play an important role in controlling cell wall extensibility by oxidative coupling (Fry, 1979; Kamisaka *et al.*, 1990; Tan *et al.*, 1991). This theory grew out of the discovery of trans,trans-diferulic acid (DFA; **33**), an oxidative coupling product of ferulic acid, as a component of wheat flour arabinoxylans (Markwalder and Neukom, 1976). It is interesting to note that no similar dimer of *p*-coumarate has, to date, been found. DFA is thought to be produced by the oxidative coupling of two adjacent feruloylated hemicellulose chains via a peroxidase/H_2O_2 system. Such a claim has been strengthened by the isolation from digests of bamboo shoot cell walls of a conjugate (**34**) in which DFA is ester linked, via each of its carboxy groups, to a hemicellulose-derived trisaccharide (Ara-Xyl$_2$) (Ishii, 1991b). Such a structure does not distinguish between intra- and interpolysaccharide cross-links. However, the suggested occurrence of an arabinoxylan–DFA–xyloglucan complex would appear to provide much stronger evidence for interpolysaccharide cross-links. Although the diferulic acid content of cell walls appears to be somewhat insignificant in terms of weight, the data of Neukom (1976) on the peroxidase-catalyzed oxidative gelling of feruloylated polysaccharides suggest that published levels of DFA in cell

33

34

walls are, theoretically, more than enough to cause appreciable cross-linking of cell wall polysaccharides (Fry, 1979).

E. Significance of Quinone Methides

The proposed peroxidase-catalyzed oxidation of feruloyl esters would theoretically produce a number of possible dimers as well as DFA. Some of these dimers—quinone methides (**35**)—will spontaneously, and possibly enzymatically, react with nucleophiles, e.g., hydroxy groups (Leary, 1980). Whereas much of the work in this area has been performed with the oxidation products of coniferyl alcohol to study lignin–carbohydrate grafting (see Section II,F), it is also the most likely explanation for the proposed existence of ether-linked phenolic acids in the plant cell wall (Whitmore, 1976; Fry, 1984; Scalbert *et al.*, 1985; Lam *et al.*, 1992). Unfortunately little attention has been paid to the reactions of phenolic acids under such oxidative conditions. The studies that do exist have started with the sodium salt of ferulic acid and have found that lactonization took place

35

(Freudenberg and Grion, 1959), which probably could not happen with ferulate esters, e.g., feruloyl–polysaccharides.

The controlling factors in such reactions appear to be the relative nucleophilicities of the reactants at a given pH (Leary, 1980) and steric hindrance (Leary *et al.*, 1977). At the pH of the cell wall (probably usually in the range pH 4–7), the reaction with carbohydrates and uronides should be favored over the reaction with phenolic groups, strongly suggesting a system whereby two phenolic moieties could possibly cross-link three or even four hemicellulose chains (Biggs and Fry, 1987).

F. Photodimerization of Hydroxycinnamoyl–Polysaccharides

Another possible cross-linking mechanism has recently been proposed with the discovery of 4,4′-dihydroxytruxillic acid (**36**) in alkali extracts from Italian ryegrass (Hartley *et al.*, 1988). 4,4′-Dihydroxytruxillic acid is a cyclodimer of *p*-coumaric acid and can be produced by the action of light on *p*-coumaric acid in the solid state (Cohen *et al.*, 1963). The dimerization can yield numerous isomeric forms: head-to-tail (ht, 4,4′-dihydroxytruxillic acid) versus head-to-head (hh, 4,4′-dihydroxytruxinic acid); syn versus anti; and cis versus trans. In all, 12 isomers are theoretically possible (Eraso and Hartley, 1990). The corresponding ferulate–ferulate and

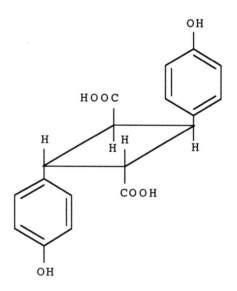

36

ferulate–*p*-coumarate dimers have also been found in alkali hydrolysates of cell walls (Hartley *et al.*, 1990a). To date, these structures have only been found in graminaceous species, with *p*-coumarate–*p*-coumarate (ht), *p*-coumarate–ferulate (ht), and ferulate–ferulate (ht) being the most common. The amounts present in cell walls vary considerably and appear to be highest in tropical (C_4) grasses. For example, there is 0.39% (cell wall basis) in bermuda grass (C_4) but only 0.11% in Italian rygrass shoots and 0.05% in barley straw (C_3) grasses. It has been suggested that they may play a more important role than DFA. The suggestion that such dimers are produced *in vivo* by photodimerization (Hartley *et al.*, 1988) has recently been supported by the finding that Fer-Ara-Xyl$_2$ (**31**) and Cou-Ara-Xyl$_2$ will also produce such structures on exposure to visible light (Hartley *et al.*, 1990b).

G. Biological Roles of Hydroxycinnamoyl–Polysaccharides

The precise role of phenolic acids in the cell wall is uncertain, although a number of theories have been proposed. It has been suggested that they play a role in inhibiting enzymatic degradation of the hemicellulose (Hartley, 1972), that they provide initiation sites for lignin synthesis (Gordon, 1975), that they act as UV receptors and control phototropism (Towers and Abeysekera, 1984), and that they are responsible for limiting wall extensibility by cross-linking (Fry, 1983).

Towers and Abeysekera (1984) proposed that the UV-mediated cis/trans isomerization of hemicellulose-bound hydroxycinnamates may affect cell wall extensibility and could be responsible for the phototropic response. However, their model system depends on the presence of another photoreceptor that will absorb in the blue region (400–450 nm) and transfer its excitation energy. To date, however, no such receptor has been demonstrated.

The light-mediated synthesis of 4,4'-dihydroxytruxillate groups offers an intriguing alternative explanation of phototropism—viz. that the light directly photodimerizes *p*-coumarate and/or ferulate residues, thereby cross-linking polysaccharides and thus restricting cell expansion on the illuminated side of a stem or coleoptile. Unfortunately, little is known about the action spectrum for photodimerization, except that incandescent light from a controlled environment chamber was more efficient than UV light (Ford and Hartley, 1989); therefore, a quantitative comparison cannot be made with the well-established action spectrum for phototropism (λ_{max} 400–500 nm).

A likely role for hydroxycinnamate groups is to provide sites through which lignin may be grafted on to polysaccharides. Recent work (Iiyama *et al.*, 1990; Lam *et al.*, 1992), using a newly developed technique for the selective cleavage of benzyl–aryl ethers by high-temperature alkaline hydrolysis, has demonstrated that the bulk of esterified ferulic acid residues in *Phalaris* and *Triticum* stem cell walls are also ether linked, therefore suggesting they act as bridge points between lignin and polysaccharides. Interestingly, *p*-coumarate does not appear to act as such a bridge.

Phenolic acids are unlikely to be major constraints in cell wall degradability and certainly the ferulate content is not well correlated with indigestibility; ferulate would also appear to be preferentially lost during rumen incubation (Chesson, 1981; Åman and Nordkvist, 1983). The close correlation between *p*-coumarate and indigestibility is likely to reflect its close association with lignin, which is more generally accepted as being the major limitation in cell wall degradation (Chesson, 1988).

Recent enzymatic studies have found *p*-coumaroyl and feruloyl esterases produced by anaerobic rumen fungi (Borneman *et al.*, 1990) and bacteria (McDermid *et al.*, 1990). However, it is still uncertain how specific they are or how important a role they play in the degradation of cell walls in the rumen.

The effects of free phenolic acids on the rumen microflora have been extensively studied. Ferulic and *p*-coumaric acids (at 5 m*M*) have been shown to be toxic to both rumen fungi (Akin and Rigsby, 1985) and bacteria (Chesson *et al.*, 1982). Interestingly, all the bacteria tested demonstrated the ability to hydrogenate the acids to phloretic (**37**) and 3-methoxy phloretic acids, which proved to be considerably less toxic. The effects of *p*-coumaroyl esters and truxillic acid have also been tested against rumen bacteria (Hartley and Akin, 1989). Butyl *p*-coumarate was

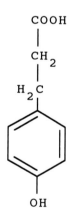

37

found to be just as, if not more, toxic than the free acid. 4,4'-Dihydroxy-truxillic acid was found to have a negligible effect. Phenolic acids (at micromolar concentrations) have also been shown to be effective at inhibiting the action of commercial preparations of hemicellulase and cellulase (Martin and Blake, 1989) and at inhibiting the production of some polysaccharidases in white- and brown-rot fungi (Highley and Micales, 1990).

Another use for phenolic acids in some plants is in allelopathy (Rice, 1984), i.e., the way in which germination and growth of one species of plant may be influenced by the presence of another. Certainly ferulic and *p*-coumaric acids (micromolar concentrations) have been shown to be toxic to pea roots, especially in a nutrient-deficient environment (Vaughan and Ord, 1990).

IV. Other Phenolic Components in the Plant Cell Wall

A. Cutin and Suberin

Cutin and suberin are lipophilic polymers, traditionally revealed by their ability to take up colored lipophilic substances such as Sudan III, and characteristically found in or on the walls of epidermal and cork cells, respectively. Cutin is present in large amounts in the thick cuticles found on the surface of many leaves and stems, but it also occurs, in smaller amounts, in the outer cell walls of young, meristematic tissue such as at the shoot apex. Cutin or a cutinlike material is also found in the outer, preepidermal cells of very young globular embryoids produced in carrot cell cultures (Sterk *et al.,* 1991). Suberin occurs in the cork produced by the normal phellogen near the stem surface in many plants (Wattendorf, 1974), and also in the wound phellem that differentiates at cut surfaces, e.g., of potato tubers, as a healing reaction. Suberin has also been found in the exodermis (Olesen, 1978), in green cotton fibers (Schmutz *et al.,* 1992), in seed coats, and in the Casparian strip in roots (van Fleet, 1950).

Alkali-catalyzed methanolysts (i.e., heating in methanolic sodium methoxide) liberates hydroxy fatty acids (as their methyl esters) from both cutinized and suberized walls. The ratio of these substances is the main chemical criterion that distinguishes cutin from suberin. The major building blocks of cutin include 16-hydroxyhexadecanoic acid, 9,16-dihydroxyhexadecanoic acid, 10,16-dihydroxyhexadecanoic acid, 18-hydroxyoctadeca-9-enoic acid, and 18-hydroxyoctadecanoic acid 9,10-epoxide. The major monomers of suberin are ω-hydroxy fatty acids (C_{16} to C_{24}), α,ω-dicarboxylic fatty acids, and a range of C_{20} to C_{30} fatty acids and fatty alcohols. In addition, recent evidence shows that ester-bonded glycerol is present in suberized and possibly also cutinized cell walls (Schmutz et al., 1992).

All cell walls that contain cutin or suberin are also found to yield phenolic material on hydrolysis or methanolysis, and it has been widely accepted that the phenolic units are an integral part of cutin and suberin (Kolattukudy, 1981). However, it has not been clearly established whether these phenolic units are covalently bonded to the aliphatic "core" of the cutin/suberin, whether they are associated with ligninlike materials that co-occur in the same cell walls, or whether they are phenolic acids esterified to the polysaccharides of the cutinized wall.

B. Soluble Extraprotoplasmic Phenolics

Analysis of the spent culture media of plant cell suspension cultures inevitably reveals the presence of soluble phenolic material that is clearly not covalently bound to cell walls. This material is chemically diverse, but includes phenylpropanoids. Feeding of trans-[^{14}C]cinnamic acid to cultured spinach cells resulted in the rapid formation of soluble conjugates of p-[^{14}C]coumarate and [^{14}C]ferulate within the cells (Fry, 1984), and low but steady levels of free p-[^{14}C]coumaric and [^{14}C]ferulic acids in the culture medium. Added [^{14}C]coumaric and [^{14}C]ferulic acids are rapidly taken up by the cells. It thus seems likely that the pool of intracellular hydroxycinnamates are in a state of dynamic interconversion, via the free acids, and that the free acids are sufficiently permeable to the plasma membrane that they can equilibrate between the cell and the medium. It is interesting to speculate that the extracellular free acids are subject to oxidative coupling reactions analogous to those reported for hydroxycinnamoyl–polysaccharides in the cell wall.

The accumulation of certain nonpolar neutral phenolics in the culture medium of spinach cells is increased by treatment with growth-promoting levels of gibberellic acid in parallel with a suppression of peroxidase secretion (Fry, 1979). It has been suggested that the low peroxidase levels prevented the efficient formation of growth-inhibiting oxidation products (Fry, 1980).

Solvent-extractable (therefore noncovalently bound) phenolic materials are also present in the exine layer of pine pollen (Schulze Osthoff and Wiermann, 1987). A related observation is also relevant: partially methylated flavonol glucosides have been localized, by immunological probes, in the cell walls of Chrysosplenium (Brisson et al., 1986).

Thus, extraprotoplasmic low molecular weight phenolic may be quite widespread in plants. Although often water soluble, these compounds may be regarded as wall components because they are contained in the water that permeates the cell wall.

Noncovalently bound cell wall phenolics may influence the ability of microorganisms to digest the polysaccharide components of the wall (see Section II). In addition, they could influence the ability of the plant's own enzymes to loosen the cell wall during growth. They could, by undergoing oxidation to yield more hydrophobic dimers or quinones, lower the effective water availability in the cell wall and thereby further interfere with the enzymatic loosening of the wall. On the other hand, they could potentially become oxidatively coupled to the covalently bound phenolic moieties of the cell wall, thus modulating the ability of the latter to participate in peroxidase-catalyzed cross-linking. They could provide a mechanism, perhaps in conjunction with ascorbate, for controlling the redox potential of the apoplastic environment: this in turn could influence the action of numerous wall enzymes. All these possibilities are highly speculative, but will perhaps encourage serious attention to be paid to these extracellular phenolics.

C. Sporopollenin

Sporopollenin is a highly resistant polymer found in the exine (outer) pollen wall of higher plants (Heslop-Harrison, 1971). Owing to its chemical resistance, sporopollenin has been very difficult to characterize. It was suggested to be a condensation product of carotenoids (Brooks and Shaw, 1968); however, recent evidence has shown that inhibitors of carotenoid biosynthesis do not block sporopollenin formation (Prahl et al., 1985), and that sporopollenin has a phenolic nature. Thus, IR spectra, saponification, KOH fusion, nitrobenzene oxidation, and treatment with AlI_3 all demonstrate the presence of phenolic material, the degradation products including p-hydroxybenzoic acid, p-coumaric acid, vanillic acid, vanillin, p-hydroxybenzaldehyde, and ferulic acid. In young tulip pollen cells, ^{14}C from [U-^{14}C]phenylalanine was efficiently incorporated into the p-hydroxybenzoic acid component released from the sporopollenin by KOH fusion (Rittscher and Wiermann, 1988). Treatment of sporopollenin from pine pollen with AlI_3 gives a high yield of p-coumaric acid, suggesting ether linkage, and subsequent saponification increased the yield, also suggesting ester linkages (Wehling et al., 1989).

V. Concluding Remarks

In conclusion, we emphasize that phenolic compounds in plant cell walls are both chemically diverse and botanically widespread. Although much of the detailed characterization of wall phenolics has concerned the chemistry of lignin in mature wood, we wish to stress that other phenolic compounds are also present in the walls of numerous growing plant tissues. These compounds often constitute only a small proportion of the weight of the primary cell wall, but the evidence suggests that they have a far-reaching qualitative importance to the life of the plant, influencing particularly the growth rate and digestibility of the tissues. Furthermore, they are dynamic components of the cell wall, undergoing postsynthetic modifications, especially oxidative and light-induced dimerization. Phenolic components of the primary cell wall are therefore important both in basic studies of plant physiology and in applied studies of plant biotechnology and agriculture. Much remains to be discovered about them, especially about their biosynthesis and *in vivo* cross-linking, and about the precise biological roles that they play.

Acknowledgment

The authors thank the Agricultural and Food Research Council for a research grant in support of their work.

References

Akin, D. E., and Rigsby, L. L. (1985). *Agron. J.* **77**, 180–182.
Akin, D. E., Hanna, W. W., Snook, M. E., Himmelsbach, D. S., Barton, F. E., and Windham, W. R. (1986). *Agron. J.* **78**, 832–837.
Akin, D. E., Rigsby, L. L., Barton, F. E., Gelfand, P., Himmelsbach, D. S., and Windham, W. R. (1987). *Food Microstructure* **6**, 103–113.
Akin, D. E., Ames-Gottfried, N., Hartley, R. D., Fulcher, R. G., and Rigsby, L. L. (1990). *Crop Sci.* **30**, 396–401.
Aldington, S., and Fry, S. C. (1992). *Adv. Bot. Res.* **19**, 1–101.
Alibert, G., and Ranjeeva, R. (1971). *FEBS Lett.* **19**, 11·14.
Aloni, R., Tollier, M. T., and Monties, B. (1990). *Plant Physiol.* **94**, 1743–1747.
Åman, P., and Nordkvist, E. (1983). *Swed. J. Agric. Res.* **13**, 61–67.
Angelini, R., Manes, F., and Federico, R. (1990). *Planta* **182**, 89–96.
Augeri, M. I., Angelini, R., and Federico, R. (1990). *J. Plant Physiol.* **136**, 690–695.
Azuma, J., Takahashi, N., and Koshijima, T. (1981). *Carbohydr. Res.* **93**, 91–104.
Azuma, J., Kato, A., Koshijima, T., and Okamura, K. (1990). *Agric. Biol. Chem.* **54**, 2181–2182.
Baayen, R. P. (1988). *Can. J. Bot.* **66**, 784–792.
Beardmore, J., Ride, J. P., and Granger, J. W. (1983). *Physiol. Plant Pathol.* **22**, 209–220.
Biggs, K. J., and Fry, S. C. (1987). *In* "Physiology of Cell Expansion during Plant Growth" (D. J. Cosgrove and D. P. Kneivel, eds.), pp. 46–57. The American Society for Plant Physiology, Rockville, Maryland.
Björkman, A. (1956). *Svensk Papperstidning* **59**, 477–485.
Borneman, W. S., Akin, D. E., and VanEsseltine, W. P. (1986). *Appl. Environ. Microbiol.* **52**, 1331–1339.

Borneman, W. S., Hartley, R. D., Morrison, W. H., Akin, D. E., and Ljungdahl, L. G. (1990). *Appl. Microbiol. Biotechnol.* **33,** 345–351.

Brisson, L., Vacha, W. E. K., and Ibrahim, R. K. (1986). *Plant Sci.* **44,** 175–181.

Brooks, J., and Shaw, G. (1968). *Nature (London)* **219,** 532–533.

Brunow, G., Sipilä, J., and Makela, T. (1989). *Holzforschung* **43,** 55–59.

Bucholtz, D. L., Cantrell, R. P., Axtell, J. D., and Lechtenberg, V. L. (1980). *Agric. Food Chem.* **28,** 1239–1241.

Chesson, A. (1981). *J. Sci. Food Agric.* **32,** 745–758.

Chesson, A. (1988). *Animal Feed Sci. Technol.* **21,** 219–228.

Chesson, A., Stewart, C. S., and Wallace, R. J. (1982). *Appl. Environ. Microbiol.* **44,** 597–603.

Clifford, M. N. (1974). *J. Chromatogr.* **94,** 321–324.

Codignola, A., Verotta, L., Spanu, P., Maffei, M., Scannerini, S., and Bonfante-Fasolo, P. (1989). *New Phytol.* **112,** 221–228.

Cohen, M. D., Schmidt, G. J. M., and Sonntag, F. I. (1963). *J. Chem. Soc.* **1963,** 2000–2013.

Conchie, J., Hay, A. J., and Lomax, J. A. (1988). *Carbohydr. Res.* **177,** 127–151.

Cone, J. W., and Engels, F. M. (1990). *J. Agric. Sci.* **114,** 207–212.

Dill, J., Salnikow, J., and Kraepelin, G. (1984). *Appl. Environ. Microbiol.* **48,** 1259–1261.

Dordick, J. S., Marletta, M. A., and Klibanov, A. M. (1986). *Proc. Natl. Acad. Sci. U.S.A.* **83,** 6255–6257.

Engels, F. M. (1989). *In* "Physicochemical Characterisation of Plant Residues for Industrial and Feed Use" (A. Chesson and E. R. Ørskov, eds.), pp. 80–87. Elsevier Applied Science, London.

Eraso, F., and Hartley, R. D. (1990). *J. Sci. Food Agric.* **51,** 163–170.

Erickson, M., and Miksche, G. E. (1974). *Holzforschung* **28,** 157–159.

Erickson, M., Miksche, G. E., and Somfai, I. (1973). *Holzforschung* **27,** 147–150.

Ericksson, O., Gôring, D. I., and Lindgren, B. O. (1980). *Wood Sci. Technol.* **14,** 267–279.

Floss, H. G. (1979). *Rec. Adv. Phytochem.* **12,** 58–89.

Ford, C. W. (1986). *Carbohydr. Res.* **147,** 101–117.

Ford, C. W. (1990). *Carbohydr. Res.* **201,** 299–309.

Ford, C. W., and Hartley, R. D. (1989). *J. Sci. Food Agric.* **46,** 301–310.

Freudenberg, K., and Friedmann, M. (1960). *Chem. Ber.* **93,** 2138–2145.

Freudenberg, K., and Grion, G. (1959). *Chem. Ber.* **32,** 1355–1363.

Freudenberg, K., and Neish, A. C. (1968). "Constitution and Biosynthesis of Lignin." Springer-Verlag, Berlin.

Freudenberg, K., and Schulter, H. (1955). *Chem. Ber.* **88,** 617–623.

Fry, S. C. (1979). *Planta* **146,** 343–351.

Fry, S. C. (1980). *Phytochemistry* **19,** 735–740.

Fry, S. C. (1982). *Biochem. J.* **203,** 493–504.

Fry, S. C. (1983). *Planta* **157,** 111–123.

Fry, S. C. (1984). *Phytochemistry* **23,** 59–64.

Fry, S. C. (1987). *Planta* **171,** 205–211.

Fukushima, K., and Terashima, N. (1990). *J. Wood Chem. Technol.* **10,** 413–433.

Fulcher, R. G., O'Brien, T. P., and Lee, J. W. (1972). *Aust. J. Biol. Sci.* **25,** 23–34.

Gaillard, B. D. E., and Richards, G. N. (1975). *Carbohydr. Res.* **42,** 135–145.

Gordon, A. J. (1975). *J. Sci. Food Agric.* **26,** 1551–1559.

Gordon, A. H., Lomax, J. A., Dalgarno, K., and Chesson, A. (1985). *J. Sci. Food Agric.* **36,** 509–519.

Grand, C., Sarni, F., and Boudet, A. M. (1985). *Planta* **163,** 232–237.

Gross, G. G. (1979). *Rec. Adv. Phytochem.* **12,** 177–220.

Gross, G. G. (1981). In "The Biochemistry of Plants" (P. K. Stumpf and E. E. Conn, eds.), Vol. 7, pp. 301–316. Academic Press, New York.

Gross, G. G., Janse, C., and Elstner, E. F. (1977). *Planta* **136,** 271–276.

Hagendoorn, M. J. M., Traas, T. P., Boon, J. J., and van der Plas, L. H. W. (1990). *J. Plant Physiol.* **137,** 72–80.

Hansen, K. R., and Havir, E. A. (1981). In "The Biochemistry of Plants" (P. K. Stumpf and E. E. Conn, eds.), Vol. 7, pp. 577–625, Academic Press, New York.

Harmatha, J., Lubke, H., Rybarik, I., and Mahdalik, M. (1978). *Coll. Czech. Chem. Commun.* **43,** 774–781.

Harris, P. J., and Hartley, R. D. (1976). *Nature (London)* **259,** 508–510.

Harris, P. J., and Hartley, R. D. (1980). *Biochem. Syst. Ecol.* **8,** 153–160.

Hartley, R. D. (1972). *J. Sci. Food Agric.* **23,** 1347–1354.

Hartley, R. D., and Akin, D. E. (1989). *J. Sci. Food Agric.* **49,** 405–411.

Hartley, R. D., and Harris, P. J. (1981). *Biochem. Syst. Ecol.* **9,** 189–203.

Hartley, R. D., and Keene, A. S. (1984). *Phytochemistry* **23,** 1305–1307.

Hartley, R. D., Whattey, F. R., and Harris, P. J. (1988). *Phytochemistry* **27,** 349–351.

Hartley, R. D., Morrison, W. H., Balza, F., and Towers, G. H. N. (1990a). *Phytochemistry* **29,** 3699–3703.

Hartley, R. D., Morrison, W. H., Himmelsbach, D. S., and Borneman, W. S. (1990b). *Phytochemistry* **29,** 3705–3709.

He, L., and Terashima, N. (1989a). *Mokuzai Gakkaishi* **35,** 116–122.

He, L., and Terashima, N. (1989b). *Mokuzai Gakkaishi* **35,** 123–129.

Heslop-Harrison, J. (1971). In "Pollen, Development and Physiology" (J. Heslop-Harrison, ed.), pp. 75–98. Butterworths, London.

Highley, T. L., and Micales, J. A. (1990). *FEMS Microbiol. Lett.* **66,** 15–22.

Higuchi, T. (1990). *Wood Sci. Technol.* **24,** 23–63.

Higuchi, T., Ito, Y., Shimada, M., and Kawamura, I. (1967). *Phytochemistry* **6,** 1551–1556.

Iiyama, K., and Lam, T. B. T. (1990). *J. Sci. Food Agric.* **51,** 481–491.

Iiyama, K., Lam, T. B. T., and Stone, B. A. (1990). *Phytochemistry* **29,** 733–737.

Ishii, T. (1991a). *Phytochemistry* **30,** 2317–2320.

Ishii, T. (1991b). *Carbohydr. Res.* **219,** 15–22.

Ishii, T., and Hiroi, T. (1990). *Carbohydr. Res.* **206,** 297–311.

Jangaard, N. O. (1974). *Phytochemistry* **13,** 1765–1768.

Jeffs, R. A., and Northcote, D. H. (1966). *Biochem. J.* **101,** 146–152.

Joseleau, J.-P., Miksche, G. E., and Yasuda, S. (1977). *Holzforschung* **31,** 19–20.

Kamisaka, S., Takeda, S., Takahashi, K., and Shibata, K. (1990). *Physiol. Plant.* **78,** 1–7.

Katayama, Y., Morohoshi, N., and Haraguchi, T. (1980). *Mokuzai Gakkaishi* **26,** 414–420.

Kato, A., Azuma, J., and Koshijima, T. (1987). *Agric. Biol. Chem.* **51,** 1692–1693.

Kato, Y., and Nevins, D. J. (1985). *Carbohydr. Res.* **137,** 139–150.

Kivaisi, A. M., op den Camp, H. J. M., Lubberding, H. L., Boon, J. J., and Vogels, G. D. (1990). *Appl. Microbiol. Biotechnol.* **33,** 93–98.

Kolattukudy, P. E. (1981). *Annu. Rev. Plant Physiol.* **32,** 539–567.

Koshijima, T., Yaku, F., and Tanaka, R. (1976). *Appl. Polymer Symp.* **28,** 1028–1039.

Lam, T. B. T., Iiyama, K., and Stone, B. A. (1990). *J. Sci. Food Agric.* **51,** 493–506.

Lam, T. B. T., Iiyama, K., and Stone, B. A. (1992). *Phytochemistry* **31,** 1179–1183.

Lapierre, C., Monties, B., and Rolando, C. (1986). *Holzforschung* **40,** 113–118.

Lapierre, C., Pollet, B., Monties, B., and Rolando, C. (1991). *Holzforschung* **45,** 61–68.

Leary, G. J. (1980). *Wood Sci. Technol.* **14,** 21–34.

Leary, G. J., Miller, I. J., Thomas, W., and Woolhouse, A. D. (1977). *J. Chem. Soc. (Perkin Trans. II)* 737–1739.

Lewis, N. G., and Yamamoto, E. (1990). *Annu. Rev. Plant Physiol. Plant Mol. Biol.* **41,** 455–496.

Lewis, N. G., Razal, R. A., and Yamamoto, E. (1987a). *Proc. Natl. Acad. Sci. U.S.A.* **84,** 7925–7927.

Lewis, N. G., Yamamoto, E., Wooten, J. B., Just, H., and Towers, G. H. N. (1987b). *Science* **237,** 1344–1346.

Lewis, N. G., Dubelstein, A., Eberhardt, T. L., Yamamoto, E., and Towers, G. H. N. (1987c). *Phytochemistry* **26,** 2729–2734.

Lewis, N. G., Razal, R. A., Yamamoto, E., Bokelman, G. H., and Wooten, J. B. (1989). *In* "Plant Cell Wall Polymers, Biosynthesis and Biodegradation" (N. G. Lewis and M. G. Paice, eds.), ACS Symposium Series, No. 399, pp. 169–181. American Chemical Society.

Marcinowski, H., and Grisebach, H. (1977). *Phytochemistry* **16,** 1665–1667.

Markwalder, H.-U., and Neukom, H. (1976). *Phytochemistry* **15,** 836–837.

Martin, S. A., and Blake, G. G. (1989). *Nutr. Rep. Int.* **40,** 685–693.

McDermid, K. P., MacKenzie, C. R., and Forsberg, C. W. (1990). *Appl. Environ. Microbiol.* **56,** 127–132.

Meyer, K., Kohler, A., and Kauss, H. (1991). *FEBS Lett.* **290,** 209–212.

Miksche, G. E., and Masuda, S. (1978). *Phytochemistry* **17,** 503–504.

Miller, A. R., and Roberts, L. L. W. (1986). *Can. J. Bot.* **64,** 2716–2718.

Moerschbacher, B. M., Kogel, K. H., Noll, U., and Reisener, H.-J. (1986). *Z. Naturforsch.* **41c,** 830–838.

Moerschbacher, B. M., Noll, U., Flott, B. E., and Reisener, H.-J. (1988). *Physiol. Mol. Plant Pathol.* **33,** 33–46.

Moerschbacher, B. M., Noll, U., Gorrichon, L., and Reisener, H.-J. (1990a). *Plant Physiol.* **93,** 465–470.

Moerschbacher, B. M., Noll, U., Ocampo, C. A., Flott, B. E., Gotthardt, U., Wüsterfeld, A., and Reisener, H.-J. (1990b). *Physiol. Plant.* **78,** 609–615.

Mollard, A., and Robert, D. (1984). *Physiol. Vég.* **22,** 3–17.

Monties, B. (1990). *In* "Dietary Fibre, Chemical and Biological Aspects" (D. A. T. Southgate, K. Waldron, I. T. Johnson, and G. R. Fenwick, eds.), Special Publication No. 83, pp. 50–55. Royal Society of Chemistry, London.

Morelli, E., Rej, R. N., Lewis, N. G., Just, G., and Towers, G. H. N. (1986). *Phytochemistry* **25,** 1701–1705.

Morrison, I. M. (1974). *Biochem. J.* **139,** 197–204.

Morrison, W. H., Scheijen, M. A., and Boon, J. J. (1991). *Animal Feed Sci. Technol.* **32,** 17–26.

Mueller-Harvey, I., Hartley, R. D., Harris, P. J., and Curzon, E. H. (1986). *Carbohydr. Res.* **148,** 71–85.

Nakamura, Y., and Higuchi, T. (1976). *Holzforschung* **30,** 187–191.

Nakamura, Y., and Higuchi, T. (1978). *Cellulose Chem. Technol.* **12,** 209–221.

Neukom, H. (1976). *Lebensm. Wiss. Technol.* **9,** 143–148.

Niemann, G. J., Baayen, R. P., and Boon, J. J. (1990a). *Ann. Bot.* **65,** 461–472.

Niemann, G. J., Baayen, R. P., and Boon, J. J. (1990b). *Netherlands J. Plant Pathol.* **96,** 133–153.

Niku-Paavola, M.-L. (1991). *Anal. Biochem.* **197,** 101–103.

Nimz, H., and Ebel, J. (1975). *Z. Naturforsch.* **30c,** 442–444.

Nishitani, K., and Nevins, D. J. (1990). *Plant Physiol.* **93,** 396–402.

O'Brien, T. P., Feder, N., and McCully, M. E. (1964). *Protoplasma* **59,** 366–373.

Olesen, P. (1978). *Protoplasma* **94,** 325–340.

Pearce, R. B., and Ride, J. P. (1982). *Physiol. Plant Pathol.* **20,** 119–123.

Pillonel, C., Mulder, M. M., Boon, J. J., Forster, B., and Binder, A. (1991). *Planta* **185,** 538–544.

Prahl, A. K., Springstubbe, H., Grumbach, K., and Wiermann, R. (1985). *Z. Naturforsch.* **40c,** 621–626.

Rice, E. L. (1984). "Allelopathy." Academic Press, New York.
Ride, J. P. (1978). *Ann. Appl. Biol.* **89**, 302–306.
Ride, J. P., and Pearce, R. B. (1979a). *Physiol. Plant Pathol.* **15**, 79–92.
Ride, J. P., and Pearce, R. B. (1979b). *Physiol. Plant Pathol.* **16**, 179–204.
Rittscher, M., and Wiermann, R. (1988). *Sex. Plant Reprod.* **1**, 132–139.
Robert, D., Mollard, A., and Barnoud, F. (1989). *Plant Physiol. Biochem.* **27**, 297–304.
Robertsen, B., and Svalheim, Ø. (1990). *Physiol. Plant.* **79**, 512–518.
Rolando, C., Laprierre, C., and Monties, B. (1991). *In* "Methods in Lignin Chemistry" (C. Dence and S. Y. Lin, eds.). Springer-Verlag, Berlin.
Rombouts, F. M., and Thibault, J. F. (1986). *Carbohydr. Res.* **154**, 177–187.
Sakagami, H., Kawazoe, Y., Komatsu, N., Simpson, A., Nonoyama, M., Konno, K., Yoshida, T., Kuroiwa, Y., and Tanuma, S.-I. (1991). *Anticancer Res.* **11**, 881–888.
Sarkanen, K. V., and Ludwig, C. H. (eds.) (1971). "Lignins. Occurence, Formation, Structure and Reactions." Wiley, New York.
Sarkanen, S., Razal, R. A., Piccariello, T., Yamamoto, E., and Lewis, N. G. (1991). *J. Biol. Chem.* **266**, 3636–3643.
Scalbert, A., Monties, B., Lallemand, J.-Y., Guittet, E., and Rolando, C. (1985). *Phytochemistry* **24**, 1359–1362.
Scalbert, A., Monties, B., and Lallemand, J. Y. (1986). *Holzforschung* **40**, 119–126.
Scheijen, M. A., and Boon, J. J. (1989). *J. Analyt. Appl. Pyrolysis* **15**, 97–120.
Schmitt, D., Pakusch, A.-E., and Matern, U. (1991). *J. Biol. Chem.* **266**, 17416–17423.
Schmutz, A., Jenny, T., Amrhein, N., and Ryser, U. (1992). *Planta* **189**, 453–460.
Schulze Osthoff, K., and Wiermann, R. (1987). *J. Plant Physiol.* **131**, 5–15.
Sherwood, R. T., and Vance, C. P. (1976). *Phytopathology* **66**, 503–510.
Shimada, M., Fukuzuka, T., and Higuchi, T. (1971). *TAPPI* **54**, 72–78.
Sipilä, J., and Brunow, G. (1991a). *Holzforschung* **45**, 275–278.
Sipilä, J., and Brunow, G. (1991b). *Holzforschung* **45** (*Sept. 1991, Suppl.*), 3–7.
Sipilä, J., and Brunow, G. (1991c). *Holzforschung* **45** (*Sept. 1991, suppl.*), 9–14.
Smith, M. M., and Hartley, R. D. (1983). *Carbohydr. Res.* **118**, 65–80.
Smith, M. M., and O'Brien, T. P. (1979). *Aust. J. Plant. Physiol.* **6**, 201–219.
Sterk, P., Booij, H., Schellekens, G. A., van Kammen, A. B., and de Vries, S. C. (1991). *Plant Cell* **3**, 907–921.
Takahashi, N., and Koshijima, T. (1988). *Wood Sci. Technol.* **22**, 231–241.
Takahashi, N., Azuma, J., and Koshijima, T. (1982). *Carbohydr. Res.* **107**, 161–168.
Tan, K. S., Hoson, T., Masuda, Y., and Kamisaka, S. (1991). *Physiol. Plant.* **83**, 397–403.
Tanaka, K., Nakatsubo, F., and Higuchi, T. (1976). *Mokuzai Gakkaishi* **22**, 589–590.
Tanaka, K., Nakatsubo, F., and Higuchi, T. (1979). *Mokuzai Gakkaishi* **25**, 653–659.
Terashima, N. (1990). *J. Pulp Paper Sci.* **16**, J150–155.
Terashima, N., Mori, I., and Kanda, T. (1975). *Phytochemistry* **14**, 1991–1992.
Terashima, N., Okada, M., and Tomimura, Y. (1979). *Mokuzai Gakkaishi* **25**, 422–426.
Terashima, N., Fukushima, K., Sano, Y., and Takabe, K. (1988). *Holzforschung* **42**, 347–350.
Tkotz, N., and Strack, D. (1980). *Z. Naturforsch.* **35**, 835–837.
Towers, G. H. N., and Abeysekera, B. (1984). *Phytochemistry* **23**, 951–952.
Umezawa, T., Davin, L. B., Yamamoto, E., Kingston, D. G. I., Lewis, N. G. (1990). *J. Chem. Soc. Chem. Commun.* **2**, 1405–1408.
Vance, C. P., Kirk, T. M., and Sherwood, R. T. (1980). *Annu. Rev. Phytopathol.* **18**, 259–288.
van Fleet, D. S. (1950). *Am. J. Bot.* **37**, 721–725.
Van Sumere, C. F. (1989). *Meth. Plant. Biochem.* **1**, 29–73.
Vaughan, D., and Ord, B. (1990). *J. Sci. Food Agric.* **52**, 289–299.

Wallace, G., Chesson, A., Lomax, J. A., and Jarvis, M. C. (1991). *Animal Feed Sci. Technol.* **32**, 193–199.

Watanabe, T., and Koshijima, T. (1988). *Agric. Biol. Chem.* **52**, 2953–2955.

Watanabe, T., Azuma, J., and Koshijima, T. (1985). *Mokuzai Gakkaishi* **31**, 52–53.

Watanabe, T., Ohnishi, J., Yamazaki, Y., Kaizu, S., and Koshijima, T. (1989). *Agric. Biol. Chem.* **53**, 2233–2252.

Wattendorf, J. (1974). *Z. Pflanzenphysiol.* **73**, 214–225.

Wehling, K., Niester, C., Boon, J. J., Willemse, M. T. M., and Wiermann, R. (1989). *Planta* **179**, 376–380.

Whiting, P., and Göring, H. (1982). *Wood Sci. Technol.* **16**, 261.

Whitmore, F. W. (1976). *Phytochemistry* **15**, 375–378.

Whitmore, F. W. (1978). *Plant Sci. Lett.* **13**, 241–245.

Willemse, M. T. M., and Emons, A. M. C. (1991). *Acta Bot. Neerl.* **40**, 115–124.

Wilson, H. A., Sawyer, J., Hatcher, P. G., and Lerch, H. E. (1989). *Phytochemistry* **28**, 1395–1400.

Yaku, F., Tuji, S., and Koshijima, T. (1979). *Holzforschung* **33**, 54–59.

Yamamoto, E., and Towers, G. H. N. (1985). *J. Plant Physiol.* **117**, 441–449.

Yamamoto, E., Inciong, M. E. J., Davin, L. B., and Lewis, N. G. (1990). *Plant Physiol.* **94**, 209–213.

Young, L. Y., and Frazer, A. C. (1987). *Geomicrobiol. J.* **5**, 261–293.

Young, M. R., Towers, G. H. N., and Neish, A. C. (1966). *Can. J. Bot.* **44**, 341–349.

Zenk, M. H. (1971). *In* "Pharmacognosy and Phytochemistry" (H. Wagner and L. Hor-hammer, eds.), pp. 314–346. Springer-Verlag, Berlin.

Zenk, M. H. (1979). *Rec. Adv. Phytochem.* **12**, 139–176.

Literature review completed December 1992.

Index

ISBN 0-12-364554-9